Plasma Physics

Plasma Physics

Volume 4 of *Modern Classical Physics*

KIP S. THORNE *and* **ROGER D. BLANDFORD**

PRINCETON UNIVERSITY PRESS

Princeton and Oxford

Published by Princeton University Press
41 William Street, Princeton, New Jersey 08540
6 Oxford Street, Woodstock, Oxfordshire OX20 1TR

press.princeton.edu

All Rights Reserved
ISBN (pbk.) 978-0-691-21550-1
ISBN (e-book) 978-0-691-21553-2

British Library Cataloging-in-Publication Data is available

Editorial: Ingrid Gnerlich and Arthur Werneck
Production Editorial: Mark Bellis
Text and Cover Design: Wanda España
Production: Jacqueline Poirier
Publicity: Matthew Taylor and Amy Stewart
Copyeditor: Cyd Westmoreland

This book has been composed in MinionPro, Whitney, and Ratio Modern by Windfall Software, Carlisle, Massachusetts, using ZzTEX

Printed on acid-free paper.

Printed in China

10 9 8 7 6 5 4 3 2 1

A NOTE TO READERS
This book is the fourth in a series of volumes that together comprise a unified work titled *Modern Classical Physics*. Each volume is designed to be read independently of the others and can be used as a textbook in an advanced undergraduate- to graduate-level course on the subject of the title or for self-study. However, as the five volumes are highly complementary to one another, we hope that reading one volume may inspire the reader to investigate others in the series—or the full, unified work—and thereby explore the rich scope of modern classical physics.

To Carolee and Liz

CONTENTS

`T2`   Track Two; see page xvi

PART VI PLASMA PHYSICS 995

BOXES

When we (the authors of this book) were students, some understanding of plasma physics was expected of us, and there were physics courses in which we could learn it. That is no longer true, but we believe it should be, and this book is designed to facilitate such courses. *Why?*

Plasmas play major roles, for example, (1) in attempts to achieve controlled thermonuclear fusion using magnetic and inertial confinement; (2) in explanations of radio wave propagation in the ionosphere and the observed behavior of the solar corona and wind; and (3) in astrophysics, where they are responsible for emission throughout the electromagnetic spectrum (e.g., from black holes, highly magnetized neutron stars, and ultrarelativistic outflows).

Plasmas exhibit an amazingly rich set of phenomena and behaviors, which enrich their roles in nature and vastly complicate the technology required for their control and manipulation, most importantly in controlled fusion. Among the most interesting phenomena are (1) the surfing of individual electrons and ions on plasma waves that consist of collective, collisionless oscillations of electrons, ions, and sometimes magnetic fields; (2) that surfing's amplification of some wave modes (feeding of energy from the surfing particles to the collective excitations) and damping of other wave modes (feeding of energy from the modes to the particles); and (3) nonlinear wave-wave coupling in which, for example, two incoming waves (collective oscillations of collisionless particles and magnetic field) interact to produce a new outgoing wave.

This rich behavior arises from a plasma's atomic-scale structure: A plasma is a gas that is significantly ionized (usually by heating or photons) and thus is composed of electrons and ions, and sometimes has an embedded or confining magnetic field. In plasmas, the mean-free paths of electrons and ions are often comparable to or even far longer than macroscopic length scales, so the plasma typically does not behave like a fluid. Its dynamics can be strongly influenced by the particles' velocity distributions. Rich dynamical behavior occurs both in physical space and in velocity space, and the two are strongly coupled and are also coupled to any embedded magnetic field.

QUANTUM PHYSICS IN THIS BOOK

This book deals primarily with *classical* plasma physics. Nevertheless, we make frequent reference to quantum mechanical concepts and phenomena, and we often use quantum concepts and techniques in the classical domain, for example, in our analyses of nonlinear wave-wave coupling. This is because classical physics arises from quantum physics as an approximation, and sometimes—especially in plasmas—the imprints left on classical physics by its quantum roots are so strong that classical phenomena are most powerfully discussed and analyzed in quantum language.

GUIDANCE FOR READERS

The amount and variety of material covered in this book may seem overwhelming. If so, keep in mind that

- *the primary goals of this book* are to teach the fundamental concepts of plasma physics, which are not so extensive that they should overwhelm, to illustrate those concepts in action, and, through our illustrations, to give the reader some physical understanding of how plasmas behave.

We do not intend to provide a mastery of the many illustrative applications contained in the book. To further help students and other readers who feel overwhelmed, we have labeled as "Track Two" sections that can be skipped on a first reading, or skipped entirely—but are sufficiently interesting and important that many readers may choose to browse or study them. Track-Two sections are labeled by the symbol **T2** .

We have aimed this book at advanced undergraduates and first- and second-year graduate students, of whom we expect only (1) a typical physics or engineering student's facility with applied mathematics, and (2) a typical undergraduate-level understanding of classical mechanics, electromagnetism, elementary thermodynamics, and quantum mechanics. We also target working scientists and engineers who want to learn or improve their understanding of plasma physics.

This book is appropriate for a one-quarter or one-semester course in plasma physics. We presume it will also be used as supplementary reading in other courses where plasmas are important—for example, in astrophysics, geophysics, and controlled fusion.

This book is the fourth of five volumes that together constitute a single treatise, *Modern Classical Physics* (or "MCP," as we shall call it). The full treatise was published in 2017 as an embarrassingly thick single book. (The electronic edition is a good deal lighter.) For readers' convenience, we have placed, at the end of this volume, the Table of Contents, Preface, and Acknowledgments of MCP. The five separate textbooks of this decomposition are

- Volume 1: *Statistical Physics,*
- Volume 2: *Optics,*
- Volume 3: *Elasticity and Fluid Dynamics,*

- Volume 4: *Plasma Physics*, and

- Volume 5: *Relativity and Cosmology*.

These individual volumes are much more suitable for human transport and for use in individual courses than their one-volume parent treatise, MCP.

The present volume is enriched by extensive cross-references to the other four volumes—cross-references that elucidate the rich interconnections of various areas of physics.

In this and the other four volumes, we have retained the chapter numbers from MCP and, for the body of each volume, MCP's pagination. In fact, the body of this volume is identical to the corresponding MCP chapters, aside from corrections of errata (which are tabulated at the MCP website http://press.princeton.edu/titles/MCP .html) and a small amount of updating that has not changed pagination. For readers' cross-referencing convenience, a list of the chapters in each of the five volumes appears immediately after this Preface.

EXERCISES

Exercises are a major component of this volume, as well as of the other four volumes of MCP. The exercises are classified into five types:

1. *Practice.* Exercises that provide practice at mathematical manipulations (e.g., of tensors).

2. *Derivation.* Exercises that fill in details of arguments skipped over in the text.

3. *Example.* Exercises that lead the reader step by step through the details of some important extension or application of the material in the text.

4. *Problem.* Exercises with few, if any, hints, in which the task of figuring out how to set up the calculation and get started on it often is as difficult as doing the calculation itself.

5. *Challenge.* Especially difficult exercises whose solution may require reading other books or articles as a foundation for getting started.

We urge readers to try working many of the exercises—especially the examples, which should be regarded as continuations of the text and which contain many of the most illuminating applications. Exercises that we regard as especially important are designated by **.

UNITS

Throughout this volume, we use SI units, as is customary today in plasma physics.

BRIEF OUTLINE OF THIS BOOK

When a plasma's dynamical timescales are sufficiently long, it behaves like a fluid and so can be understood and analyzed by the techniques of fluid dynamics (Volume 3 of MCP). If this fluid-like plasma has an embedded magnetic field, the plasma's (usually high) electric conductivity will strongly couple that field to the fluid. The study of such a magnetized fluid is called *magnetohydrodynamics,* or *MHD* for short, and is the subject of MCP Chap. 19. In MCP, we formally classify MHD and Chap. 19 as part of fluid dynamics, because they also accurately describe other electrically conducting, magnetized fluids (e.g., liquid metals like mercury and liquid sodium). However, by far the most important and widespread application of MHD today is to plasmas, so we have included Chap. 19 in this volume instead of in our fluid dynamics book, Volume 3 of MCP.

In Chap. 19, we derive and elucidate the basic equations and principles of MHD, and we then illustrate them for a nondynamical plasma (*magnetostatics*) in two ways: (1) We analyze the steady flow of plasma along magnetic ducts and describe applications to electric power generation and to spacecraft propulsion. (2) We describe a *tokamak,* the currently most promising configuration for magnetic confinement of a hot plasma in controlled-fusion R&D. Turning to dynamical plasmas, we use MHD to analyze the stability of various magnetostatic equilibria, most importantly the tokamak and other, simpler configurations for magnetic confinement of plasmas. We also discuss the physics of some of the many unstable modes of oscillation that plague magnetic confinement configurations. To illustrate the application of MHD to geophysics and planetary science, we discuss the generation of Earth's magnetic field by dynamo flows in its liquid core; and we analyze magnetosonic waves propagating in a magnetized plasma, such as the interstellar medium.

Important aspects of a plasma's dynamics can be understood by what is happening microscopically, with individual electrons and ions, and groups of them. In Chap. 20, we study the particle kinetics of plasmas. Among the important particle-kinetic phenomena we analyze in Chap. 20 are: (1) Debye shielding—the manner in which the electric field of an individual electron or ion is shielded by collective redistribution of other electrons and ions, so the field dies out much faster than $1/r$. (2) The remarkably long mean free paths of electrons and ions when the only impediment to their constant-velocity motion is Coulomb scattering, with the consequence that the primary impediment to unimpeded motion is scattering off plasma waves (wave-particle interactions, treated in Chap. 23). (3) The motions of individual particles (electrons or ions), and collections of them, in a magnetic field that may be inhomogeneous and time varying, and implications of these motions for behaviors of the plasma.

When a plasma's dynamical timescales are sufficiently short, and the velocities of its electrons (and those of its ions) are not greatly spread out (so the plasma is "cold"), then the negatively charged electrons are coupled together and behave like one fluid, and the positively charged ions are coupled together and behave like a second fluid. These two fluids, interacting with each other and with a magnetic field, can be analyzed using a *two-fluid formalism* that we develop in Chap. 21. We illustrate this formalism by its most important application: to the wide variety of waves that can propagate in a cold plasma like our ionosphere. We also use the two-fluid formalism to illustrate how easily waves can feed off an ordered relative motion of ions and electrons and can grow in strength by extracting kinetic energy from the ordered motions: the *two-stream instability*.

For warm plasmas (with wide spreads of electron and ion velocities), the velocity distributions can undergo interesting and complex dynamics, which couples to the particles' spatial dynamics and to the magnetic field, if present. In other words, the rich dynamics occurs in the six-dimensional phase space of particle positions and velocities, and so is best analyzed using a *kinetic theory formalism* that we develop and illustrate in Chaps. 22 and 23.

In Chap. 22, we develop this kinetic-theory formalism and then linearize it in the electron and ion distribution functions and the magnetic field. For simplicity, we focus largely but not entirely on an unmagnetized plasma, applying it most importantly to *Langmuir waves*—longitudinal oscillations of the electron distribution with the restoring force partially due to thermal pressure and partially electrostatic; and *ion-acoustic* waves—the analog, for ions, of Langmuir waves. We discover, in our linearized kinetic-theory analysis, the damping of these wave modes by electrons or ions that surf on the waves (*Landau damping*) and, under some circumstances (such as the two-stream instability), the feeding of the surfing particles' kinetic energy into the waves so the waves grow instead of damp. We explore the instabilities that result from this surfing in several ways, including a study of particles trapped in the waves, and *Nyquist's method* of analyzing the waves' dispersion relation. And we illustrate the power of Nyquist's method by exhibiting its application to a system far from plasma physics: the stability of a feedback-control system (e.g., an automobile's cruise control). We conclude this chapter with a discussion of N-particle distribution functions, which, despite their apparent formality, can lead to usable answers for practical problems, such as calculating the Coulomb correction to the equation of state of a plasma.

In Chap. 23, on the nonlinear dynamics of plasmas, we restore the nonlinear terms to the kinetic-theory formalism, one after another, and discover their effects. The first nonlinear term reveals (not surprisingly) the back action of Langmuir and ion-acoustic waves on the surfing particles, to reduce or increase the particles' kinetic energy at the same rate as the wave gains or loses energy. The formalism at this order is called *quasilinear theory*. The next nonlinear term gives rise to the nonlinear interaction of two waves to resonantly produce a third wave. This is called *three-wave mixing*, and it is a ubiquitous phenomenon that occurs in many other areas of science and technology, perhaps most importantly today, for light waves that interact in a nonlinear crystal (see Chap. 10 of MCP Volume 2)—a major foundation for nonlinear-optics technology. Although we derive our weakly nonlinear kinetic-theory formalism from the equations of classical plasma physics, we discover in Chap. 23 that the formalism can be written much more elegantly and powerfully in the language of the quantum theory of interacting bosonic plasmons (the quanta associated with the plasma waves), electrons, and ions. In the quantum formalism, we easily identify a term, missed classically, that describes a high-speed charged particle's spontaneous Cerenkov-type emission of Langmuir plasmons (and ion-acoustic plasmons). By comparing the quantum theory's stimulated emission term with that computed from the classical theory, we infer the quantitative strength of this spontaneous emission and discover that it can be as important as the purely classical processes we have been studying. This analysis exhibits the deep and powerful relationships between quantum physics and classical physics that occur widely elsewhere in science and technology. We conclude this chapter with a discussion of collisionless shock waves and their deep relationship to solitons.

In three short appendixes, we present sections from other volumes of MCP that underpin portions of this volume:

- Appendix A, Evolution of Vorticity—A good starting point for our analysis of the evolution of the magnetic field in Chap. 19 on magnetohydrodynamics.

- Appendix B, Geometric Optics—An essential foundation for some of our analysis of waves in plasmas in Chaps. 21, 22, and 23.

- Appendix C, Distribution Function and Mean Occupation Number—The link between more traditional kinetic theory and its application to plasmas in Chaps. 22 and 23.

Volume 4: Plasma Physics

Volume 5: Relativity and Cosmology

19

Magnetohydrodynamics

. . . it is only the plasma itself which does not 'understand' how beautiful the theories are and absolutely refuses to obey them.

HANNES ALFVÉN (1970)

19.1 Overview

In preceding chapters we have described the consequences of incorporating viscosity and thermal conductivity into the description of a fluid. We now turn to our final embellishment of fluid mechanics, in which the fluid is electrically conducting and moves in a magnetic field. The study of flows of this type is known as *magnetohydrodynamics*, or MHD for short. In our discussion, we eschew full generality and with one exception just use the basic Euler equation (no viscosity, no heat diffusion, etc.) augmented by magnetic terms. This approach suffices to highlight peculiarly magnetic effects and is adequate for many applications.

The simplest example of an electrically conducting fluid is a liquid metal, for example, mercury or liquid sodium. However, the major application of MHD is in plasma physics—discussed in Part VI. (A plasma is a hot, ionized gas containing free electrons and ions.) It is by no means obvious that plasmas can be regarded as fluids, since the mean free paths for Coulomb-force collisions between a plasma's electrons and ions are macroscopically long. However, as we shall learn in Sec. 20.5, collective interactions between large numbers of plasma particles can isotropize the particles' velocity distributions in some local mean reference frame, thereby making it sensible to describe the plasma macroscopically by a mean density, velocity, and pressure. These mean quantities can then be shown to obey the same conservation laws of mass, momentum, and energy as we derived for fluids in Chap. 13. As a result, a fluid description of a plasma is often reasonably accurate. We defer to Part VI further discussion of this point, asking the reader to take it on trust for the moment. In MHD, we also implicitly assume that the average velocity of the ions is nearly the same as the average velocity of the electrons. This is usually a good approximation; if it were not so, then the plasma would carry an unreasonably large current density.

Two serious technological applications of MHD may become very important in the future. In the first, strong magnetic fields are used to confine rings or columns of hot plasma that (it is hoped) will be held in place long enough for thermonuclear fusion to occur and for net power to be generated. In the second, which is directed toward a similar goal, liquid metals or plasmas are driven through a magnetic field to generate electricity. The study of magnetohydrodynamics is also motivated by its widespread application to the description of space (in the solar system) and astrophysical plasmas (beyond the solar system). We illustrate the principles of MHD using examples drawn from all these areas.

After deriving the basic equations of MHD (Sec. 19.2), we elucidate magnetostatic (also called "hydromagnetic") equilibria by describing a *tokamak* (Sec. 19.3). This is currently the most popular scheme for the magnetic confinement of hot plasma. In our second application (Sec. 19.4) we describe the flow of conducting liquid metals or plasma along magnetized ducts and outline its potential as a practical means of electrical power generation and spacecraft propulsion. We then return to the question of magnetostatic confinement of hot plasma and focus on the stability of equilibria (Sec. 19.5). This issue of stability has occupied a central place in our development of fluid mechanics, and it will not come as a surprise to learn that it has dominated research on thermonuclear fusion in plasmas. When a magnetic field plays a role in the equilibrium (e.g., for magnetic confinement of a plasma), the field also makes possible new modes of oscillation, and some of these MHD modes can be unstable to exponential growth. Many magnetic-confinement geometries exhibit such instabilities. We demonstrate this qualitatively by considering the physical action of the magnetic field, and also formally by using variational methods.

In Sec. 19.6, we turn to a geophysical problem, the origin of Earth's magnetic field. It is generally believed that complex fluid motions in Earth's liquid core are responsible for regenerating the field through dynamo action. We use a simple model to illustrate this process.

When magnetic forces are added to fluid mechanics, a new class of waves, called magnetosonic waves, can propagate. We conclude our discussion of MHD in Sec. 19.7 by deriving the properties of these wave modes in a homogeneous plasma and discussing how they control the propagation of cosmic rays in the interplanetary and interstellar media.

As in previous chapters, we encourage our readers to view films; on magnetohydrodynamics, for example, Shercliff (1965).

19.2 19.2 Basic Equations of MHD

The equations of MHD describe the motion of a conducting fluid in a magnetic field. This fluid is usually either a liquid metal or a plasma. In both cases, the conductivity, strictly speaking, should be regarded as a tensor (Sec. 20.6.3) if the electrons' cyclotron frequency (Sec. 20.6.1) exceeds their collision frequency (the inverse of the mean time between collisions; Sec. 20.4.1). (If there are several collisions per cyclotron orbit, then the influence of the magnetic field on the transport coefficients will be minimal.) However, to keep the mathematics simple, we treat the conductivity as a constant scalar, κ_e. In fact, it turns out that for many of our applications, it is adequate to take the conductivity as infinite, and it does not matter whether that infinity is a scalar or a tensor!

Two key physical effects occur in MHD, and understanding them well is key to developing physical intuition. The first effect arises when a good conductor moves into a magnetic field (Fig. 19.1a). Electric current is induced in the conductor, which, by Lenz's law, creates its own magnetic field. This induced magnetic field tends to cancel the original, externally supported field, thereby in effect excluding the magnetic field lines from the conductor. Conversely, when the magnetic field penetrates the conductor and the conductor is moved out of the field, the induced field reinforces the applied field. The net result is that the lines of force appear to be dragged along with the conductor—they "go with the flow." Naturally, if the conductor is a fluid with complex motions, the ensuing magnetic field distribution can become quite complex, and the current builds up until its growth is balanced by Ohmic dissipation.

The second key effect is dynamical. When currents are induced by a motion of a conducting fluid through a magnetic field, a Lorentz (or $\mathbf{j} \times \mathbf{B}$) force acts on the fluid and modifies its motion (Fig. 19.1b). In MHD, the motion modifies the field, and the field, in turn, reacts back and modifies the motion. This behavior makes the theory highly nonlinear.

Before deriving the governing equations of MHD, we should consider the choice of primary variables. In electromagnetic theory, we specify the spatial and temporal variation of either the electromagnetic field or its source, the electric charge density and current density. One choice is computable (at least in principle) from the other

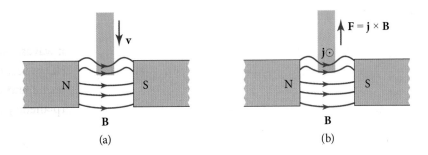

FIGURE 19.1 The two key physical effects that occur in MHD. (a) A moving conductor modifies the magnetic field by dragging the field lines with it. When the conductivity is infinite, the field lines are frozen in the moving conductor. (b) When electric current, flowing in the conductor, crosses magnetic field lines, a Lorentz force is generated that accelerates the fluid.

two key physical effects in MHD

using Maxwell's equations, augmented by suitable boundary conditions. So it is with MHD, and the choice depends on convenience. It turns out that for the majority of applications, it is most instructive to deal with the magnetic field as primary, and to use Maxwell's equations

in MHD the magnetic field is the primary variable

Maxwell's equations

$$\nabla \cdot \mathbf{E} = \frac{\rho_e}{\epsilon_0}, \tag{19.1a}$$

$$\nabla \cdot \mathbf{B} = 0, \tag{19.1b}$$

$$\nabla \times \mathbf{E} = -\frac{\partial \mathbf{B}}{\partial t}, \tag{19.1c}$$

$$\nabla \times \mathbf{B} = \mu_0 \mathbf{j} + \mu_0 \epsilon_0 \frac{\partial \mathbf{E}}{\partial t} \tag{19.1d}$$

to express the electric field $\mathbf{E}$, the current density $\mathbf{j}$, and the charge density ρ_e in terms of the magnetic field (next subsection).

19.2.1

19.2.1 Maxwell's Equations in the MHD Approximation

As normally formulated, Ohm's law is valid only in the rest frame of the conductor. In particular, for a conducting fluid, Ohm's law relates the current density $\mathbf{j}'$ measured in the fluid's local rest frame to the electric field $\mathbf{E}'$ measured there:

$$\mathbf{j}' = \kappa_e \mathbf{E}', \tag{19.2}$$

where κ_e is the scalar electric conductivity. Because the fluid is generally accelerated, $d\mathbf{v}/dt \neq 0$, its local rest frame is generally not inertial. Since it would produce a terrible headache to have to transform time and again from some inertial frame to the continually changing local rest frame when applying Ohm's law, it is preferable to reformulate Ohm's law in terms of the fields $\mathbf{E}$, $\mathbf{B}$, and $\mathbf{j}$ measured in an inertial frame. To facilitate this (and for completeness), we explore the frame dependence of all our electromagnetic quantities $\mathbf{E}$, $\mathbf{B}$, $\mathbf{j}$, and ρ_e.

Throughout our development of magnetohydrodynamics, we assume that the fluid moves with a nonrelativistic speed $v \ll c$ relative to our chosen reference frame. We can then express the rest-frame electric field in terms of the inertial-frame electric and magnetic fields as

$$\mathbf{E}' = \mathbf{E} + \mathbf{v} \times \mathbf{B}; \quad E' = |\mathbf{E}'| \ll E, \quad \text{so } \mathbf{E} \simeq -\mathbf{v} \times \mathbf{B}. \tag{19.3a}$$

In the first equation we have set the Lorentz factor $\gamma \equiv 1/\sqrt{1 - v^2/c^2}$ to unity, consistent with our nonrelativistic approximation. The second equation follows from the high conductivity of the fluid, which guarantees that current will quickly flow in whatever manner it must to annihilate any electric field $\mathbf{E}'$ that might be formed in the fluid's local rest frame. By contrast with the extreme frame dependence (19.3a) of the electric field, the magnetic field is essentially the same in the fluid's local rest frame as in the laboratory. More specifically, the analog of Eq. (19.3a) is $\mathbf{B}' = \mathbf{B} - (\mathbf{v}/c^2) \times \mathbf{E}$; and since $E \sim vB$, the second term is of magnitude $(v/c)^2 B$, which is negligible, giving

$$\mathbf{B}' \simeq \mathbf{B}. \tag{19.3b}$$

Because $\mathbf{E}$ is highly frame dependent, so is its divergence, the electric charge density ρ_e. In the laboratory frame, where $E \sim vB$, Gauss's and Ampère's laws [Eqs. (19.1a,d)] imply that $\rho_e \sim \epsilon_0 vB/L \sim (v/c^2)j$, where L is the lengthscale on which $\mathbf{E}$ and $\mathbf{B}$ vary; and the relation $E' \ll E$ with Gauss's law implies $|\rho_e'| \ll |\rho_e|$:

$$\rho_e \sim j\, v/c^2, \qquad |\rho_e'| \ll |\rho_e|. \tag{19.3c}$$

By transforming the current density between frames and approximating $\gamma \simeq 1$, we obtain $\mathbf{j}' = \mathbf{j} + \rho_e \mathbf{v} = \mathbf{j} + \mathrm{O}(v/c)^2 j$; so in the nonrelativistic limit (first order in v/c) we can ignore the charge density and write

$$\mathbf{j}' = \mathbf{j}. \tag{19.3d}$$

To recapitulate, in nonrelativistic magnetohydrodynamic flows, the magnetic field and current density are frame independent up to fractional corrections of order $(v/c)^2$, while the electric field and charge density are highly frame dependent and are generally small in the sense that $E/c \sim (v/c)B \ll B$ and $\rho_e \sim (v/c^2)j \ll j/c$ [in Gaussian cgs units we have $E \sim (v/c)B \ll B$ and $\rho_e c \sim (v/c)j \ll j$].

in MHD, magnetic field and current density are approximately frame independent; electric field and charge density are small and frame dependent

Combining Eqs. (19.2), (19.3a), and (19.3d), we obtain the nonrelativistic form of Ohm's law in terms of quantities measured in our chosen inertial, laboratory frame:

$$\mathbf{j} = \kappa_e(\mathbf{E} + \mathbf{v} \times \mathbf{B}). \tag{19.4}$$

Ohm's law

We are now ready to derive explicit equations for the (inertial-frame) electric field and current density in terms of the (inertial-frame) magnetic field. In our derivation, we denote by L the lengthscale on which the magnetic field changes.

We begin with Ampère's law written as $\nabla \times \mathbf{B} - \mu_0 \mathbf{j} = \mu_0 \epsilon_0 \partial \mathbf{E}/\partial t = (1/c^2)\partial \mathbf{E}/\partial t$, and we notice that the time derivative of $\mathbf{E}$ is of order $Ev/L \sim Bv^2/L$ (since $E \sim vB$). Therefore, the right-hand side is $\mathrm{O}[Bv^2/(c^2 L)]$ and thus can be neglected compared to the $\mathrm{O}(B/L)$ term on the left, yielding:

$$\boxed{\mathbf{j} = \frac{1}{\mu_0} \nabla \times \mathbf{B}.} \tag{19.5a}$$

current density in terms of magnetic field

We next insert this expression for $\mathbf{j}$ into the inertial-frame Ohm's law (19.4), thereby obtaining

$$\boxed{\mathbf{E} = -\mathbf{v} \times \mathbf{B} + \frac{1}{\kappa_e \mu_0} \nabla \times \mathbf{B}.} \tag{19.5b}$$

electric field in terms of magnetic field

If we happen to be interested in the charge density (which is rare in MHD), we can compute it by taking the divergence of this electric field:

charge density in terms of magnetic field

$$\rho_e = -\epsilon_0 \nabla \cdot (\mathbf{v} \times \mathbf{B}).$$

(19.5c)

Equations (19.5) express all the secondary electromagnetic variables in terms of our primary one, $\mathbf{B}$. This has been possible because of the high electric conductivity κ_e and our choice to confine ourselves to nonrelativistic (low-velocity) situations; it would not be possible otherwise.

We next derive an evolution law for the magnetic field by taking the curl of Eq. (19.5b), using Maxwell's equation $\nabla \times \mathbf{E} = -\partial \mathbf{B}/\partial t$ and the vector identity $\nabla \times (\nabla \times \mathbf{B}) = \nabla(\nabla \cdot \mathbf{B}) - \nabla^2 \mathbf{B}$, and using $\nabla \cdot \mathbf{B} = 0$. The result is

evolution law for magnetic field

$$\frac{\partial \mathbf{B}}{\partial t} = \nabla \times (\mathbf{v} \times \mathbf{B}) + \left(\frac{1}{\mu_0 \kappa_e} \right) \nabla^2 \mathbf{B},$$

(19.6)

which, using Eqs. (14.4) and (14.5) with $\boldsymbol{\omega}$ replaced by $\mathbf{B}$, can also be written as

$$\frac{D\mathbf{B}}{Dt} = -\mathbf{B}\nabla \cdot \mathbf{v} + \left(\frac{1}{\mu_0 \kappa_e} \right) \nabla^2 \mathbf{B},$$

(19.7)

where D/Dt is the fluid derivative defined in Eq. (14.5). When the flow is incompressible (as it often will be), the $\nabla \cdot \mathbf{v}$ term vanishes.

Equation (19.6)—or equivalently, Eq. (19.7)—is called the *induction equation* and describes the temporal evolution of the magnetic field. It is the same in form as the propagation law for vorticity $\boldsymbol{\omega}$ in a flow with $\nabla P \times \nabla \rho = 0$ [Eq. (14.3), or (14.6) with $\boldsymbol{\omega}\nabla \cdot \mathbf{v}$ added in the compressible case]. The $\nabla \times (\mathbf{v} \times \mathbf{B})$ term in Eq. (19.6) dominates when the conductivity is large and can be regarded as describing the freezing of magnetic field lines in the fluid in the same way as the $\nabla \times (\mathbf{v} \times \boldsymbol{\omega})$ term describes the freezing of vortex lines in a fluid with small viscosity ν (Fig. 19.2). By analogy with Eq. (14.10), when flux-freezing dominates, the fluid derivative of $\mathbf{B}/\rho$ can be written as

for large conductivity: freezing of magnetic field into the fluid

$$\frac{D}{Dt}\left(\frac{\mathbf{B}}{\rho} \right) \equiv \frac{d}{dt}\left(\frac{\mathbf{B}}{\rho} \right) - \left(\frac{\mathbf{B}}{\rho} \cdot \nabla \right) \mathbf{v} = 0,$$

(19.8)

where ρ is mass density (not to be confused with charge density ρ_e). Equation (19.8) states that $\mathbf{B}/\rho$ evolves in the same manner as the separation $\Delta \mathbf{x}$ between two points in the fluid (cf. Fig. 14.4 and associated discussion).

The term $[1/(\mu_0 \kappa_e)]\nabla^2 \mathbf{B}$ in the B-field evolution equation (19.6) or (19.7) is analogous to the vorticity diffusion term $\nu\nabla^2\boldsymbol{\omega}$ in the vorticity evolution equation (14.3) or (14.6). Therefore, when κ_e is not too large, magnetic field lines will diffuse through the fluid. The effective diffusion coefficient (analogous to ν) is

magnetic diffusion coefficient

$$D_M = 1/(\mu_0 \kappa_e).$$

(19.9a)

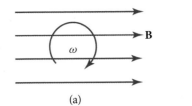
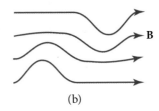

(a)   (b)

FIGURE 19.2 Pictorial representation of the evolution of the magnetic field in a fluid endowed with infinite electrical conductivity. (a) A uniform magnetic field at time $t = 0$ in a vortex. (b) At a later time, when the fluid has rotated through $\sim 30°$, the circulation has stretched and distorted the magnetic field.

Earth's magnetic field provides an example of field diffusion. That field is believed to be supported by electric currents flowing in Earth's iron core. Now, we can estimate the electric conductivity of iron under these conditions and from it deduce a value for the diffusivity, $D_M \sim 1 \, \text{m}^2 \, \text{s}^{-1}$. The size of Earth's core is $L \sim 10^4 \, \text{km}$, so if there were no fluid motions, then we would expect the magnetic field to diffuse out of the core and escape from Earth in a time

$$\boxed{\tau_M \sim \frac{L^2}{D_M}}$$ (19.9b)   **magnetic decay time**

$\sim$3 million years, which is much shorter than the age of Earth, $\sim$5 billion years. The reason for this discrepancy, as we discuss in Sec. 19.6, is that there are internal circulatory motions in the liquid core that are capable of regenerating the magnetic field through dynamo action.

Although Eq. (19.6) describes a genuine diffusion of the magnetic field, to compute with confidence the resulting magnetic decay time, one must solve the complete boundary value problem. To give a simple illustration, suppose that a poor conductor (e.g., a weakly ionized column of plasma) is surrounded by an excellent conductor (e.g., the metal walls of the container in which the plasma is contained), and that magnetic field lines supported by wall currents thread the plasma. The magnetic field will only diminish after the wall currents undergo Ohmic dissipation, which can take much longer than the diffusion time for the plasma column alone.

It is customary to introduce a dimensionless number called the *magnetic Reynolds number*, R_M, directly analogous to the fluid Reynolds number Re, to describe the relative importance of flux freezing and diffusion. The fluid Reynolds number can be regarded as the ratio of the magnitude of the vorticity-freezing term, $\boldsymbol{\nabla} \times (\mathbf{v} \times \boldsymbol{\omega}) \sim (V/L)\omega$, in the vorticity evolution equation, $\partial\boldsymbol{\omega}/\partial t = \boldsymbol{\nabla} \times (\mathbf{v} \times \boldsymbol{\omega}) + \nu\nabla^2\boldsymbol{\omega}$, to the magnitude of the diffusion term, $\nu\nabla^2\boldsymbol{\omega} \sim (\nu/L^2)\omega$: $\text{Re} = (V/L)(\nu/L^2)^{-1} = VL/\nu$. Here V is a characteristic speed, and L a characteristic lengthscale of the flow. Similarly, the magnetic Reynolds number is the ratio of the magnitude of the

TABLE 19.1: Characteristic magnetic diffusivities D_M, decay times τ_M, and magnetic Reynolds numbers R_M for some common MHD flows with characteristic length scales L and velocities V

Substance	L (m)	V (m s^{-1})	D_M (m^2 s^{-1})	τ_M (s)	R_M
Mercury	0.1	0.1	1	0.01	0.01
Liquid sodium	0.1	0.1	0.1	0.1	0.1
Laboratory plasma	1	100	10	0.1	10
Earth's core	10^7	0.1	1	10^{14}	10^6
Interstellar gas	10^{17}	10^3	10^3	10^{31}	10^{17}

magnetic-flux-freezing term, $\mathbf{\nabla} \times (\mathbf{v} \times \mathbf{B}) \sim (V/L)B$, to the magnitude of the magnetic-flux-diffusion term, $D_M \nabla^2 \mathbf{B} = [1/(\mu_o \kappa_e)]\nabla^2 \mathbf{B} \sim B/(\mu_o \kappa_e L^2)$, in the induction equation (19.6):

magnetic Reynolds number and magnetic field freezing

$$R_M = \frac{V/L}{D_M/L^2} = \frac{VL}{D_M} = \mu_0 \kappa_e V L. \tag{19.9c}$$

When $R_M \gg 1$, the field lines are effectively frozen in the fluid; when $R_M \ll 1$, Ohmic dissipation is dominant, and the field lines easily diffuse through the fluid.

Magnetic Reynolds numbers and diffusion times for some typical MHD flows are given in Table 19.1. For most laboratory conditions, R_M is modest, which means that electric resistivity $1/\kappa_e$ is significant, and the magnetic diffusivity D_M is rarely negligible. By contrast, in space physics and astrophysics, R_M is usually very large, $R_M \gg 1$, so the resistivity can be ignored almost always and everywhere. This limiting case, when the electric conductivity is treated as infinite, is often called *perfect MHD*.

perfect MHD: infinite conductivity and magnetic field freezing

The phrase "almost always and everywhere" needs clarification. Just as for large-Reynolds-number fluid flows, so also here, boundary layers and discontinuities can be formed, in which the gradients of physical quantities are automatically large enough to make $R_M \sim 1$ locally. An important example discussed in Sec. 19.6.3 is *magnetic reconnection*. This occurs when regions magnetized along different directions are juxtaposed, for example, when the solar wind encounters Earth's magnetosphere. In such discontinuities and boundary layers, the current density is high, and magnetic diffusion and Ohmic dissipation are important. As in ordinary fluid mechanics, these dissipative layers and discontinuities can control the character of the overall flow despite occupying a negligible fraction of the total volume.

magnetic reconnection and its influence

19.2.2 Momentum and Energy Conservation

19.2.2

The fluid dynamical aspects of MHD are handled by adding an electromagnetic force term to the Euler or Navier-Stokes equation. The magnetic force density $\mathbf{j} \times \mathbf{B}$ is the sum of the Lorentz forces acting on all the fluid's charged particles in a unit volume.

There is also an electric force density $\rho_e \mathbf{E}$, but this is smaller than $\mathbf{j} \times \mathbf{B}$ by a factor $O(v^2/c^2)$ by virtue of Eqs. (19.5), so we ignore it. When $\mathbf{j} \times \mathbf{B}$ is added to the Euler equation (13.44) (or equivalently, to the Navier-Stokes equation with the viscosity neglected as unimportant in the situations we shall study), it takes the following form:

$$\rho \frac{d\mathbf{v}}{dt} = \rho\mathbf{g} - \nabla P + \mathbf{j} \times \mathbf{B} = \rho\mathbf{g} - \nabla P + \frac{(\nabla \times \mathbf{B}) \times \mathbf{B}}{\mu_0}.$$ (19.10)

MHD equation of motion for fluid

Here , 951 we have used expression (19.5a) for the current density in terms of the magnetic field. This is our basic MHD force equation. In Sec. 20.6.2 we will generalize it to situations where, due to electron cyclotron motion, the pressure P is anisotropic.

Like all other force densities in this equation, the magnetic one $\mathbf{j} \times \mathbf{B}$ can be expressed as minus the divergence of a stress tensor, the magnetic portion of the Maxwell stress tensor:

$$\mathbf{T}_M = \frac{B^2 \mathbf{g}}{2\mu_0} - \frac{\mathbf{B} \otimes \mathbf{B}}{\mu_0};$$ (19.11)

magnetic stress tensor

see Ex. 19.1. By virtue of $\mathbf{j} \times \mathbf{B} = -\nabla \cdot \mathbf{T}_M$ and other relations explored in Sec. 13.5 and Box 13.4, we can convert the force-balance equation (19.10) into the conservation law for momentum [generalization of Eq. (13.42)]:

momentum conservation

$$\frac{\partial(\rho\mathbf{v})}{\partial t} + \nabla \cdot (P\mathbf{g} + \rho\mathbf{v} \otimes \mathbf{v} + \mathbf{T}_g + \mathbf{T}_M) = 0.$$ (19.12)

Here $\mathbf{T}_g$ is the gravitational stress tensor [Eq. (1) of Box 13.4], which resembles the magnetic one:

$$\mathbf{T}_g = -\frac{g^2 \mathbf{g}}{8\pi G} + \frac{\mathbf{g} \otimes \mathbf{g}}{4\pi G};$$ (19.13)

it is generally unimportant in laboratory plasmas but can be quite important in and near stars and black holes.

The two terms in the magnetic Maxwell stress tensor [Eq. (19.11)] can be identified as the "push" of an isotropic magnetic pressure of $B^2/(2\mu_0)$ that acts just like the gas pressure P, and the "pull" of a tension B^2/μ_0 that acts parallel to the magnetic field. The combination of the tension and the isotropic pressure give a net tension $B^2/(2\mu_0)$ along the field and a net pressure $B^2/(2\mu_0)$ perpendicular to the field lines (Ex. 1.14).

The magnetic force density

magnetic force density

$$\mathbf{f}_m = -\nabla \cdot \mathbf{T}_M = \mathbf{j} \times \mathbf{B} = \frac{(\nabla \times \mathbf{B}) \times \mathbf{B}}{\mu_0}$$ (19.14)

can be rewritten, using standard vector identities, as

$$\mathbf{f}_m = -\nabla\left(\frac{B^2}{2\mu_0}\right) + \frac{(\mathbf{B} \cdot \nabla)\mathbf{B}}{\mu_0} = -\left[\nabla\left(\frac{B^2}{2\mu_0}\right)\right]_\perp + \left[\frac{(\mathbf{B} \cdot \nabla)\mathbf{B}}{\mu_0}\right]_\perp.$$ (19.15)

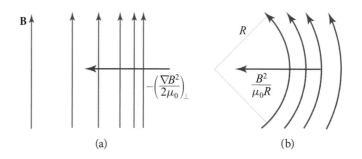

FIGURE 19.3 Contributions to the electromagnetic force density acting on a conducting fluid in a nonuniform magnetic field. A magnetic-pressure force density $-[\nabla B^2/(2\mu_0)]_\perp$ acts perpendicularly to the field. And a magnetic-curvature force density $[(\mathbf{B} \cdot \nabla)\mathbf{B}/\mu_0]_\perp$, which is also perpendicular to the magnetic field and lies in the plane of the field's bend, points toward its center of curvature. The magnitude of this curvature force density is $B^2/(\mu_0 R)$, where R is the radius of curvature.

Here "$\perp$" means keep only the components perpendicular to the magnetic field; the fact that $\mathbf{f}_m = \mathbf{j} \times \mathbf{B}$ guarantees that the net force parallel to $\mathbf{B}$ must vanish, so we can throw away the component along $\mathbf{B}$ in each term. This transversality of $\mathbf{f}_m$ means that the magnetic force neither inhibits nor promotes motion of the fluid along the magnetic field. Instead, fluid elements are free to slide along the field like beads that slide without friction along a magnetic "wire."

The "$\perp$" expressions in Eq. (19.15) indicate that the magnetic force density has two parts: first, the negative of the 2-dimensional gradient of the magnetic pressure $B^2/(2\mu_0)$ orthogonal to $\mathbf{B}$ (Fig. 19.3a), and second, an orthogonal *curvature force* $(\mathbf{B} \cdot \nabla)\mathbf{B}/\mu_0$, which has magnitude $B^2/(\mu_0 R)$, where R is the radius of curvature of a field line. This curvature force acts toward the field line's center of curvature (Fig. 19.3b) and is the magnetic-field-line analog of the force that acts on a curved wire or curved string under tension.

Just as the magnetic force density dominates and the electric force is negligible $[O(v^2/c^2)]$ in our nonrelativistic situation, so also the electromagnetic contribution to the energy density is predominantly due to the magnetic term $U_M = B^2/(2\mu_0)$ with negligible electric contribution. The electromagnetic energy flux is just the Poynting flux $\mathbf{F}_M = \mathbf{E} \times \mathbf{B}/\mu_0$, with $\mathbf{E}$ given by Eq. (19.5b). Inserting these expressions into the law of energy conservation (13.58) (and continuing to neglect viscosity), we obtain

energy conservation

$$\frac{\partial}{\partial t}\left[\left(\frac{1}{2}v^2 + u + \Phi\right)\rho + \frac{B^2}{2\mu_0}\right] + \nabla \cdot \left[\left(\frac{1}{2}v^2 + h + \Phi\right)\rho\mathbf{v} + \frac{\mathbf{E} \times \mathbf{B}}{\mu_0}\right] = 0.$$

(19.16)

When the fluid's self-gravity is important, we must augment this equation with the gravitational energy density and flux, as discussed in Box 13.4.

As in Sec. 13.7.4, we can combine this energy conservation law with mass conservation and the first law of thermodynamics to obtain an equation for the evolution of entropy: Eqs. (13.75) and (13.76) are modified to read

$$\frac{\partial(\rho s)}{\partial t} + \nabla \cdot (\rho s \mathbf{v}) = \rho \frac{ds}{dt} = \frac{j^2}{\kappa_e T}.$$

(19.17)

entropy evolution; Ohmic dissipation

Thus, just as viscosity increases entropy through viscous dissipation, and thermal conductivity increases entropy through diffusive heat flow [Eqs. (13.75) and (13.76)], so also electrical resistivity (formally, κ_e^{-1}) increases entropy through Ohmic dissipation. From Eq. (19.17) we see that our fourth transport coefficient κ_e, like our previous three (the two coefficients of viscosity $\eta \equiv \rho \nu$ and ζ and the thermal conductivity κ), is constrained to be positive by the second law of thermodynamics.

Exercise 19.1 *Derivation: Basic Equations of MHD*

EXERCISES

(a) Verify that $-\nabla \cdot \mathbf{T}_M = \mathbf{j} \times \mathbf{B}$, where $\mathbf{T}_M$ is the magnetic stress tensor (19.11).

(b) Take the scalar product of the fluid velocity $\mathbf{v}$ with the equation of motion (19.10) and combine with mass conservation to obtain the energy conservation equation (19.16).

(c) Combine energy conservation (19.16) with the first law of thermodynamics and mass conservation to obtain Eq. (19.17) for the evolution of the entropy.

19.2.3 Boundary Conditions

19.2.3

The equations of MHD must be supplemented by boundary conditions at two different types of interfaces. The first is a *contact discontinuity* (i.e., the interface between two distinct media that do not mix; e.g., the surface of a liquid metal or a rigid wall of a plasma containment device). The second is a shock front that is being crossed by the fluid. Here the boundary is between shocked and unshocked fluid.

types of interfaces: contact discontinuity and shock front

We can derive the boundary conditions by transforming into a primed frame in which the interface is instantaneously at rest (not to be confused with the fluid's local rest frame) and then transforming back into our original unprimed inertial frame. In the primed frame, we resolve the velocity and magnetic and electric vectors into components normal and tangential to the surface. If $\mathbf{n}$ is a unit vector normal to the surface, then the normal and tangential components of velocity in either frame are

$$v_n = \mathbf{n} \cdot \mathbf{v}, \quad \mathbf{v}_t = \mathbf{v} - (\mathbf{n} \cdot \mathbf{v})\mathbf{n},$$

(19.18)

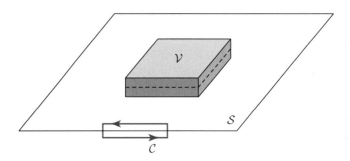

FIGURE 19.4 Elementary pill box $\mathcal{V}$ and elementary circuit $\mathcal{C}$ used in deriving the MHD junction conditions at a surface $\mathcal{S}$.

and similarly for the **E** and **B**. At a contact discontinuity, we have

boundary conditions at interfaces: Eqs. (19.19)

$$v'_n = v_n - v_{sn} = 0 \tag{19.19a}$$

on both sides of the interface surface; here v_{sn} is the normal velocity of the surface. At a shock front, mass flux across the surface is conserved [cf. Eq. (17.29a)]:

junction condition for mass flux

$$[\rho v'_n] = [\rho(v_n - v_{sn})] = 0. \tag{19.19b}$$

Here, as in Sec. 17.5, we use the notation $[X]$ to signify the difference in some quantity X across the interface, that is, the *junction condition* for X.

When we consider the magnetic field, it does not matter which frame we use, since **B** is unchanged to the Galilean order at which we are working. Let us construct a thin "pill box" $\mathcal{V}$ (Fig. 19.4) and integrate the equation $\nabla \cdot \mathbf{B} = 0$ over its volume, invoke the divergence theorem, and let the box thickness diminish to zero; thereby we see that

electromagnetic junction conditions

$$[B_n] = 0. \tag{19.19c}$$

By contrast, the tangential component of the magnetic field can be discontinuous across an interface because of surface currents: by integrating $\nabla \times \mathbf{B} = \mu_0 \mathbf{j}$ across the shock front, we can deduce that

$$[\mathbf{B}_t] = -\mu_0 \mathbf{n} \times \mathbf{J}, \tag{19.19d}$$

where **J** is the surface current density.

We deduce the junction condition on the electric field by integrating Maxwell's equation $\nabla \times \mathbf{E} = -\partial \mathbf{B}/\partial t$ over the area bounded by the circuit $\mathcal{C}$ in Fig. 19.4 and using Stokes' theorem, letting the two short legs of the circuit vanish. We thereby obtain

$$[\mathbf{E}'_t] = [\mathbf{E}_t] + [(\mathbf{v}_s \times \mathbf{B})_t] = 0, \tag{19.19e}$$

where $\mathbf{v}_s$ is the velocity of a frame that moves with the surface. Note that only the normal component of the velocity contributes to this expression, so we can replace $\mathbf{v}_s$ by $v_{sn}\mathbf{n}$. The normal component of the electric field, like the tangential component of the magnetic field, can be discontinuous, as there may be surface charge at the interface.

There are also dynamical junction conditions that can be deduced by integrating the laws of momentum conservation (19.12) and energy conservation (19.16) over the pill box and using Gauss's theorem to convert the volume integral of a divergence to a surface integral. The results, naturally, are the requirements that the normal fluxes of momentum $\mathbf{T} \cdot \mathbf{n}$ and energy $\mathbf{F} \cdot \mathbf{n}$ be continuous across the surface. Here $\mathbf{T}$ is the total stress [i.e., the quantity inside the divergence in Eq. (19.12)], and $\mathbf{F}$ is the total energy flux [i.e., the quantity inside the divergence in Eq. (19.16)]; see Eqs. (17.29)–(17.31) and associated discussion. The normal and tangential components of $[\mathbf{T} \cdot \mathbf{n}] = 0$ read

dynamical junction conditions

$$\left[P + \rho(v_n - v_{sn})^2 + \frac{B_t^2}{2\mu_0} \right] = 0, \tag{19.19f}$$

$$\left[\rho(v_n - v_{sn})(\mathbf{v}_t - \mathbf{v}_{st}) - \frac{B_n \mathbf{B}_t}{\mu_0} \right] = 0, \tag{19.19g}$$

where we have omitted the gravitational stress, since it will always be continuous in situations studied in this chapter (no surface layers of mass). Similarly, continuity of the energy flux $[\mathbf{F} \cdot \mathbf{n}] = 0$ reads

$$\left[\left(\frac{1}{2}v^2 + h \right) \rho(v_n - v_{sn}) + \frac{\mathbf{n} \cdot [(\mathbf{E} + \mathbf{v}_s \times \mathbf{B}) \times \mathbf{B}]}{\mu_0} \right] = 0. \tag{19.19h}$$

When the interface has plasma on one side and a vacuum magnetic field on the other, as in devices for magnetic confinement of plasmas (Sec. 19.3), the vacuum electromagnetic field, like that in the plasma, has small time derivatives: $\partial/\partial t \sim v\partial/\partial x^j$. As a result, the vacuum displacement current $\epsilon_0 \partial \mathbf{E}/\partial t$ is very small, and the vacuum Maxwell equations reduce to the same form as those in the MHD plasma but with ρ_e and $\mathbf{j}$ zero. As a result, the boundary conditions at the vacuum-plasma interface (Ex. 19.2) are those discussed above [Eqs. (19.19)], but with ρ and P vanishing on the vacuum side.

EXERCISES

Exercise 19.2 *Example and Derivation: Perfect MHD Boundary Conditions at a Fluid-Vacuum Interface*
When analyzing the stability of configurations for magnetic confinement of a plasma (Sec. 19.5), one needs boundary conditions at the plasma-vacuum interface for the special case of perfect MHD (electrical conductivity idealized as arbitrarily large).

Denote by a tilde ($\tilde{\mathbf{B}}$ and $\tilde{\mathbf{E}}$) the magnetic and electric fields in the vacuum, and reserve non-tilde symbols for quantities on the plasma side of the interface.

(a) Show that the normal-force boundary condition (19.19f) reduces to an equation for the vacuum region's tangential magnetic field:

$$\frac{\tilde{B}_t^2}{2\mu_0} = P + \frac{B_t^2}{2\mu_0}. \tag{19.20a}$$

(b) By combining Eqs. (19.19c) and (19.19g) and noting that $v_n - v_{sn}$ must vanish (why?), and assuming that the magnetic confinement entails surface currents on the interface, show that the normal component of the magnetic field must vanish on both sides of the interface:

$$\tilde{B}_n = B_n = 0. \tag{19.20b}$$

(c) When analyzing energy flow across the interface, it is necessary to know the tangential electric field. On the plasma side $\tilde{\mathbf{E}}_t$ is a secondary quantity fixed by projecting tangentially the relation $\mathbf{E} + \mathbf{v} \times \mathbf{B} = 0$. On the vacuum side it is fixed by the boundary condition (19.19e). By combining these two relations, show that

$$\tilde{\mathbf{E}}_t + \mathbf{v}_{sn} \times \tilde{\mathbf{B}}_t = 0. \tag{19.20c}$$

Exercise 19.3 *Problem: Diffusion of Magnetic Field*
Consider an infinitely long cylinder of plasma with constant electric conductivity, surrounded by vacuum. Assume that the cylinder initially is magnetized uniformly parallel to its length, and assume that the field decays quickly enough that the plasma's inertia keeps it from moving much during the decay (so $\mathbf{v} \simeq 0$).

(a) Show that the reduction of magnetic energy as the field decays is compensated by the Ohmic heating of the plasma plus energy lost to outgoing electromagnetic waves (which will be negligible if the decay is slow).

(b) Compute the approximate magnetic profile after the field has decayed to a small fraction of its original value. Your answer should be expressible in terms of a Bessel function.

Exercise 19.4 *Example: Shock with Transverse Magnetic Field*
Consider a normal shock wave ($\mathbf{v}$ perpendicular to the shock front), in which the magnetic field is parallel to the shock front, analyzed in the shock front's rest frame.

(a) Show that the junction conditions across the shock are the vanishing of all the following quantities:

$$[\rho v] = [P + \rho v^2 + B^2/(2\mu_0)] = [h + v^2/2 + B^2/(\mu_0 \rho)] = [vB] = 0. \tag{19.21}$$

(b) Specialize to a fluid with equation of state $P \propto \rho^\gamma$. Show that these junction conditions predict no compression, $[\rho] = 0$, if the upstream velocity is $v_1 = C_f \equiv \sqrt{(B_1^2/\mu_0 + \gamma P_1)/\rho_1}$. For v_1 greater than this C_f, the fluid gets compressed.

(c) Explain why the result in part (b) means that the speed of sound perpendicular to the magnetic field must be C_f. As we shall see in Sec. 19.7.2, this indeed is the case: C_f is the speed [Eq. (19.77)] of a *fast magnetosonic wave,* the only kind of sound wave that can propagate perpendicular to **B**.

Exercise 19.5 *Problem: Earth's Bow Shock*
The solar wind is a supersonic, hydromagnetic flow of plasma originating in the solar corona. At the radius of Earth's orbit, the wind's density is $\rho \sim 6 \times 10^{-21}\,\mathrm{kg\,m^{-3}}$, its velocity is $v \sim 400\,\mathrm{km\,s^{-1}}$, its temperature is $T \sim 10^5\,\mathrm{K}$, and its magnetic field strength is $B \sim 1\,\mathrm{nT}$.

(a) By balancing the wind's momentum flux with the magnetic pressure exerted by Earth's dipole magnetic field, estimate the radius above Earth at which the solar wind passes through a bow shock (Fig. 17.2).

(b) Consider a strong perpendicular shock at which the magnetic field is parallel to the shock front. Show that the magnetic field strength will increase by the same ratio as the density, when crossing the shock front. Do you expect the compression to increase or decrease as the strength of the field is increased, keeping all of the other flow variables constant?

19.2.4 Magnetic Field and Vorticity

We have already remarked on how the magnetic field and the vorticity are both axial vectors that can be written as the curl of a polar vector and that they satisfy similar transport equations. It is not surprising that they are physically intimately related. To explore this relationship in full detail would take us beyond the scope of this book. However, we can illustrate their interaction by showing how they can create each other. In brief: vorticity can twist a magnetic field, amplifying it; and an already twisted field, trying to untwist itself, can create vorticity.

vorticity-magnetic-field interactions

First, consider a simple vortex through which passes a uniform magnetic field (Fig. 19.2a). If the magnetic Reynolds number is large enough, then the magnetic field is carried with the flow and is wound up like spaghetti on the end of a fork (Fig. 19.2b, continued for a longer time). This process increases the magnetic energy in the vortex, though not the mean flux of the magnetic field. This amplification continues until either the field gradient is large enough that the field decays through Ohmic dissipation, or the field strength is large enough to react back on the flow and stop it from spinning.

Second, consider an irrotational flow containing a twisted magnetic field (Fig. 19.5a). Provided that the magnetic Reynolds number is sufficiently large, the magnetic stress, attempting to untwist the field, will act on the flow and induce vorticity (Fig. 19.5b). We can describe this formally by taking the curl of the equation

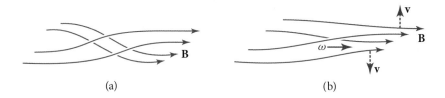

(a)   (b)

FIGURE 19.5 (a) A twisted magnetic field is frozen in an irrotational flow. (b) The field tries to untwist and in the process creates vorticity.

of motion (19.10). Assuming, for simplicity, that the density ρ is constant and the electric conductivity is infinite, we obtain

$$\frac{\partial \boldsymbol{\omega}}{\partial t} - \boldsymbol{\nabla} \times (\mathbf{v} \times \boldsymbol{\omega}) = \frac{\boldsymbol{\nabla} \times [(\boldsymbol{\nabla} \times \mathbf{B}) \times \mathbf{B}]}{\mu_0 \rho}. \tag{19.22}$$

The term on the right-hand side of this equation changes the number of vortex lines threading the fluid, just like the $-\boldsymbol{\nabla} P \times \boldsymbol{\nabla} \rho / \rho^2$ term on the right-hand side of Eq. (14.3). However, because the divergence of the vorticity is zero, any fresh vortex lines that are made must be created as continuous curves that grow out of points or lines where the vorticity vanishes.

19.3	**19.3 Magnetostatic Equilibria**
19.3.1	19.3.1 Controlled Thermonuclear Fusion

GLOBAL POWER DEMAND

motivation for controlled fusion program

We start this section with an oversimplified discussion of the underlying problem. Earth's population has quadrupled over the past century to its present (2016) value of 7.4 billion and is still rising. We consume $\sim 16\,\mathrm{TW}$ of power more or less equally for manufacture, transportation, and domestic use. The power is derived from oil ($\sim 5\,\mathrm{TW}$), coal ($\sim 4\,\mathrm{TW}$), gas ($\sim 4\,\mathrm{TW}$), nuclear (fission) reactors ($\sim 1\,\mathrm{TW}$), hydroelectric turbines ($\sim 1\,\mathrm{TW}$), and alternative sources, such as solar, wind, wave, and biomass ($\sim 1\,\mathrm{TW}$). The average power consumption is $\sim 2\,\mathrm{kW}$ per person, with Canada and the United States in the lead, consuming $\sim 10\,\mathrm{kW}$ per person. Despite conservation efforts, the demand for energy still appears to be rising.

Meanwhile, the burning of coal, oil, and gas produces carbon dioxide at a rate of $\sim 1\,\mathrm{Gg\,s^{-1}}$, about 10% of the total transfer rate in Earth's biomass–atmosphere–ocean *carbon cycle*. This disturbance of the carbon-cycle equilibrium has led to an increase in the atmospheric concentration of carbon dioxide by about a third over the past century and it is currently growing at an average rate of about a half percent per year. There is strong evidence to link this increase in carbon dioxide and other greenhouse gases to climate change, as exemplified by an increase in the globally averaged mean temperature of $\sim 1\,\mathrm{K}$ over the past century. Given the long time constants associated with the three components of the carbon cycle, future projections of climate change

958   Chapter 19. Magnetohydrodynamics

are alarming. These considerations strongly motivate the rapid deployment of low-carbon sources of power—renewables and nuclear—and conservation.

THERMONUCLEAR FUSION

For more than 60 years, plasma physicists have striven to address this problem by releasing nuclear energy in a controlled, peaceful manner through confining plasma at a temperature in excess of 100 million degrees using strong magnetic fields. In the most widely studied scheme, deuterium and tritium combine according to the reaction

$$d + t \rightarrow \alpha + n + 22.4 \text{ MeV}. \tag{19.23}$$

d, t fusion reaction

The energy release is equivalent to $\sim$400 TJ kg^{-1}. The fast neutrons can be absorbed in a surrounding blanket of lithium, and the heat can then be used to drive a generator.

PLASMA CONFINEMENT

At first this task seemed quite simple. However, it eventually became clear that it is very difficult to confine hot plasma with a magnetic field, because most confinement geometries are unstable. In this book we restrict our attention to a few simple confinement devices, emphasizing the one that is the basis of most modern efforts, the *tokamak*.[1] In this section, we treat equilibrium configurations; in Sec. 19.5, we consider their stability.

In our discussions of both equilibrium and stability, we treat the plasma as a magnetized fluid in the MHD approximation. At first sight, treating the plasma as a fluid might seem rather unrealistic, because we are dealing with a dilute gas of ions and electrons that undergo infrequent Coulomb collisions. However, as we discuss in Sec. 20.5.2 and justify in Chaps. 22 and 23, collective effects produce a sufficiently high effective collision frequency to make the plasma behave like a fluid, so MHD is usually a good approximation for describing these equilibria and their rather slow temporal evolution.

Let us examine some numbers that characterize the regime in which a successful controlled-fusion device must operate.

PLASMA PRESSURE

The ratio of plasma pressure to magnetic pressure

$$\beta \equiv \frac{P}{B^2/(2\mu_0)} \tag{19.24}$$

pressure ratio β for controlled fusion

plays a key role. For the magnetic field to have any chance of confining the plasma, its pressure must exceed that of the plasma (i.e., β must be less than one). The most successful designs achieve $\beta \sim 0.2$. The largest field strengths that can be safely

---

1. Originally proposed in the Soviet Union by Andrei Sakharov and Igor Tamm in 1950. The word is a Russian abbreviation for "toroidal magnetic field."

sustained in the laboratory are $B \sim 10\,\text{T} = 100\,\text{kG}$, so $\beta \lesssim 0.2$ limits the gas pressure to $P \lesssim 10^7\,\text{Pa} \sim 100$ atmospheres.

LAWSON CRITERION

Plasma fusion can only be economically feasible if more power is released by nuclear reactions than is lost to radiative cooling. Both heating and cooling are proportional to the square of the number density of hydrogen ions, n^2. However, while the radiative cooling rate increases comparatively slowly with temperature, the nuclear reaction rate increases very rapidly. (This is because, as the mean energy of the ions increases, the number of ions in the Maxwellian tail of the distribution function that are energetic enough to penetrate the Coulomb barrier increases exponentially.) Thus for the rate of heat production to greatly exceed the rate of cooling, the temperature need only be modestly higher than that required for the rates to be equal—which is a minimum temperature essentially fixed by atomic and nuclear physics. In the case of a d-t plasma, this is $T_{\min} \sim 10^8\,\text{K}$. The maximum hydrogen density that can be confined is therefore $n_{\max} = P/(2k_B T_{\min}) \sim 3 \times 10^{21}\,\text{m}^{-3}$. (The factor 2 comes from the electrons, which produce the same pressure as the ions.)

Now, if a volume V of plasma is confined at a given number density n and temperature $T_{\min}$ for a time τ, then the amount of nuclear energy generated will be proportional to $n^2 V \tau$, while the energy to heat the plasma up to $T_{\min}$ is $\propto n V$. Therefore, there is a minimum value of the product $n\tau$ that must be attained before net energy is produced. This condition is known as the *Lawson criterion*. Numerically, the plasma must be confined for

$$\tau \sim (n/10^{20}\,\text{m}^{-3})^{-1}\,\text{s}, \tag{19.25}$$

typically ~ 30 ms. The sound speed at these temperatures is $\sim 1 \times 10^6\,\text{m s}^{-1}$, and so an unconfined plasma would hit the few-meter-sized walls of the vessel in which it is held in a few μs. Therefore, the magnetic confinement must be effective for typically 10^4–10^5 dynamical timescales (sound-crossing times). It is necessary that the plasma be confined and confined well if we want to build a viable fusion reactor.

19.3.2 Z-Pinch

Before discussing plasma confinement by tokamaks, we describe a simpler confinement geometry known as the *Z-pinch* and often called the Bennett pinch (Fig. 19.6a). In a Z-pinch, electric current is induced to flow along a cylinder of plasma. This current creates a toroidal magnetic field whose tension prevents the plasma from expanding radially, much like hoops on a barrel prevent it from exploding. Let us assume that the cylinder has a radius R and is surrounded by vacuum.

Now, in static equilibrium we must balance the plasma pressure gradient by a Lorentz force:

$$\nabla P = \mathbf{j} \times \mathbf{B}. \tag{19.26}$$

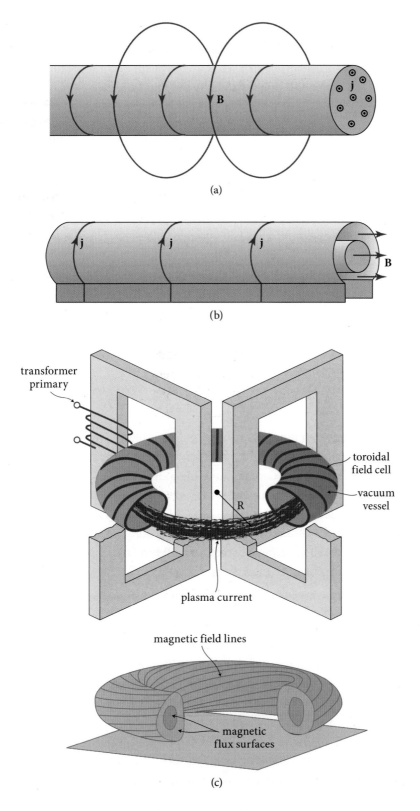

FIGURE 19.6 (a) The Z-pinch. (b) The Θ-pinch. (c) The tokamak.

(Gravitational forces can safely be ignored.) Equation (19.26) implies immediately that $\mathbf{B} \cdot \nabla P = \mathbf{j} \cdot \nabla P = 0$, so both the magnetic field and the current density lie on constant pressure (or *isobaric*) surfaces. An equivalent version of the force-balance equation (19.26), obtained using Eq. (19.15) and Fig. 19.3, is

$$\frac{d}{d\varpi}\left(P + \frac{B^2}{2\mu_0}\right) = -\frac{B^2}{\mu_0 \varpi}, \tag{19.27}$$

where ϖ is the radial cylindrical coordinate. Equation (19.27) exhibits the balance between the gradient of plasma and magnetic pressure on the left, and the magnetic tension (the "hoop force") on the right. Treating it as a differential equation for B^2 and integrating it, assuming that P falls to zero at the surface of the column, we obtain for the surface magnetic field

$$B^2(R) = \frac{4\mu_0}{R^2} \int_0^R P \varpi \, d\varpi. \tag{19.28}$$

We can reexpress the surface toroidal field in terms of the total current flowing along the plasma as $B(R) = \mu_0 I/(2\pi R)$ (Ampère's law); and assuming that the plasma is primarily hydrogen (so its ion density n and electron density are equal), we can write the pressure as $P = 2n k_B T$. Inserting these expressions into Eq. (19.28), integrating, and solving for the current, we obtain

$$I = \left(\frac{16\pi N k_B T}{\mu_0}\right)^{1/2}, \tag{19.29}$$

where N is the number of ions per unit length. For a column of plasma with diameter $2R \sim 1$ m, hydrogen density $n \sim 10^{20}$ m^{-3}, and temperature $T \sim 10^8$ K, Eq. (19.29) indicates that currents ~ 1 MA are required for confinement.

The most promising Z-pinch experiments to date have been carried out at Sandia National Laboratories in Albuquerque, New Mexico. The experimenters have impulsively compressed a column of gas into a surprisingly stable cylinder of diameter ~ 1 mm for a time ~ 100 ns, using a current of ~ 30 MA from giant capaciter banks. A transient field of ~ 1 kT was created.

19.3.3 Θ-Pinch

There is a complementary equilibrium for a cylindrical plasma, in which the magnetic field lies parallel to the axis and the current density encircles the cylinder (Fig. 19.6b). This configuration is called the Θ-*pinch*. It is usually established by making a cylindrical metal tube with a small gap, so that current can flow around it as shown in the figure. The tube is filled with cold plasma, and then the current is turned on quickly, producing a quickly growing longitudinal field in the tube (as in a solenoid). Since the plasma is highly conducting, the field lines cannot penetrate the plasma column but instead exert a pressure on its surface, causing it to shrink radially and rapidly. The plasma heats up due to both the radial work done on it and Ohmic heating. Equilibrium is established when the plasma's pressure P balances the magnetic pressure $B^2/(2\mu_0)$ at the plasma's surface.

Despite early promise, interest in Θ-pinches has waned, and mirror machines (Sec. 19.5.3) have become more popular.

19.3.4 Tokamak

19.3.4

tokamak configuration for plasma confinement

One of the problems with the Θ- and Z-pinches (and we shall find other problems below!) is that they have ends through which plasma can escape. This is readily addressed by replacing the cylinder with a torus. The most stable geometry, called the *tokamak*, combines features of both Z- and Θ-pinches; see Fig. 19.6c. If we introduce spherical coordinates (r, θ, ϕ), then magnetic field lines and currents that lie in an r-θ plane (orthogonal to $\mathbf{e}_\phi$) are called *poloidal*, whereas their ϕ components are called *toroidal*. In a tokamak, the toroidal magnetic field is created by external poloidal current windings. However, the poloidal field is mostly created as a consequence of toroidal current induced to flow in the plasma torus. The resulting net field lines wrap around the plasma torus in a helical manner, defining a magnetic surface on which the pressure is constant. The number of poloidal transits around the torus during one toroidal transit is denoted $\iota/(2\pi)$; ι is called the *rotational transform* and is a property of the magnetic surface on which the field line resides. If $\iota/(2\pi)$ is a rational number, then the field line will close after a finite number of circuits. However, in general, $\iota/(2\pi)$ will not be rational, so a single field line will cover the whole magnetic surface ergodically. This allows the plasma to spread over the whole surface rapidly. The rotational transform is a measure of the toroidal current flowing inside the magnetic surface and of course increases as we move out from the innermost magnetic surface, while the pressure decreases.

The best performers to date (2016) include the MIT Alcator C-Mod ($n \sim 2 \times 10^{20}\,\text{m}^{-3}$, $T \sim 35\,\text{MK}$, $B \sim 6\,\text{T}$, $\tau \sim 2\,\text{s}$), and the larger Joint European Torus or JET (Keilhacker and the JET team, 1998). JET generated 16 MW of nuclear power with $\tau \sim 1\,\text{s}$ by burning d-t fuel, but its input power was $\sim$25 MW and so, despite being a major step forward, JET fell short of achieving "break even."

JET

The largest device, currently under construction, is the International Thermonuclear Experimental Reactor (ITER) (whose acronym means "journey" in Latin). ITER is a tokamak-based experimental fusion reactor being constructed in France by a large international consortium (see http://www.iter.org/). The outer diameter of the device is $\sim$20 m, and the maximum magnetic field produced by its superconducting magnets will be $\sim$14 T. Its goal is to use d-t fuel to convert an input power of $\sim$50 MW into an output power of $\sim$500 MW, sustained for $\sim$3,000 s.[2] However, many engineering, managerial, financial, and political challenges remain to be addressed before mass production of economically viable, durable, and safe fusion reactors can begin.

ITER

---

2. Note that it would require of order 10,000 facilities with ITER's projected peak power operating continuously to supply, say, one-third of the current global power demand.

Exercise 19.6 *Problem: Strength of Magnetic Field in a Magnetic Confinement Device*
The currents that are sources for strong magnetic fields have to be held in place by solid conductors. Estimate the limiting field that can be sustained using normal construction materials.

Exercise 19.7 *Problem: Force-Free Equilibria*
In an equilibrium state of a very low-β plasma, the plasma's pressure force density $-\nabla P$ is ignorably small, and so the Lorentz force density $\mathbf{j} \times \mathbf{B}$ must vanish [Eq. (19.10)]. Such a plasma is said to be "force-free." As a result, the current density is parallel to the magnetic field, so $\nabla \times \mathbf{B} = \alpha \mathbf{B}$. Show that α must be constant along a field line, and that if the field lines eventually travel everywhere, then α must be constant everywhere.

Exercise 19.8 *Example and Challenge: Spheromak*
Another magnetic confinement device which brings out some important principles is the *spheromak*. Spheromaks can be made in the laboratory (Bellan, 2000) and have also been proposed as the basis of a fusion reactor.[3] It is simplest to consider a spheromak in the limit when the plasma pressure is ignorable (low β) and the magnetic field distribution is force-free (Ex. 19.7). We just describe the simplest example of this regime.

(a) As in the previous exercise, assume that α is constant everywhere and, without loss of generality, set it equal to unity. Show that the magnetic field—and also the current density and vector potential, adopting the Coulomb gauge—satisfy the vector Helmholtz equation: $\nabla^2 \mathbf{B} + \alpha^2 \mathbf{B} = 0$.

(b) Introduce a scalar χ such that $\mathbf{B} = \alpha \mathbf{r} \times \nabla \chi + \nabla \times (\mathbf{r} \times \nabla \chi)$, with $\mathbf{r}$ the radial vector pointing out of the spheromak's center, and show that χ satisfies the scalar Helmholtz equation: $\nabla^2 \chi + \alpha^2 \chi = 0$.

(c) The Helmholtz equation in part (b) separates in spherical coordinates (r, θ, ϕ). Show that it has a nonsingular solution $\chi = j_l(\alpha r) Y_{lm}(\theta, \phi)$, where $j_l(\alpha r)$ is a spherical Bessel function, and $Y_{lm}(\theta, \phi)$ is a spherical harmonic. Evaluate this for the simplest example, the spheromak, with $l = 2$ and $m = 0$.

(d) Calculate expressions for the associated magnetic field in part (c) and either sketch or plot it.

(e) Show that the magnetic field's radial component vanishes on the surface of a sphere of radius equal to the first zero of j_l. Hence explain why a spheromak may be confined in a conducting sphere or, alternatively, by a current-free field that is uniform at large distance.

---

3. Spheromak configurations of magnetic fields have even been associated with ball lightning!

Exercise 19.9 *Problem: Magnetic Helicity*

(a) A physical quantity that turns out to be useful in describing the evolution of magnetic fields in confinement devices is *magnetic helicity*. This is defined by $H = \int dV\, \mathbf{A} \cdot \mathbf{B}$, where $\mathbf{A}$ is the vector potential, and the integral should be performed over all the space visited by the field lines to remove a dependence on the electromagnetic gauge. Compute the partial derivative of $\mathbf{A}$ and $\mathbf{B}$ with respect to time to show that H is conserved if $\mathbf{E} \cdot \mathbf{B} = 0$.

(b) The helicity H primarily measures the topological linkage of the magnetic field lines. Therefore, it should not be a surprise that H turns out to be relatively well-preserved, even when the plasma is losing energy through resistivity and radiation. To discuss magnetic helicity properly would take us too far into the domain of classical electromagnetic theory, but some indication of its value follows from computing it for the case of two rings of magnetic field containing fluxes Φ_1 and Φ_2. Start with the rings quite separate, and show that $H = 0$. Then allow the rings to be linked while not sharing any magnetic field lines. Show that now $H = 2\Phi_1\Phi_2$ and that $dH/dt = -2\int dV\, \mathbf{E} \cdot \mathbf{B}$.[4]

19.4 Hydromagnetic Flows

We now introduce fluid motions into our applications of magnetohydrodynamics. Specifically, we explore a simple class of *stationary hydromagnetic flows*: the flow of an electrically conducting fluid along a duct of constant cross section perpendicular to a uniform magnetic field B_0 (see Fig. 19.7). This is sometimes known as *Hartmann flow*. The duct has two insulating walls (top and bottom, as shown in the figure), separated by a distance $2a$ that is much smaller than the separation of short side walls, which are electrically conducting.

To relate Hartmann flow to magnetic-free Poiseuille flow (viscous, laminar flow between plates; Ex. 13.18), we reinstate the viscous force in the equation of motion. For simplicity we assume that the time-independent flow ($\partial \mathbf{v}/\partial t = 0$) has traveled sufficiently far down the duct (x direction) to have reached a z-independent form, so $\mathbf{v} \cdot \nabla \mathbf{v} = 0$ and $\mathbf{v} = \mathbf{v}(y, z)$. We also assume that gravitational forces are unimportant. Then the flow's equation of motion takes the form

$$\nabla P = \mathbf{j} \times \mathbf{B} + \eta \nabla^2 \mathbf{v}, \tag{19.30}$$

where $\eta = \rho \nu$ is the coefficient of dynamical viscosity. The magnetic (Lorentz) force $\mathbf{j} \times \mathbf{B}$ alters the balance between the Poiseuille flow's viscous force $\eta \nabla^2 \mathbf{v}$ and the

---

4. For further discussion, see Bellan (2000).

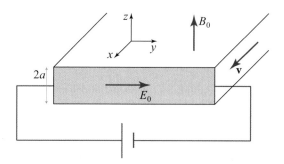

FIGURE 19.7 Hartmann flow with average speed v along a duct of thickness $2a$, perpendicular to an applied magnetic field of strength B_0. The short side walls are conducting and the two long horizontal walls are electrically insulating.

four versions of Hartmann flow

pressure gradient ∇P. The details of that altered balance and the resulting magnetic-influenced flow depend on how the walls are connected electrically. Let us consider four possibilities chosen to bring out the essential physics.

ELECTROMAGNETIC BRAKE

We short circuit the electrodes, so a current $\mathbf{j}$ can flow (Fig. 19.8a). The magnetic field lines are partially dragged by the fluid, bending them (as embodied in $\nabla \times \mathbf{B} = \mu_0 \mathbf{j}$), so they can exert a decelerating tension force $\mathbf{j} \times \mathbf{B} = (\nabla \times \mathbf{B}) \times \mathbf{B}/\mu_0 = \mathbf{B} \cdot \nabla \mathbf{B}/\mu_0$ on the flow (Fig. 19.3b). This configuration is an electromagnetic brake. The pressure gradient, which is trying to accelerate the fluid, is balanced by the magnetic tension and viscosity. The work being done (per unit volume) by the pressure gradient, $\mathbf{v} \cdot (-\nabla P)$, is converted into heat through viscous and Ohmic dissipation.

MHD POWER GENERATOR

The MHD power generator is similar to the electromagnetic brake except that an external load is added to the circuit (Fig. 19.8b). Useful power can be extracted from the flow. Variants of this configuration were developed in the 1970s–1990s in many countries, but they are currently not seen to be economically competitive with other power-generation methods.

FLOW METER

When the electrodes in a flow meter are on an open circuit, the induced electric field produces a measurable potential difference across the duct (Fig. 19.8c). This voltage will increase monotonically with the rate of flow of fluid through the duct and therefore can provide a measurement of the flow.

ELECTROMAGNETIC PUMP

Finally, we can attach a battery to the electrodes and allow a current to flow (Figs. 19.7 and 19.8d). This produces a Lorentz force which either accelerates or decelerates the

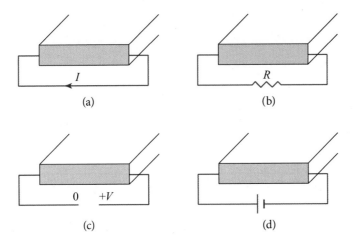

FIGURE 19.8 Four variations on Hartmann flow: (a) Electromagnetic brake. (b) MHD power generator. (c) Flow meter. (d) Electromagnetic pump.

flow, depending on the direction of the magnetic field. This method can be used to pump liquid sodium coolant around a nuclear reactor. It has also been proposed as a means of spacecraft propulsion in interplanetary space.

We consider in more detail two limiting cases of the electromagnetic pump. When there is a constant pressure gradient $Q = -dP/dx$ but no magnetic field, a flow with modest Reynolds number will be approximately laminar with velocity profile (Ex. 13.18):

$$v_x(z) = \frac{Qa^2}{2\eta}\left[1 - \left(\frac{z}{a}\right)^2\right],\tag{19.31}$$

where a is the half-width of the channel. This flow is the 1-dimensional version of the Poiseuille flow in a pipe, such as a blood artery, which we studied in Sec. 13.7.6 [cf. Eq. (13.82a)]. Now suppose that uniform electric and magnetic fields E_0 and B_0 are applied along the $\mathbf{e}_y$ and $\mathbf{e}_z$ directions, respectively (Fig. 19.7). The resulting magnetic force $\mathbf{j} \times \mathbf{B}$ can either reinforce or oppose the fluid's motion. When the applied magnetic field is small, $B_0 \ll E_0/v_x$, the effect of the magnetic force will be similar to that of the pressure gradient, and Eq. (19.31) must be modified by replacing $Q \equiv -dP/dx$ by $-dP/dx + j_y B_z = -dP/dy + \kappa_e E_0 B_0$. [Here $j_y = \kappa_e(E_y - v_x B_z) \simeq \kappa_e E_0$.]

If the strength of the magnetic field is increased sufficiently, then the magnetic force will dominate the viscous force, except in thin boundary layers near the walls. Outside the boundary layers, in the bulk of the flow, the velocity will adjust so that the electric field vanishes in the rest frame of the fluid (i.e., $v_x = E_0/B_0$). In the boundary layers v_x drops sharply from E_0/B_0 to zero at the walls, and correspondingly, a strong viscous force $\eta\nabla^2\mathbf{v}$ develops. Since the pressure gradient ∇P must be essentially the same in the boundary layer as in the adjacent bulk flow and thus cannot balance this

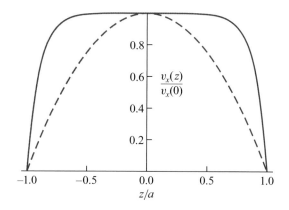

FIGURE 19.9 Velocity profiles [Eq. (19.35)] for flow in an electromagnetic pump of width $2a$ with small and large Hartmann number scaled to the velocity at the center of the channel. The dashed curve shows the almost parabolic profile for Ha $= 0.1$ [Eq. (19.31)]. The solid curve shows the almost flat-topped profile for Ha $= 10$.

large viscous force, it must be balanced instead by the magnetic force: $\mathbf{j} \times \mathbf{B} + \eta \nabla^2 \mathbf{v} = 0$ [Eq. (19.30)], with $\mathbf{j} = \kappa_e (\mathbf{E} + \mathbf{v} \times \mathbf{B}) \sim \kappa_e v_x B_0 \mathbf{e}_y$. We thereby see that the thickness of the boundary layer is given by

$$\delta_H \sim \left(\frac{\eta}{\kappa_e B_0^2} \right)^{1/2}. \tag{19.32}$$

This suggests a new dimensionless number to characterize the flow,

Hartmann number

$$\boxed{\text{Ha} = \frac{a}{\delta_H} = B_0 a \left(\frac{\kappa_e}{\eta} \right)^{1/2},} \tag{19.33}$$

called the *Hartmann number*. The square of the Hartmann number, Ha^2, is essentially the ratio of the magnetic force, $|\mathbf{j} \times \mathbf{B}| \sim \kappa_e v_x B_0^2$, to the viscous force, $\sim \eta v_x / a^2$, assuming a lengthscale a rather than δ_H for variations of the velocity.

The detailed velocity profile $v_x(z)$ away from the vertical side walls is computed in Ex. 19.10 and is shown for low and high Hartmann numbers in Fig. 19.9. Notice that at low Hartmann numbers, the plotted profile is nearly parabolic as expected, and at high Hartmann numbers it consists of boundary layers at $z \sim -a$ and $z \sim a$, and a uniform flow in between.

In Exs. 19.11 and 19.12 we explore two important astrophysical examples of hydromagnetic flow: the magnetosphere of a rotating, magnetized star or other body, and the solar wind.

EXERCISES

Exercise 19.10 *Example: Hartmann Flow*
Compute the velocity profile of a conducting fluid in a duct of thickness $2a$ perpendicular to externally generated, uniform electric and magnetic fields ($E_0 \mathbf{e}_y$

and $B_0\mathbf{e}_z$) as shown in Fig. 19.7. Away from the vertical sides of the duct, the velocity v_x is just a function of z, and the pressure can be written in the form $P = -Qx + p(z)$, where Q is the longitudinal pressure gradient.

(a) Show that the velocity field satisfies the differential equation

$$\frac{d^2v_x}{dz^2} - \frac{\kappa_e B_0^2}{\eta} v_x = -\frac{(Q + \kappa_e B_0 E_0)}{\eta}. \qquad (19.34)$$

(b) Impose suitable boundary conditions at the bottom and top walls of the channel, and solve this differential equation to obtain the following velocity field:

$$v_x = \frac{Q + \kappa_e B_0 E_0}{\kappa_e B_0^2} \left[1 - \frac{\cosh(\mathrm{Ha}\, z/a)}{\cosh(\mathrm{Ha})} \right], \qquad (19.35)$$

where Ha is the Hartmann number (see Fig. 19.9).

Exercise 19.11 **Example: Rotating Magnetospheres*
Many self-gravitating cosmic bodies are both spinning and magnetized. Examples are Earth, the Sun, black holes surrounded by highly conducting accretion disks (which hold a magnetic field on the hole), neutron stars (pulsars), and magnetic white dwarfs. As a consequence of the magnetic field's spin-induced motion, large electric fields are produced outside the rotating body. The divergence of these electric fields must be balanced by free electric charge, which implies that the region around the body cannot be a vacuum. It is usually filled with plasma and is called a *magnetosphere*. MHD provides a convenient formalism for describing the structure of this magnetosphere. Magnetospheres are found around most planets and stars. Magnetospheres surrounding neutron stars and black holes are believed to be responsible for the emissions from pulsars and quasars.

As a model of a rotating magnetosphere, consider a magnetized and infinitely conducting star, spinning with angular frequency Ω_*. Suppose that the magnetic field is stationary and axisymmetric with respect to the spin axis and that the magnetosphere, like the star, is perfectly conducting.

(a) Show that the azimuthal component E_ϕ of the magnetospheric electric field must vanish if the magnetic field is to be stationary. Hence show that there exists a function $\mathbf{\Omega}(\mathbf{r})$ that must be parallel to $\mathbf{\Omega}_*$ and must satisfy

$$\mathbf{E} = -(\mathbf{\Omega} \times \mathbf{r}) \times \mathbf{B}. \qquad (19.36)$$

Show that if the motion of the magnetosphere's conducting fluid is simply a rotation, then its angular velocity must be $\mathbf{\Omega}$.

(b) Use the induction equation (magnetic-field transport law) to show that

$$(\mathbf{B} \cdot \nabla)\mathbf{\Omega} = 0. \qquad (19.37)$$

(c) Use the boundary condition at the surface of the star to show that the magnetosphere corotates with the star (i.e., $\mathbf{\Omega} = \mathbf{\Omega}_*$). This is known as *Ferraro's law of isorotation.*

Exercise 19.12 *Example: Solar Wind*

The solar wind is a magnetized outflow of plasma that emerges from the solar corona. We make a simple model of it by generalizing the results from the previous exercise that emphasize its hydromagnetic features while ignoring gravity and thermal pressure (which are also important in practice). We consider stationary, axisymmetric motion in the equatorial plane, and idealize the magnetic field as having the form $B_r(r)$, $B_\phi(r)$. (If this were true at all latitudes, the Sun would have to contain magnetic monopoles!)

(a) Use the results from the previous exercise plus the perfect MHD relation, $\mathbf{E} = -\mathbf{v} \times \mathbf{B}$, to argue that the velocity field can be written in the form

$$\mathbf{v} = \frac{\kappa \mathbf{B}}{\rho} + (\mathbf{\Omega} \times \mathbf{r}), \tag{19.38}$$

where κ and $\mathbf{\Omega}$ are constant along a field line. Interpret this relation kinematically.

(b) Resolve the velocity and the magnetic field into radial and azimuthal components, v_r, v_ϕ, and B_r, B_ϕ, and show that $\rho v_r r^2$ and $B_r r^2$ are constant.

(c) Use the induction equation to show that

$$\frac{v_r}{v_\phi - \Omega r} = \frac{B_r}{B_\phi}. \tag{19.39}$$

(d) Use the equation of motion to show that the specific angular momentum, including both the mechanical and the magnetic contributions,

$$\Lambda = r v_\phi - \frac{r B_r B_\phi}{\mu_0 \rho v_r}, \tag{19.40}$$

is constant.

(e) Combine Eqs. (19.39) and (19.40) to argue that

$$v_\phi = \frac{\Omega r [M_A^2 \Lambda / (\Omega r^2) - 1]}{M_A^2 - 1}, \tag{19.41}$$

where

$$M_A = \frac{v_r (\mu_0 \rho)^{1/2}}{B_r} \tag{19.42}$$

is the Alfvén Mach number [cf. Eq. (19.73)]. Show that the solar wind must pass through a critical point (Sec. 17.3.2) where its radial speed equals the Alfvén speed.

(f) Suppose that the critical point in part (e) is located at 20 solar radii and that $v_r = 100 \text{ km s}^{-1}$, $v_\phi = 20 \text{ km s}^{-1}$, and $B_r = 400 \text{ nT}$ there. Calculate values for Λ, B_ϕ, ρ and use these to estimate the fraction of the solar mass and the solar angular momentum that could have been carried off by the solar wind over its ~5 Gyr lifetime. Comment on your answer.

(g) Suppose that there is no poloidal current density so that $B_\phi \propto r^{-1}$, and deduce values for the velocity and the magnetic field in the neighborhood of Earth where $r \sim 200$ solar radii. Sketch the magnetic field and the path of the solar wind as it flows from r_c to Earth.

The solar mass, radius and rotation period are $\sim 2 \times 10^{30}$ kg, $\sim 7 \times 10^8$ m, and ~ 25 d.

19.5 Stability of Magnetostatic Equilibria

19.5

Having used the MHD equation of motion to analyze some simple flows, we return to the problem of magnetic confinement and demonstrate a procedure to analyze the stability of the confinement's magnetostatic equilibria. We first perform a straightforward linear perturbation analysis about equilibrium, obtaining an eigenequation for the perturbation's oscillation frequencies ω. For sufficiently simple equilibria, this eigenequation can be solved analytically, but most equilibria are too complex for this approach, so the eigenequation must be solved numerically or by other approximation techniques. This is rather similar to the task one faces in attempting to solve the Schrödinger equation for multi-electron atoms. It will not be a surprise to learn that variational methods are especially practical and useful, and we develop a suitable formalism for them.

We develop the perturbation theory, eigenequation, and variational formalism in some detail not only because of their importance for the stability of magnetostatic equilibria, but also because essentially the same techniques (with different equations) are used in studying the stability of other equilibria. One example is the oscillations and stability of stars, in which the magnetic field is unimportant, while self-gravity is crucial [see, e.g., Shapiro and Teukolsky (1983, Chap. 6), and Sec. 16.2.4 of this book, on helioseismology]. Another example is the oscillations and stability of elastostatic equilibria, in which **B** is absent but shear stresses are important (Secs. 12.3 and 12.4).

19.5.1 Linear Perturbation Theory

19.5.1

Consider a perfectly conducting, isentropic fluid at rest in equilibrium, with pressure gradients that balance magnetic forces—for example, the Z-pinch, Θ-pinch, and tokamak configurations of Fig. 19.6. For simplicity, we ignore gravity. (This is usually justified in laboratory situations.) The equation of equilibrium then reduces to

$$\nabla P = \mathbf{j} \times \mathbf{B} \qquad (19.43)$$

[Eq. (19.10)].

equation of magnetostatic equilibrium without gravity

We now perturb slightly about our chosen equilibrium and ignore the (usually negligible) effects of viscosity and magnetic-field diffusion, so $\eta = \rho\nu \simeq 0$ and $\kappa_e \simeq \infty$. It is useful and conventional to describe the perturbations in terms of two different types of quantities: (i) The change in a quantity (e.g., the fluid density) moving with the fluid, which is called a *Lagrangian* perturbation and is denoted by the symbol Δ (e.g., the lagrangian density perturbation $\Delta\rho$). (ii) The change at a fixed location in

Lagrangian and Eulerian perturbations and how they are related space, which is called an *Eulerian* perturbation and is denoted by the symbol δ (e.g., the Eulerian density perturbation $\delta\rho$). The fundamental variable used in the theory is the fluid's *Lagrangian displacement* $\Delta\mathbf{x} \equiv \boldsymbol{\xi}(\mathbf{x}, t)$ (i.e., the change in a fluid element's location). A fluid element whose location is $\mathbf{x}$ in the unperturbed equilibrium is moved to location $\mathbf{x} + \boldsymbol{\xi}(\mathbf{x}, t)$ by the perturbations. From their definitions, one can see that the Lagrangian and Eulerian perturbations are related by

$$\boxed{\Delta = \delta + \boldsymbol{\xi} \cdot \nabla \quad (\text{e.g., } \Delta\rho = \delta\rho + \boldsymbol{\xi} \cdot \nabla\rho).} \tag{19.44}$$

Now consider the transport law for the magnetic field in the limit of infinite electrical conductivity, $\partial\mathbf{B}/\partial t = \nabla \times (\mathbf{v} \times \mathbf{B})$ [Eq. (19.6)]. To linear order, the fluid velocity is $\mathbf{v} = \partial\boldsymbol{\xi}/\partial t$. Inserting this into the transport law and setting the full magnetic field at fixed $\mathbf{x}$ and t equal to the equilibrium field plus its Eulerian perturbation $\mathbf{B} \rightarrow \mathbf{B} + \delta\mathbf{B}$, we obtain $\partial\delta\mathbf{B}/\partial t = \nabla \times [(\partial\boldsymbol{\xi}/\partial t) \times (\mathbf{B} + \delta\mathbf{B})]$. Linearizing in the perturbation, and integrating in time, we obtain for the Eulerian perturbation of the magnetic field:

$$\delta\mathbf{B} = \nabla \times (\boldsymbol{\xi} \times \mathbf{B}). \tag{19.45a}$$

Since the current and the field are related, in general, by the linear equation $\mathbf{j} = \nabla \times \mathbf{B}/\mu_0$, their Eulerian perturbations are related in the same way:

$$\delta\mathbf{j} = \nabla \times \delta\mathbf{B}/\mu_0. \tag{19.45b}$$

In the equation of mass conservation, $\partial\rho/\partial t + \nabla \cdot (\rho\mathbf{v}) = 0$, we replace the density by its equilibrium value plus its Eulerian perturbation, $\rho \rightarrow \rho + \delta\rho$, and replace $\mathbf{v}$ by $\partial\boldsymbol{\xi}/\partial t$. We then linearize in the perturbation to obtain

$$\delta\rho + \rho\nabla \cdot \boldsymbol{\xi} + \boldsymbol{\xi} \cdot \nabla\rho = 0. \tag{19.45c}$$

The lagrangian density perturbation, obtained from this via Eq. (19.44), is

$$\Delta\rho = -\rho\nabla \cdot \boldsymbol{\xi}. \tag{19.45d}$$

We assume that, as it moves, the fluid gets compressed or expanded adiabatically (no Ohmic or viscous heating, or radiative cooling). Then the Lagrangian change of pressure ΔP in each fluid element (moving with the fluid) is related to the Lagrangian change of density by

$$\Delta P = \left(\frac{\partial P}{\partial \rho}\right)_s \Delta\rho = \frac{\gamma P}{\rho}\Delta\rho = -\gamma P\nabla \cdot \boldsymbol{\xi}, \tag{19.45e}$$

where γ is the fluid's adiabatic index (ratio of specific heats), which might or might not be independent of position in the equilibrium configuration. Correspondingly, the Eulerian perturbation of the pressure (perturbation at fixed location) is

$$\delta P = \Delta P - (\boldsymbol{\xi} \cdot \nabla)P = -\gamma P(\nabla \cdot \boldsymbol{\xi}) - (\boldsymbol{\xi} \cdot \nabla)P. \tag{19.45f}$$

This is the pressure perturbation that appears in the fluid's equation of motion.

By replacing $\mathbf{v} \to \partial \boldsymbol{\xi}/\partial t$, $P \to P + \delta P$, $\mathbf{B} \to \mathbf{B} + \delta \mathbf{B}$, and $\mathbf{j} \to \mathbf{j} + \delta \mathbf{j}$ in the fluid's equation of motion (19.10) and neglecting gravity, and by then linearizing in the perturbation, we obtain

$$\rho \frac{\partial^2 \boldsymbol{\xi}}{\partial t^2} = \mathbf{j} \times \delta \mathbf{B} + \delta \mathbf{j} \times \mathbf{B} - \boldsymbol{\nabla} \delta P \equiv \hat{\mathbf{F}}[\boldsymbol{\xi}]. \qquad (19.46)$$

dynamical equation for adiabatic perturbations of a magnetostatic equilibrium

Here $\hat{\mathbf{F}}$ is a real, linear differential operator, whose form one can deduce by substituting expressions (19.45a), (19.45b), and (19.45f) for $\delta \mathbf{B}$, $\delta \mathbf{j}$, and δP, and by substituting $\boldsymbol{\nabla} \times \mathbf{B}/\mu_0$ for $\mathbf{j}$. Performing these substitutions and carefully rearranging the terms, we eventually convert the operator $\hat{\mathbf{F}}$ into the following form, expressed in slot-naming index notation:

$$\hat{F}_i[\boldsymbol{\xi}] = \left\{ \left[(\gamma - 1)P + \frac{B^2}{2\mu_0} \right] \xi_{k;k} + \frac{B_j B_k}{\mu_0} \xi_{j;k} \right\}_{;i}$$
$$+ \left[\left(P + \frac{B^2}{2\mu_0} \right) \xi_{j;i} + \frac{B_j B_k}{\mu_0} \xi_{i;k} + \frac{B_i B_j}{\mu_0} \xi_{k;k} \right]_{;j}. \qquad (19.47)$$

Honestly! Here the semicolons denote gradients (partial derivatives in Cartesian coordinates; connection coefficients (Sec. 11.8) are required in curvilinear coordinates).

We write the operator $\hat{F}_i$ in the explicit form (19.47) because of its power for demonstrating that $\hat{F}_i$ is self-adjoint (Hermitian, with real variables rather than complex): by introducing the Kronecker-delta components of the metric, $g_{ij} = \delta_{ij}$, we can easily rewrite Eq. (19.47) in the form

force operator $\hat{F}_i[\boldsymbol{\xi}]$ is self-adjoint

$$\hat{F}_i[\boldsymbol{\xi}] = (T_{ijkl}\xi_{k;l})_{;j}, \qquad (19.48)$$

where T_{ijkl} are the components of a fourth-rank tensor that is symmetric under interchange of its first and second pairs of indices: $T_{ijkl} = T_{klij}$.

Now, our magnetic-confinement equilibrium configuration (e.g., Fig. 19.6) typically consists of a plasma-filled interior region $\mathcal{V}$ surrounded by a vacuum magnetic field (which in turn may be surrounded by a wall). Our MHD equations with force operator $\hat{\mathbf{F}}$ are valid only in the plasma region $\mathcal{V}$, and not in vacuum, where Maxwell's equations with small displacement current prevail. We use Eq. (19.48) to prove that $\hat{\mathbf{F}}$ is a self-adjoint operator when integrated over $\mathcal{V}$, with the appropriate boundary conditions at the vacuum interface.

Specifically, we contract a vector field $\boldsymbol{\zeta}$ into $\hat{\mathbf{F}}[\boldsymbol{\xi}]$, integrate over $\mathcal{V}$, and perform two integrations by parts to obtain

$$\int_{\mathcal{V}} \boldsymbol{\zeta} \cdot \mathbf{F}[\boldsymbol{\xi}] dV = \int_{\mathcal{V}} \zeta_i (T_{ijkl}\xi_{k;l})_{;j} dV = -\int_{\mathcal{V}} T_{ijkl}\zeta_{i;j}\xi_{k;l} dV = \int_{\mathcal{V}} \xi_i (T_{ijkl}\zeta_{k;l})_{;j} dV$$
$$= \int_{\mathcal{V}} \boldsymbol{\xi} \cdot \mathbf{F}[\boldsymbol{\zeta}] \, dV. \qquad (19.49)$$

Here we have discarded the integrals of two divergences, which by Gauss's theorem can be expressed as surface integrals at the fluid-vacuum interface $\partial\mathcal{V}$. Those unwanted surface integrals vanish if $\boldsymbol{\xi}$ and $\boldsymbol{\zeta}$ satisfy

$$T_{ijkl}\xi_{k;l}\zeta_i n_j = 0, \quad \text{and} \quad T_{ijkl}\zeta_{k;l}\xi_i n_j = 0, \tag{19.50}$$

with n_j the normal to the interface ∂V.

boundary conditions for perturbations

Now, $\boldsymbol{\xi}$ and $\boldsymbol{\zeta}$ are physical displacements of the MHD fluid, and as such, they must satisfy the appropriate boundary conditions at the boundary $\partial\mathcal{V}$ of the plasma region $\mathcal{V}$. In the simplest, idealized case, the conducting fluid would extend far beyond the region where the disturbances have appreciable amplitude, so $\boldsymbol{\xi} = \boldsymbol{\zeta} = 0$ at the distant boundary, and Eqs. (19.50) are satisfied. More reasonably, the fluid might butt up against rigid walls at $\partial\mathcal{V}$, where the normal components of $\boldsymbol{\xi}$ and $\boldsymbol{\zeta}$ vanish, guaranteeing again that Eqs. (19.50) are satisfied. This configuration is fine for liquid mercury or sodium, but not for a hot plasma, which would quickly destroy the walls. For confinement by a surrounding vacuum magnetic field, no current flows outside $\mathcal{V}$, and the displacement current is negligible, so $\nabla \times \delta\mathbf{B} = 0$ there. By combining this with the rest of Maxwell's equations and paying careful attention to the motion of the interface and boundary conditions [Eqs. (19.20)] there, one can show, once again, that Eqs. (19.50) are satisfied (see, e.g., Goedbloed and Poedts, 2004, Sec. 6.6.2). Therefore, in all these cases Eq. (19.49) is also true, which demonstrates the self-adjointness (Hermiticity) of $\hat{\mathbf{F}}$.[5] We use this property below.

Returning to our perturbed MHD system, we seek its normal modes by assuming a harmonic time dependence, $\boldsymbol{\xi} \propto e^{-i\omega t}$. The first-order equation of motion (19.46) then becomes

Sturm-Liouville type eigenequation for perturbations

$$\boxed{\hat{\mathbf{F}}[\boldsymbol{\xi}] + \rho\omega^2\boldsymbol{\xi} = 0.} \tag{19.51}$$

This is an eigenequation for the fluid's Lagrangian displacement $\boldsymbol{\xi}$, with eigenvalue ω^2. It must be augmented by the boundary conditions (19.20) at the edge $\partial\mathcal{V}$ of the fluid.

By virtue of the elegant, self-adjoint mathematical form (19.48) of the differential operator $\hat{\mathbf{F}}$, our eigenequation (19.51) is of a very special and powerful type, called *Sturm-Liouville*; see any good text on mathematical physics (e.g., Mathews and Walker, 1970; Arfken, Weber, and Harris, 2013; Hassani, 2013). From the general (rather simple) theory of Sturm-Liouville equations, we can infer that all the eigenvalues ω^2 are real, so the normal modes are purely oscillatory ($\omega^2 > 0$, $\boldsymbol{\xi} \propto e^{\pm i|\omega|t}$) or are purely exponentially growing or decaying ($\omega^2 < 0$, $\boldsymbol{\xi} \propto e^{\pm|\omega|t}$). Exponentially growing modes represent instabilities. Sturm-Liouville theory also implies that all

properties of eigenfrequencies and eigenfunctions

---

5. Self-adjointness can also be deduced from energy conservation without getting entangled in detailed boundary conditions (see, e.g., Bellan, 2006, Sec. 10.4.2).

eigenfunctions [labeled by indices "(n)"] with different eigenfrequencies are orthogonal to one another, in the sense that $\int_{\mathcal{V}} \rho \boldsymbol{\xi}^{(n)} \cdot \boldsymbol{\xi}^{(m)} dV = 0$.

Exercise 19.13 *Derivation: Properties of Eigenmodes*

Derive the properties of the eigenvalues and eigenfunctions for perturbations of an MHD equilibrium that are asserted in the last paragraph of Sec. 19.5.1, namely, the following.

(a) For each normal mode the eigenfrequency ω_n is either real or imaginary.

(b) Eigenfunctions $\boldsymbol{\xi}^{(m)}$ and $\boldsymbol{\xi}^{(n)}$ that have different eigenvalues $\omega_m \neq \omega_n$ are orthogonal to each other: $\int \rho\, \boldsymbol{\xi}^{(m)} \cdot \boldsymbol{\xi}^{(m)} dV = 0$.

19.5.2 Z-Pinch: Sausage and Kink Instabilities

We illustrate MHD stability theory using a simple, analytically tractable example: a variant of the Z-pinch configuration of Fig. 19.6a. We consider a long, cylindrical column of a conducting, incompressible liquid (e.g., mercury) with column radius R and fluid density ρ. The column carries a current I longitudinally along its surface (rather than in its interior as in Fig. 19.6a), so $\mathbf{j} = [I/(2\pi R)]\delta(\varpi - R)\mathbf{e}_z$, and the liquid is confined by the resulting external toroidal magnetic field $B_\phi \equiv B$. The interior of the plasma is field free and at constant pressure P_0. From $\nabla \times \mathbf{B} = \mu_0 \mathbf{j}$, we deduce that the exterior magnetic field is

$$B_\phi \equiv B = \frac{\mu_0 I}{2\pi \varpi} \quad \text{at } \varpi \geq R. \tag{19.52}$$

Here (ϖ, ϕ, z) are the usual cylindrical coordinates. This is a variant of the Z-pinch, because the z-directed current on the column's surface creates the external toroidal field B, which pinches the column until its internal pressure is balanced by the field's pressure:

$$P_0 = \left(\frac{B^2}{2\mu_0}\right)_{\varpi = R}. \tag{19.53}$$

It is quicker and more illuminating to analyze the stability of this Z-pinch equilibrium directly instead of by evaluating $\hat{\mathbf{F}}$, and the outcome is the same. (For a treatment based on $\hat{\mathbf{F}}$, see Ex. 19.16.) Treating only the most elementary case, we consider small, axisymmetric perturbations with an assumed variation $\boldsymbol{\xi} \propto e^{i(kz-\omega t)}\mathbf{f}(\varpi)$ for some function $\mathbf{f}$. As the magnetic field interior to the column vanishes, the equation of motion $\rho d\mathbf{v}/dt = -\nabla(P + \delta P)$ becomes

$$-\omega^2 \rho \xi_\varpi = -\delta P', \quad -\omega^2 \rho \xi_z = -ik\delta P, \tag{19.54a}$$

where the prime denotes differentiation with respect to radius ϖ. Combining these two equations, we obtain

$$\xi_z' = ik\xi_\varpi. \tag{19.54b}$$

Because the fluid is incompressible, it satisfies $\nabla \cdot \boldsymbol{\xi} = 0$:

$$\varpi^{-1}(\varpi \xi_\varpi)' + ik\xi_z = 0, \tag{19.54c}$$

which, with Eq. (19.54b), leads to

$$\xi_z'' + \frac{\xi_z'}{\varpi} - k^2\xi_z = 0. \tag{19.54d}$$

The solution of this equation that is regular at $\varpi = 0$ is

$$\xi_z = AI_0(k\varpi) \quad \text{at } \varpi \leq R, \tag{19.54e}$$

where A is a constant, and $I_n(x)$ is the modified Bessel function: $I_n(x) = i^{-n}J_n(ix)$. From Eq. (19.54b) and $dI_0(x)/dx = I_1(x)$, we obtain

$$\xi_\varpi = -iAI_1(k\varpi). \tag{19.54f}$$

Next we consider the region exterior to the fluid column. As this is vacuum, it must be current free; and as we are dealing with a purely axisymmetric perturbation, the ϖ component of Maxwell's equation $\nabla \times \delta\mathbf{B} = 0$ (with negligible displacement current) reads

$$\frac{\partial \delta B_\phi}{\partial z} = ik\delta B_\phi = 0. \tag{19.54g}$$

The ϕ component of the magnetic perturbation therefore vanishes outside the column.

The interior and exterior solutions must be connected by the law of force balance, that is, by the boundary condition (19.19f) [or equivalently, Eq. (19.20a) with the tildes removed] at the plasma's surface. Allowing for the displacement of the surface and retaining only linear terms, this becomes

$$P_0 + \Delta P = P_0 + (\boldsymbol{\xi} \cdot \nabla)P_0 + \delta P = \frac{(B + \Delta B_\phi)^2}{2\mu_0} = \frac{B^2}{2\mu_0} + \frac{B}{\mu_0}(\boldsymbol{\xi} \cdot \nabla)B + \frac{B\delta B_\phi}{\mu_0}, \tag{19.54h}$$

where all quantities are evaluated at $\varpi = R$. The equilibrium force-balance condition gives us $P_0 = B^2/(2\mu_0)$ [Eq. (19.53)] and $\nabla P_0 = 0$. In addition, we have shown that $\delta B_\phi = 0$. Therefore, Eq. (19.54h) becomes simply

$$\delta P = \frac{BB'}{\mu_0}\xi_\varpi. \tag{19.54i}$$

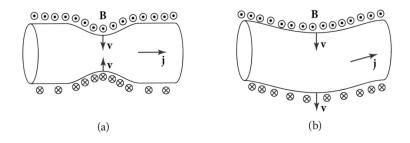

(a)   (b)

FIGURE 19.10 Physical interpretation of (a) sausage and (b) kink instabilities.

Substituting δP from Eqs. (19.54a) and (19.54e), B from Eq. (19.52), and ξ_ϖ from Eq. (19.54f), we obtain the dispersion relation

$$\omega^2 = \frac{-\mu_0 I^2}{4\pi^2 R^4 \rho} \frac{kR I_1(kR)}{I_0(kR)}$$

$$\simeq \frac{-\mu_0 I^2}{8\pi^2 R^2 \rho} k^2; \quad \text{for } k \ll R^{-1}$$

$$\simeq \frac{-\mu_0 I^2}{4\pi^2 R^3 \rho} k; \quad \text{for } k \gg R^{-1}, \tag{19.55}$$

where we have used $I_0(x) \sim 1$, $I_1(x) \sim x/2$ as $x \to 0$ and $I_1(x)/I_0(x) \to 1$ as $x \to \infty$.

Because I_0 and I_1 are positive for all $kR > 0$, for every wave number k this dispersion relation shows that ω^2 is negative. Therefore, ω is imaginary, the perturbation grows exponentially with time, and so *the Z-pinch configuration is dynamically unstable.* If we define a characteristic Alfvén speed by $a = B(R)/(\mu_0 \rho)^{1/2}$ [see Eq. (19.73)], then we see that the growth time for modes with wavelengths comparable to the column diameter is roughly an Alfvén crossing time $2R/a$. This is fast!

This instability is sometimes called a *sausage instability,* because its eigenfunction $\xi_\varpi \propto e^{ikz}$ consists of oscillatory pinches of the column's radius that resemble the pinches between sausages in a link. This sausage instability has a simple physical interpretation (Fig. 19.10a), one that illustrates the power of the concepts of flux freezing and magnetic tension for developing intuition. If we imagine an inward radial motion of the fluid, then the toroidal loops of magnetic field will be carried inward, too, and will therefore shrink. As the external field is unperturbed, $\delta B_\phi = 0$, we have $B_\phi \propto 1/\varpi$, whence the surface field at the inward perturbation increases, leading to a larger "hoop" stress or, equivalently, a larger $\mathbf{j} \times \mathbf{B}$ Lorentz force, which accelerates the inward perturbation (see Fig. 19.10a).

So far, we have only considered axisymmetric perturbations. We can generalize our analysis by allowing the perturbations to vary as $\boldsymbol{\xi} \propto \exp(im\phi)$. (Our sausage instability corresponds to $m = 0$.) Modes with $m \geq 1$, like $m = 0$, are also generically unstable. The $m = 1$ modes are known as *kink* modes. In this case the column bends,

sausage instability of Z-pinch

kink instability of Z-pinch

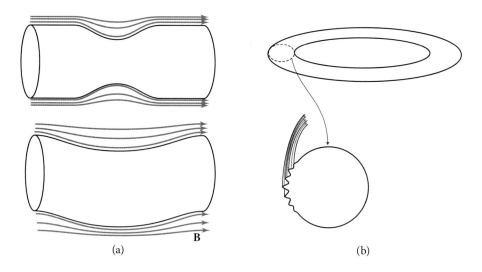

FIGURE 19.11 (a) Stabilizing magnetic fields for Θ-pinch configuration. (b) Flute instability for Θ-pinch configuration made into a torus.

and the field strength is intensified along the inner face of the bend and reduced on the outer face, thereby amplifying the instability (Fig. 19.10b). The incorporation of compressibility, as is appropriate for plasma instead of mercury, introduces only algebraic complexity; the conclusions are unchanged. The column is still highly unstable. It remains so if we distribute the longitudinal current throughout the column's interior, thereby adding a magnetic field to the interior, as in Fig. 19.6a. MHD instabilities such as these have bedeviled attempts to confine plasma long enough to bring about nuclear fusion. Indeed, considerations of MHD stability were one of the primary motivations for the tokamak, the most consistently successful of experimental fusion devices.

19.5.3

19.5.3 The Θ-Pinch and Its Toroidal Analog; Flute Instability; Motivation for Tokamak

full stability of Θ-pinch

By contrast with the extreme instability of the Z-pinch configuration, the Θ-pinch configuration (Sec. 19.3.3 and Fig. 19.6b) is fully stable against MHD perturbations! (See Ex. 19.14.) This is easily understood physically (Fig. 19.11a). When the plasma cylinder is pinched or bent, at outward displaced regions of the cylinder, the external longitudinal magnetic field lines are pushed closer together, thereby strengthening the magnetic field and its pressure and thence creating an inward restoring force. Similarly, at inward displaced regions, the field lines are pulled apart, weakening their pressure and creating an outward restoring force.

toroidal Θ-pinch: flute instability

Unfortunately, the Θ-pinch configuration cannot confine plasma without closing its ends. The ends can be partially closed by a pair of magnetic mirrors, but there remain losses out the ends that cause problems. The ends can be fully closed by bending the column into a closed torus, but, sadly, the resulting toroidal Θ-pinch configuration exhibits a new MHD "flute" instability (Fig. 19.11b).

This flute instability arises on and near the outermost edge of the torus. That edge is curved around the torus's symmetry axis in just the same way as the face of the Z-pinch configuration is curved around its symmetry axis—with the magnetic field in both cases forming closed loops around the axis. As a result, this outer edge is subject to a sausage-type instability, similar to that of the Z-pinch. The resulting corrugations are translated along the torus, so they resemble the flutes on a Greek column that supports an architectural arch or roof (hence the name "flute instability"). This fluting can also be understood as a flux-tube interchange instability (Ex. 19.15).

The flute instability can be counteracted by endowing the torus with an internal magnetic field that twists (shears) as one goes radially inward (the tokamak configuration of Fig. 19.6c). Adjacent magnetic surfaces (isobars), with their different field directions, counteract each other's MHD instabilities. The component of the magnetic field along the plasma torus provides a pressure that stabilizes against sausage instabilities and a tension that stabilizes against kink-type instabilities; the component around the torus's guiding circle acts to stabilize its flute modes. In addition, the formation of image currents in the conducting walls of a tokamak vessel can also have a stabilizing influence.

how tokamak configuration counteracts instabilities

Exercise 19.14 *Problem and Challenge: Stability of Θ-Pinch*
Derive the dispersion relation $\omega^2(k)$ for axisymmetric perturbations of the Θ-pinch configuration when the magnetic field is confined to the cylinder's exterior, and conclude from it that the Θ-pinch is stable against axisymmetric perturbations. [Hint: The analysis of the interior of the cylinder is the same as for the Z-pinch analyzed in the text.] Repeat your analysis for a general, variable-separated perturbation of the form $\boldsymbol{\xi} \propto e^{i(m\theta+kz-\omega t)}$, and thereby conclude that the Θ-pinch is fully MHD stable.

Exercise 19.15 *Example: Flute Instability Understood by Flux-Tube Interchange*
Carry out an analysis of the flute instability patterned after that for rotating Couette flow (first long paragraph of Sec. 14.6.3) and that for convection in stars (Fig. 18.5): Imagine exchanging two plasma-filled magnetic flux tubes that reside near the outermost edge of the torus. Argue that the one displaced outward experiences an unbalanced outward force, and the one displaced inward experiences an unbalanced inward force. [Hint: (i) To simplify the analysis, let the equilibrium magnetic field rise from zero continuously in the outer layers of the torus rather than arising discontinuously at its surface, and consider flux tubes in that outer, continuous region. (ii) Argue that the unbalanced force per unit length on a displaced flux tube is $[-\boldsymbol{\nabla}(P + B^2/(2\mu_0)) - (B_{\text{tube}}^2/\varpi)\mathbf{e}_\varpi]A$. Here A is the tube's cross sectional area, ϖ is distance from the torus's symmetry axis, $\mathbf{e}_\varpi$ is the unit vector pointing away from the symmetry axis, B_{tube} is the field strength inside the displaced tube, and $\boldsymbol{\nabla}(P + B^2/(2\mu_0))$ is the net surrounding pressure gradient at the tube's

displaced location.] Argue further that the innermost edge of the torus is stable against flux-tube interchange, so the flute instability is confined to the torus's outer face.

19.5.4

19.5.4 Energy Principle and Virial Theorems T2

For the perturbation eigenequation (19.46) and its boundary conditions, analytical or even numerical solutions are only readily obtained in the most simple of geometries and for the simplest fluids. However, as the eigenequation is self-adjoint, it is possible to write down a variational principle and use it to derive approximate stability criteria. This variational principle has been a powerful tool for analyzing the stability of tokamak and other magnetostatic configurations.

To derive the variational principle, we begin by multiplying the fluid velocity $\dot{\boldsymbol{\xi}} = \partial\boldsymbol{\xi}/\partial t$ into the eigenequation (equation of motion) $\rho\,\partial^2\boldsymbol{\xi}/\partial t^2 = \hat{\mathbf{F}}[\boldsymbol{\xi}]$. We then integrate over the plasma-filled region $\mathcal{V}$, and use the self-adjointness of $\hat{\mathbf{F}}$ to write $\int_{\mathcal{V}} dV\,\dot{\boldsymbol{\xi}} \cdot \hat{\mathbf{F}}[\boldsymbol{\xi}] = \frac{1}{2}\int_{\mathcal{V}} dV\,(\dot{\boldsymbol{\xi}} \cdot \hat{\mathbf{F}}[\boldsymbol{\xi}] + \boldsymbol{\xi} \cdot \hat{\mathbf{F}}[\dot{\boldsymbol{\xi}}])$. We thereby obtain

conserved energy for adiabatic perturbations of magnetostatic configurations

$$\frac{dE}{dt} = 0, \quad \text{where } E = T + W, \tag{19.56a}$$

$$T = \int_{\mathcal{V}} dV\,\frac{1}{2}\rho\dot{\boldsymbol{\xi}}^2, \quad \text{and} \quad W = W[\boldsymbol{\xi}] \equiv -\frac{1}{2}\int_{\mathcal{V}} dV\,\boldsymbol{\xi} \cdot \hat{\mathbf{F}}[\boldsymbol{\xi}]. \tag{19.56b}$$

The integrals T and W are the perturbation's kinetic and potential energy, and $E = T + W$ is the conserved total energy.

Any solution of the equation of motion $\partial^2\boldsymbol{\xi}/\partial t^2 = \hat{\mathbf{F}}[\boldsymbol{\xi}]$ can be expanded in terms of a complete set of normal modes $\boldsymbol{\xi}^{(n)}(\mathbf{x})$ with eigenfrequencies ω_n: $\boldsymbol{\xi} = \sum_n A_n \boldsymbol{\xi}^{(n)} e^{-i\omega_n t}$. Because $\hat{\mathbf{F}}$ is a real, self-adjoint operator, these normal modes can all be chosen to be real and orthogonal, even when some of their frequencies are degenerate. As the perturbation evolves, its energy sloshes back and forth between kinetic T and potential W, so time averages of T and W are equal: $\overline{T} = \overline{W}$. This implies, for each normal mode,

$$\omega_n^2 = \frac{W[\boldsymbol{\xi}^{(n)}]}{\int_{\mathcal{V}} dV\,\frac{1}{2}\rho\boldsymbol{\xi}^{(n)2}}. \tag{19.57}$$

energy principle (Rayleigh principle) for stability

As the denominator is positive definite, we conclude that *a magnetostatic equilibrium is stable against small perturbations if and only if the potential energy $W[\boldsymbol{\xi}]$ is a positive-definite functional of the perturbation $\boldsymbol{\xi}$.* This is sometimes called the *Rayleigh principle* for a general Sturm-Liouville problem. In the MHD context, it is known as the *energy principle.*

action principle for eigenfrequencies

It is straightforward to verify, by virtue of the self-adjointness of $\hat{\mathbf{F}}[\boldsymbol{\xi}]$, that expression (19.57) serves as an action principle for the eigenfrequencies: If one inserts into Eq. (19.57) a trial function $\boldsymbol{\xi}_{\text{trial}}$ in place of $\boldsymbol{\xi}^{(n)}$, then the resulting value of the equation will be stationary under small variations of $\boldsymbol{\xi}_{\text{trial}}$ if and only if $\boldsymbol{\xi}_{\text{trial}}$ is equal to some eigenfunction $\boldsymbol{\xi}^{(n)}$; and the stationary value of Eq. (19.57) is that eigenfunction's squared eigenfrequency ω_n^2. This action principle is most useful for estimating the lowest few squared frequencies ω_n^2. Because first-order differences between $\boldsymbol{\xi}_{\text{trial}}$ and $\boldsymbol{\xi}^{(n)}$ produce second-order errors in ω_n^2, relatively crude trial eigenfunctions can furnish surprisingly accurate eigenvalues.

Whatever may be our chosen trial function $\boldsymbol{\xi}_{\text{trial}}$, the computed value of the action (19.57) will always be larger than ω_0^2, the squared eigenfrequency of the most unstable mode. Therefore, if we compute a negative value of Eq. (19.57) using some trial eigenfunction, we know that the equilibrium must be even more unstable.

The MHD energy principle and action principle are special cases of the general conservation law and action principle for Sturm-Liouville differential equations (see, e.g., Mathews and Walker, 1970; Arfken, Weber, and Harris, 2013; Hassani, 2013). For further insights into the energy and action principles, see the original MHD paper by Bernstein et al. (1958), in which these ideas were developed; also see Mikhailovskii (1998), Goedbloed and Poedts (2004, Chap. 6), and Bellan (2006, Chap. 10).

Exercise 19.16 *Example: Reformulation of the Energy Principle;*
Application to Z-Pinch T2
The form of the potential energy functional derived in the text [Eq. (19.47)] is optimal for demonstrating that the operator $\hat{\mathbf{F}}$ is self-adjoint. However, there are several simpler, equivalent forms that are more convenient for practical use.

(a) Use Eq. (19.46) to show that

$$\boldsymbol{\xi} \cdot \hat{\mathbf{F}}[\boldsymbol{\xi}] = \mathbf{j} \times \mathbf{b} \cdot \boldsymbol{\xi} - b^2/\mu_0 - \gamma P (\nabla \cdot \boldsymbol{\xi})^2 - (\nabla \cdot \boldsymbol{\xi})(\boldsymbol{\xi} \cdot \nabla) P$$
$$+ \nabla \cdot [(\boldsymbol{\xi} \times \mathbf{B}) \times \mathbf{b}/\mu_0 + \gamma P \boldsymbol{\xi}(\nabla \cdot \boldsymbol{\xi}) + \boldsymbol{\xi}(\boldsymbol{\xi} \cdot \nabla) P], \quad (19.58)$$

where $\mathbf{b} \equiv \delta \mathbf{B}$ is the Eulerian perturbation of the magnetic field.

(b) Insert Eq. (19.58) into expression (19.56b) for the potential energy $W[\boldsymbol{\xi}]$ and convert the volume integral of the divergence into a surface integral. Then impose the boundary condition of a vanishing normal component of the magnetic field at $\partial \mathcal{V}$ [Eq. (19.20b)] to show that

$$W[\boldsymbol{\xi}] = \frac{1}{2} \int_{\mathcal{V}} dV \left[-\mathbf{j} \times \mathbf{b} \cdot \boldsymbol{\xi} + b^2/\mu_0 + \gamma P (\nabla \cdot \boldsymbol{\xi})^2 + (\nabla \cdot \boldsymbol{\xi})(\boldsymbol{\xi} \cdot \nabla) P \right]$$
$$- \frac{1}{2} \int_{\partial \mathcal{V}} d\boldsymbol{\Sigma} \cdot \boldsymbol{\xi} \left[\gamma P (\nabla \cdot \boldsymbol{\xi}) + \boldsymbol{\xi} \cdot \nabla P - \mathbf{B} \cdot \mathbf{b}/\mu_0 \right]. \quad (19.59)$$

(c) Consider axisymmetric perturbations of the cylindrical Z-pinch of an incompressible fluid, as discussed in Sec. 19.5.2, and argue that the surface integral vanishes.

(d) Adopt a simple trial eigenfunction, and obtain a variational estimate of the growth rate of the sausage instability's fastest growing mode.

Exercise 19.17 *Problem: Potential Energy in Its Most Physically Interpretable Form* T2

(a) Show that the potential energy (19.59) can be transformed into the following form:

$$W[\boldsymbol{\xi}] = \frac{1}{2} \int_{\mathcal{V}} dV \left[-\mathbf{j} \times \mathbf{b} \cdot \boldsymbol{\xi} + \frac{\mathbf{b}^2}{\mu_0} + \delta P \frac{\Delta \rho}{\rho} \right]$$

$$+ \frac{1}{2} \int_{\partial \mathcal{V}} d\boldsymbol{\Sigma} \cdot \boldsymbol{\nabla} \left(\frac{\tilde{B}^2}{2\mu_0} - P - \frac{B^2}{2\mu_0} \right) \xi_n^2 + \frac{1}{2} \int_{\text{vacuum}} dV \frac{\tilde{\mathbf{b}}^2}{\mu_0}. \quad (19.60)$$

Here symbols without tildes represent quantities in the plasma region, and those with tildes are in the vacuum region; ξ_n is the component of the fluid displacement orthogonal to the vacuum-plasma interface.

(b) Explain the physical interpretation of each term in the expression for the potential energy in part (a). Notice that, although our original expression for the potential energy, $W = -\frac{1}{2} \int_{\mathcal{V}} \boldsymbol{\xi} \cdot \mathbf{F}[\boldsymbol{\xi}] dV$, entails an integral only over the plasma region, it actually includes the vacuum region's magnetic energy.

Exercise 19.18 *Example: Virial Theorems* T2

Additional mathematical tools that are useful in analyzing MHD equilibria and their stability—and are also useful in astrophysics—are the *virial theorems*. In this exercise and the next, you will deduce time-dependent and time-averaged virial theorems for any system for which the law of momentum conservation can be written in the form

$$\frac{\partial (\rho v_j)}{\partial t} + T_{jk;k} = 0. \quad (19.61)$$

Here ρ is mass density, v_j is the material's velocity, ρv_j is momentum density, and T_{jk} is the stress tensor. We have met this formulation of momentum conservation in elastodynamics [Eq. (12.2b)], in fluid mechanics with self-gravity [Eq. (2) of Box 13.4], and in magnetohydrodynamics with self-gravity [Eq. (19.12)].

The virial theorems involve integrals over any region $\mathcal{V}$ for which there is no mass flux or momentum flux across its boundary: $\rho v_j n_j = T_{jk} n_k = 0$ everywhere on $\partial \mathcal{V}$, where n_j is the normal to the boundary.[6] For simplicity we use Cartesian coordinates, so there are no connection coefficients to worry about, and momentum conservation becomes $\partial (\rho v_j) + \partial T_{jk}/\partial x_k = 0$.

(a) Show that mass conservation, $\partial \rho/\partial t + \boldsymbol{\nabla} \cdot (\rho \mathbf{v}) = 0$, implies that for any field f (scalar, vector, or tensor), we have

$$\frac{d}{dt} \int_{\mathcal{V}} \rho f dV = \int_{\mathcal{V}} \rho \frac{df}{dt} dV. \quad (19.62)$$

---

6. If self-gravity is included, then the boundary will have to be at spatial infinity (i.e., no boundary), as $T_{jk}^{\text{grav}} n_k$ cannot vanish everywhere on a finite enclosing wall.

(b) Use Eq. (19.62) to show that

$$\frac{d^2 I_{jk}}{dt^2} = 2 \int_{\mathcal{V}} T_{jk} dV,$$ (19.63a)

where I_{jk} is the second moment of the mass distribution:

$$I_{jk} = \int_{\mathcal{V}} \rho x_j x_k dV$$ (19.63b)

with x_j the distance from a chosen origin. This is the *time-dependent tensor viral theorem*. (Note that the system's mass quadrupole moment is the trace-free part of I_{jk}, $\mathcal{I}_{jk} = I_{jk} - \frac{1}{3} I g_{jk}$, and the system's moment of inertia tensor is $\mathfrak{I}_{jk} = I_{jk} - I g_{jk}$, where $I = I_{jj}$ is the trace of I_{jk}.)

(c) If the time integral of $d^2 I_{jk}/dt^2$ vanishes, then the time-averaged stress tensor satisfies

$$\int_{\mathcal{V}} \bar{T}_{jk} dV = 0.$$ (19.64)

This is the *time-averaged tensor virial theorem*. Under what circumstances is this theorem true?

Exercise 19.19 *Example: Scalar Virial Theorems* T2

(a) By taking the trace of the time-dependent tensorial virial theorem and specializing to an MHD plasma with (or without) self-gravity, show that

$$\frac{1}{2} \frac{d^2 I}{dt^2} = 2 E_{\text{kin}} + E_{\text{mag}} + E_{\text{grav}} + 3 E_P,$$ (19.65a)

where I is the trace of I_{jk}, E_{kin} is the system's kinetic energy, E_{mag} is its magnetic energy, E_P is the volume integral of its pressure, and E_{grav} is its gravitational self-energy:

$$I = \int_{\mathcal{V}} \rho \mathbf{x}^2 dV, \quad E_{\text{kin}} = \int_{\mathcal{V}} \frac{1}{2} \rho \mathbf{v}^2 dV, \quad E_{\text{mag}} = \int \frac{\mathbf{B}^2}{2\mu_0}, \quad E_P = \int_{\mathcal{V}} P dV,$$

$$E_{\text{grav}} = \frac{1}{2} \int_{\mathcal{V}} \rho \Phi = -\int_{\mathcal{V}} \frac{1}{8\pi G} (\nabla \Phi)^2 = -\frac{1}{2} \int_{\mathcal{V}} \int_{\mathcal{V}} G \frac{\rho(\mathbf{x})\rho(\mathbf{x}')}{|\mathbf{x} - \mathbf{x}'|} dV' dV, \text{ (19.65b)}$$

with Φ the gravitational potential energy [cf. Eq. (13.62)].

(b) When the time integral of $d^2 I/dt^2$ vanishes, then the time average of the right-hand side of Eq. (19.65a) vanishes:

$$2\bar{E}_{\text{kin}} + \bar{E}_{\text{mag}} + \bar{E}_{\text{grav}} + 3\bar{E}_P = 0.$$ (19.66)

This is the *time-averaged scalar virial theorem*. Give examples of circumstances in which it holds.

(c) Equation (19.66) is a continuum analog of the better-known scalar virial theorem, $2\bar{E}_{\text{kin}} + \bar{E}_{\text{grav}} = 0$, for a system consisting of particles that interact via their self-gravity—for example, the solar system (see, e.g., Goldstein, Poole, and Safko, 2002, Sec. 3.4).

(d) As a simple but important application of the time-averaged scalar virial theorem, show—neglecting self-gravity—that it is impossible for internal currents in a plasma to produce a magnetic field that confines the plasma to a finite volume: external currents (e.g., in solenoids) are necessary.

(e) For applications to the oscillation and stability of self-gravitating systems, see Chandrasekhar (1961, Sec. 118).

19.6   **19.6 Dynamos and Reconnection of Magnetic Field Lines**   T2

dynamo

As we have already remarked, the timescale for Earth's magnetic field to decay is estimated to be roughly 1 million years. Since Earth is far older than that, some process inside Earth must be regenerating the magnetic field. This process is known as a *dynamo*. Generally speaking, in a dynamo the motion of the fluid stretches the magnetic field lines, thereby increasing their magnetic energy density, which compensates for the decrease in magnetic energy associated with Ohmic decay. The details of how this happens inside Earth are not well understood. However, some general principles of dynamo action have been formulated, and their application to the Sun is somewhat better understood (Exs. 19.20 and 19.21).

19.6.1   19.6.1   Cowling's Theorem   T2

It is simple to demonstrate the impossibility of dynamo action in any time-independent, axisymmetric flow. Suppose that there were such a dynamo configuration, and the time-independent, poloidal (meridional) field had—for concreteness—the form sketched in Fig. 19.12. Then there must be at least one neutral point marked $\mathcal{P}$ (actually a circle about the symmetry axis), where the poloidal field vanishes. However, the curl of the magnetic field does not vanish at $\mathcal{P}$, so a toroidal current j_ϕ must exist there. Now, in the presence of finite resistivity, there must also be a toroidal electric field at $\mathcal{P}$, since

$$j_\phi = \kappa_e[E_\phi + (\mathbf{v}_P \times \mathbf{B}_P)_\phi] = \kappa_e E_\phi. \tag{19.67}$$

Here $\mathbf{v}_P$ and $\mathbf{B}_P$ are the poloidal components of $\mathbf{v}$ and $\mathbf{B}$. The nonzero E_ϕ in turn implies, via $\nabla \times \mathbf{E} = -\partial \mathbf{B}/\partial t$, that the amount of poloidal magnetic flux threading the circle at $\mathcal{P}$ must change with time, violating our original supposition that the magnetic field distribution is stationary.

Cowling's theorem for dynamos

We therefore conclude that any time-independent, self-perpetuating dynamo must be more complicated than a purely axisymmetric magnetic field. This result is known as *Cowling's theorem.*

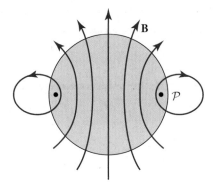

FIGURE 19.12 Impossibility of an axisymmetric dynamo.

19.6.2 Kinematic Dynamos T2

19.6.2

The simplest types of dynamo to consider are those in which we specify a particular velocity field and allow the magnetic field to evolve according to the transport law (19.6). Under certain circumstances, this can produce dynamo action. Note that we do not consider in our discussion the dynamical effect of the magnetic field on the velocity field.

The simplest type of motion is one in which a *dynamo cycle* occurs. In this cycle, there is one mechanism for creating a toroidal magnetic field from a poloidal field and a separate mechanism for regenerating the poloidal field. The first mechanism is usually differential rotation. The second is plausibly magnetic buoyancy, in which a toroidal magnetized loop is lighter than its surroundings and therefore rises in the gravitational field. As the loop rises, Coriolis forces twist the flow, causing a poloidal magnetic field to appear, which completes the dynamo cycle.

dynamo cycle

Small-scale, turbulent velocity fields may also be responsible for dynamo action. In this case, it can be shown on general symmetry grounds that the velocity field must contain *hydrodynamic helicity*—a nonzero mean value of $\mathbf{v} \cdot \boldsymbol{\omega}$ [which is an obvious analog of magnetic helicity (Ex. 19.9), the volume integral or mean value of $\mathbf{A} \cdot \mathbf{B} = \mathbf{A} \cdot (\boldsymbol{\nabla} \times \mathbf{A})$].

dynamo action in turbulence

hydrodynamic helicity

If the magnetic field strength grows, then its dynamical effect will eventually react back on the flow and modify the velocity field. A full description of a dynamo must include this back reaction. Dynamos are a prime target for numerical simulations of MHD, and in recent years, significant progress has been made using these simulations to understand the terrestrial dynamo and other specialized problems.

Exercise 19.20 *Problem: Differential Rotation in the Solar Dynamo* T2

EXERCISES

This problem shows how differential rotation leads to the production of a toroidal magnetic field from a poloidal field.

(a) Verify that for a fluid undergoing differential rotation around a symmetry axis with angular velocity $\Omega(r, \theta)$, the ϕ component of the induction equation reads

$$\frac{\partial B_\phi}{\partial t} = \sin\theta \left(B_\theta \frac{\partial \Omega}{\partial \theta} + B_r r \frac{\partial \Omega}{\partial r} \right),$$ (19.68)

where θ is the co-latitude. (The resistive term can be ignored.)

(b) It is observed that the angular velocity on the solar surface is largest at the equator and decreases monotonically toward the poles. Analysis of solar oscillations (Sec. 16.2.4) has shown that this variation $\Omega(\theta)$ continues inward through the convection zone (cf. Sec. 18.5). Suppose that the field of the Sun is roughly poloidal. Sketch the appearance of the toroidal field generated by the poloidal field.

Exercise 19.21 *Problem: Buoyancy in the Solar Dynamo* T2

Consider a slender flux tube with width much less than its length which, in turn, is much less than the external pressure scale height H. Also assume that the magnetic field is directed along the tube so there is negligible current along the tube.

(a) Show that the requirement of magnetostatic equilibrium implies that inside the flux tube

$$\nabla \left(P + \frac{B^2}{2\mu_0} \right) \simeq 0.$$ (19.69)

(b) Consider a segment of this flux tube that is horizontal and has length ℓ. Holding both ends fixed, bend it vertically upward so that the radius of curvature of its center line is $R \gg \ell$. Assume that the fluid is isentropic with adiabatic index γ. By balancing magnetic tension against buoyancy, show that magnetostatic equilibrium is possible if $R \simeq 2\gamma H$. Do you think this equilibrium could be stable?

(c) In the solar convection zone (cf. Sec. 18.5), small entropy differences are important in driving the convective circulation. Following Ex. 19.20(b), suppose that a length of toroidal field is carried upward by a convecting "blob." Consider the action of the Coriolis force due to the Sun's rotation (cf. Sec. 14.2.1) on a single blob, and argue that it will rotate. What will this do to the magnetic field? Sketch the generation of a large-scale poloidal field from a toroidal field through the combined effects of many blobs. What do you expect to observe when a flux tube breaks through the solar surface (known as the photosphere)?

The combined effect of differential rotation, magnetic stress, and buoyancy, as outlined in Exs. 19.20 and 19.21, is thought to play an important role in sustaining the solar dynamo cycle. See Goedbloed and Poedts (2004, Secs. 8.2, 8.3) for further insights.

19.6.3

19.6.3 Magnetic Reconnection T2

So far, our discussion of the evolution of the magnetic field has centered on the induction equation (19.6) (the magnetic transport law). We have characterized the

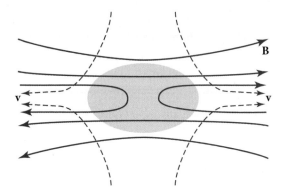

FIGURE 19.13 Magnetic reconnection. In the shaded reconnection region, Ohmic diffusion is important and allows magnetic field lines to "exchange partners," changing the overall field topology. Magnetic field components perpendicular to the plane of the illustration do not develop large gradients and so do not inhibit the reconnection process.

magnetized fluid by a magnetic Reynolds number using some characteristic length L associated with the flow, and have found that, when $R_M \gg 1$, Ohmic dissipation and field-line diffusion in the transport law are unimportant. This is reminiscent of the procedure we followed when discussing vorticity. However, for vorticity we discovered a very important exception to an uncritical neglect of viscosity, dissipation, and vortex-line diffusion at large Reynolds numbers, namely, boundary layers (Sec. 14.4). In particular, we found that large-Reynolds-number flows near solid surfaces develop large velocity gradients on account of the no-slip boundary condition, and that the local Reynolds number can thereby decrease to near unity, allowing viscous stress to change the character of the flow completely. Something similar, called *magnetic reconnection,* can happen in hydromagnetic flows with large R_M, even without the presence of solid surfaces.

Consider two oppositely magnetized regions of conducting fluid moving toward each other (the upper and lower regions in Fig. 19.13). Mutual magnetic attraction of the two regions occurs, as magnetic energy would be reduced if the two sets of field lines were superposed. However, strict flux freezing prevents superposition. Something has to give! What happens is a compromise. The attraction causes large magnetic gradients to develop, accompanied by a buildup of large current densities, until Ohmic diffusion ultimately allows the magnetic field lines to slip sideways through the fluid and to *reconnect* with the field in the other region (the sharply curved field lines in Fig. 19.13).

how magnetic reconnection occurs at high R_M

This reconnection mechanism can be clearly observed at work in tokamaks and in Earth's magnetopause, where the solar wind's magnetic field meets Earth's magnetosphere. However, the details of the reconnection mechanism are quite complex, involving plasma instabilities, anisotropic electrical conductivity, and shock fronts.

Large, inductive electric fields can also develop when the magnetic geometry undergoes rapid change. This can happen in the reversing magnetic field in Earth's *magnetotail*,[7] leading to the acceleration of charged particles that impact Earth during a *magnetic substorm*. Like dynamo action, reconnection has a major role in determining how magnetic fields actually behave in both laboratory and space plasmas.

For further detail on the physics and observations of reconnection, see, for example, Birn and Priest (2007) and Forbes and Priest (2007).

19.7   19.7 Magnetosonic Waves and the Scattering of Cosmic Rays

In Sec. 19.5, we discussed global perturbations of a nonuniform magnetostatic plasma and described how they may be unstable. We now consider a different, particularly simple example of dynamical perturbations: planar, monochromatic, propagating

magnetosonic waves

waves in a uniform, magnetized, conducting medium. These are called *magnetosonic waves*. They can be thought of as sound waves that are driven not just by gas pressure but also by magnetic pressure and tension.

Although magnetosonic waves have been studied experimentally under laboratory conditions, there the magnetic Reynolds numbers are generally quite small, so the waves damp quickly by Ohmic dissipation. No such problem arises in space plasmas, where magnetosonic waves are routinely studied by the many spacecraft that monitor the solar wind and its interaction with planetary magnetospheres. It appears that these modes perform an important function in space plasmas: they control the transport of cosmic rays. Let us describe some properties of cosmic rays before giving a formal derivation of the magnetosonic-wave dispersion relation.

19.7.1   19.7.1 Cosmic Rays

cosmic-ray spectrum

Cosmic rays are the high-energy particles, primarily protons, that bombard Earth's magnetosphere from outer space. They range in energy from $\sim 1\,\mathrm{MeV}$ to $\sim 3 \times 10^{11}\,\mathrm{GeV} = 0.3\,\mathrm{ZeV} \sim 50\,\mathrm{J}$. (The highest cosmic-ray energy ever measured was $\sim 50\,\mathrm{J}$. Thus, naturally occurring particle accelerators are far more impressive than their terrestrial counterparts, which can only reach $\sim 10\,\mathrm{TeV} = 10^4\,\mathrm{GeV}$!) Most subrelativistic particles originate in the solar system. Their relativistic counterparts, up to energies $\sim 100\,\mathrm{TeV}$, are believed to come mostly from interstellar space, where they are accelerated by expanding shock waves created by supernova explosions (cf. Sec. 17.6.3). The origin of the highest energy particles, greater than $\sim 100\,\mathrm{TeV}$, is an intriguing mystery.

isotropy

The distribution of cosmic-ray arrival directions at Earth is inferred to be quite isotropic (to better than 1 part in 10^4 at an energy of 10 GeV). This is somewhat surprising, because their sources, both in and beyond the solar system, are believed to be distributed anisotropically, so the isotropy needs to be explained. Part of the reason for the isotropy is that the interplanetary and interstellar media are magnetized, and

---

7. The magnetotail is the region containing trailing field lines on the night side of Earth.

the particles gyrate around the magnetic field with the gyro (relativistic cyclotron) frequency $\omega_G = eBc^2/\mathcal{E}$, where $\mathcal{E}$ is the (relativistic) particle energy including rest mass, and B is the magnetic field strength. The gyro (Larmor) radii of the non-relativistic particle orbits are typically small compared to the size of the solar system, and those of the relativistic particles are typically small compared to characteristic lengthscales in the interstellar medium. Therefore, this gyrational motion can effectively erase any azimuthal asymmetry around the field direction. However, this does not stop the particles from streaming away from their sources along the magnetic field, thereby producing anisotropy at Earth. So something else must be impeding this along-line flow, by scattering the particles and causing them to effectively diffuse along and across the field through interplanetary and interstellar space.

evidence for scattering

As we verify in Sec. 20.4, Coulomb collisions are quite ineffective (even if they were effective, they would cause huge cosmic-ray energy losses, in violation of observations). We therefore seek some means of changing a cosmic ray's momentum without altering its energy significantly. This is reminiscent of the scattering of electrons in metals, where it is phonons (elastic waves in the crystal lattice) that are responsible for much of the scattering. It turns out that in the interstellar medium, magnetosonic waves can play a role analogous to phonons and can scatter the cosmic rays. As an aid to understanding this phenomenon, we now derive the waves' dispersion relation.

19.7.2 Magnetosonic Dispersion Relation

19.7.2

Our procedure for deriving the dispersion relation (last paragraph of Sec. 7.2.1) should be familiar by now. We consider a uniform, isentropic, magnetized fluid at rest; we perform a linear perturbation and seek monochromatic plane-wave solutions varying as $e^{i(\mathbf{k}\cdot\mathbf{x}-\omega t)}$. We ignore gravity and dissipative processes (specifically, viscosity, thermal conductivity, and electrical resistivity), as well as gradients in the equilibrium, which can all be important in some circumstances.

It is convenient to use the velocity perturbation $\delta\mathbf{v}$ as the independent variable. The perturbed and linearized equation of motion (19.10) then takes the form

$$-i\rho\omega\delta\mathbf{v} = -iC^2\mathbf{k}\delta\rho + \frac{(i\mathbf{k} \times \delta\mathbf{B}) \times \mathbf{B}}{\mu_0}. \tag{19.70}$$

Here C is the sound speed [$C^2 = (\partial P/\partial\rho)_s = \gamma P/\rho$, not to be confused with the speed of light c], and $\delta P = C^2\delta\rho$ is the Eulerian pressure perturbation for our homogeneous equilibrium.[8] (Note that $\nabla P = \nabla\rho = 0$, so Eulerian and Lagrangian perturbations are the same.) The perturbed equation of mass conservation, $\partial\rho/\partial t + \nabla \cdot (\rho\mathbf{v}) = 0$, becomes

$$\omega\delta\rho = \rho\mathbf{k} \cdot \delta\mathbf{v}, \tag{19.71}$$

---

8. Note that we are assuming that (i) the equilibrium pressure tensor is isotropic and (ii) the perturbations to the pressure are also isotropic. This is unlikely to be the case for a collisionless plasma, so our treatment there must be modified. See Sec. 20.6.2 and Ex. 21.1.

and the MHD law of magnetic-field transport with dissipation ignored, $\partial \mathbf{B}/\partial t = \nabla \times (\mathbf{v} \times \mathbf{B})$, becomes

$$\omega \delta \mathbf{B} = -\mathbf{k} \times (\delta \mathbf{v} \times \mathbf{B}).\tag{19.72}$$

Alfvén velocity

We introduce the Alfvén velocity

$$\boxed{\mathbf{a} \equiv \frac{\mathbf{B}}{(\mu_0 \rho)^{1/2}},}\tag{19.73}$$

and insert $\delta \rho$ [Eq. (19.71)] and $\delta \mathbf{B}$ [Eq. (19.72)] into Eq. (19.70) to obtain

eigenequation for magnetosonic waves

$$\{\mathbf{k} \times [\mathbf{k} \times (\delta \mathbf{v} \times \mathbf{a})]\} \times \mathbf{a} + C^2(\mathbf{k} \cdot \delta \mathbf{v})\mathbf{k} = \omega^2 \delta \mathbf{v}.\tag{19.74}$$

Equation (19.74) is an eigenequation for the wave's squared frequency ω^2 and eigendirection $\delta \mathbf{v}$. The straightforward way to solve it is to rewrite it in the standard matrix form $M_{ij}\delta v_j = \omega^2 \delta v_i$ and then use standard matrix (determinant) methods. It is quicker, however, to seek the three eigendirections $\delta \mathbf{v}$ and eigenfrequencies ω one by one, by projection along preferred directions.

Alfvén mode and its dispersion relation

We first seek a solution to Eq. (19.74) for which $\delta \mathbf{v}$ is orthogonal to the plane formed by the unperturbed magnetic field and the wave vector, $\delta \mathbf{v} = \mathbf{a} \times \mathbf{k}$ up to a multiplicative constant. Inserting this $\delta \mathbf{v}$ into Eq. (19.74), we obtain the dispersion relation

$$\boxed{\omega = \pm \mathbf{a} \cdot \mathbf{k}; \qquad \frac{\omega}{k} = \pm a \cos\theta,}\tag{19.75}$$

where θ is the angle between $\mathbf{k}$ and the unperturbed field, and $a \equiv |\mathbf{a}| = B/(\mu_o \rho)^{1/2}$ is the Alfvén speed. This type of wave is known as the *intermediate magnetosonic mode* and also as the *Alfvén mode*. Its phase speed $\omega/k = a\cos\theta$ is plotted as the larger figure-8 curve in Fig. 19.14. The velocity and magnetic perturbations $\delta \mathbf{v}$ and $\delta \mathbf{B}$ are both along the direction $\mathbf{a} \times \mathbf{k}$, so the Alfvén wave is fully transverse. There is no compression ($\delta \rho = 0$), which accounts for the absence of the sound speed C in the dispersion relation. This Alfvén mode has a simple physical interpretation in the limiting case when $\mathbf{k}$ is parallel to $\mathbf{B}$. We can think of the magnetic field lines as strings with tension B^2/μ_0 and inertia ρ, which are plucked transversely. Their transverse oscillations then propagate with speed $\sqrt{\text{tension/inertia}} = B/\sqrt{\mu_0\rho} = a$. For details and delicacies, see Ex. 21.8.

Alfvén-mode properties

The dispersion relations for the other two modes can be deduced by projecting the eigenequation (19.74) successively along $\mathbf{k}$ and along $\mathbf{a}$ to obtain two scalar equations:

$$k^2(\mathbf{k} \cdot \mathbf{a})(\mathbf{a} \cdot \delta \mathbf{v}) = [(a^2 + C^2)k^2 - \omega^2](\mathbf{k} \cdot \delta \mathbf{v}),$$
$$C^2(\mathbf{k} \cdot \mathbf{a})(\mathbf{k} \cdot \delta \mathbf{v}) = \omega^2(\mathbf{a} \cdot \delta \mathbf{v}).\tag{19.76}$$

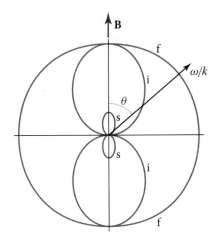

FIGURE 19.14 Phase-velocity surfaces for the three types of magnetosonic modes: fast (f), intermediate (i), and slow (s). The three curves are polar plots of the wave phase velocity ω/k in units of the Alfvén speed $a = B/\sqrt{\mu_0 \rho}$. In the particular example shown, the sound speed C is half the Alfvén speed.

Combining these equations, we obtain the dispersion relation

$$\left(\frac{\omega}{k}\right)^2 = \frac{1}{2}(a^2 + C^2)\left[1 \pm \left(1 - \frac{4C^2 a^2 \cos^2\theta}{(a^2 + C^2)^2}\right)^{1/2}\right]. \qquad (19.77)$$

dispersion relation for fast ($+$) and slow ($-$) magnetosonic modes

(By inserting this dispersion relation, with the upper or lower sign, back into Eqs. (19.76), we can deduce the mode's eigendirection $\delta\mathbf{v}$.) This dispersion relation tells us that ω^2 is positive, so no unstable modes exist, which seems reasonable, as there is no source of free energy. (The same is true, of course, for the Alfvén mode.)

These waves are compressive, with the gas being moved by a combination of gas pressure and magnetic pressure and tension. The modes can be seen to be non-dispersive, which is also to be expected, as we have introduced neither a characteristic timescale nor a characteristic length into the problem.

The mode with the plus signs in Eq. (19.77) is called the *fast magnetosonic mode;* its phase speed is depicted by the outer, quasi-circular curve in Fig. 19.14. A good approximation to its phase speed when $a \gg C$ or $a \ll C$ is $\omega/k \simeq \pm(a^2 + C^2)^{1/2}$. When propagating perpendicularly to $\mathbf{B}$, the fast mode can be regarded as simply a longitudinal sound wave in which the gas pressure is augmented by the magnetic pressure $B^2/(2\mu_0)$ (adopting a specific-heat ratio γ for the magnetic field of 2, as $B \propto \rho$ and so $P_{\text{mag}} \propto \rho^2$ under perpendicular compression).

The mode with the minus signs in Eq. (19.77) is called the *slow magnetosonic mode.* Its phase speed (depicted by the inner figure-8 curve in Fig. 19.14) can be approximated by $\omega/k = \pm aC\cos\theta/(a^2 + C^2)^{1/2}$ when $a \gg C$ or $a \ll C$. Note that

fast magnetosonic-mode properties

slow magnetosonic-mode properties

slow mode—like the intermediate mode but unlike the fast mode—is incapable of propagating perpendicularly to the unperturbed magnetic field; see Fig. 19.14. In the limit of vanishing Alfvén speed or vanishing sound speed, the slow mode ceases to exist for all directions of propagation.

In Chap. 21, we will discover that MHD is a good approximation to the behavior of plasmas only at frequencies below the "ion cyclotron frequency," which is a rather low frequency. For this reason, magnetosonic modes are usually regarded as low-frequency modes.

19.7.3 Scattering of Cosmic Rays by Alfvén Waves

19.7.3

mechanism for Alfvén waves to scatter cosmic-ray particles

Now let us return to the issue of cosmic-ray propagation, which motivated our investigation of magnetosonic modes. Let us consider 100-GeV particles in the interstellar medium. The electron (and ion, mostly proton) density and magnetic field strength are typically $n \sim -3 \times 10^4 \, \text{m}^{-3}$ and $B \sim 300 \, \text{pT}$. The Alfvén speed is then $a \sim 30 \, \text{km s}^{-1}$, much slower than the speeds of the cosmic rays. When analyzing cosmic-ray propagation, a magnetosonic wave can therefore be treated as essentially a magnetostatic perturbation. A relativistic cosmic ray of energy $\mathcal{E}$ has a gyro (relativistic Larmor) radius of $r_G = \mathcal{E}/eBc$, in this case $\sim 3 \times 10^{12}$ m. Cosmic rays will be unaffected by waves with wavelength either much greater than or much less than r_G. However, magnetosonic waves (especially Alfvén waves, which are responsible for most of the scattering), with wavelength matched to the gyro radius, will be able to change the particle's *pitch angle* α (the angle its momentum makes with the mean magnetic field direction). See Sec. 23.4.1 for some details. If the Alfvén waves in this wavelength range have rms dimensionless amplitude $\delta B/B \ll 1$, then the particle's pitch angle will change by an amount $\delta \alpha \sim \delta B/B$ for every wavelength. Now, if the wave spectrum is broadband, individual waves can be treated as uncorrelated, so the particle pitch angle changes stochastically. In other words, the particle diffuses in pitch angle. The effective diffusion coefficient is

$$D_\alpha \sim \left(\frac{\delta B}{B}\right)^2 \omega_G, \tag{19.78}$$

where $\omega_G = c/r_G$ is the gyro frequency (relativistic analog of cyclotron frequency ω_c). The particle is therefore scattered by roughly a radian in pitch angle every time it traverses a distance $\ell \sim (B/\delta B)^2 r_G$. This is effectively the particle's collisional mean free path. Associated with this mean free path is a spatial diffusion coefficient

$$D_{\mathbf{x}} \sim \frac{\ell c}{3}. \tag{19.79}$$

estimated and measured cosmic-ray anisotropy

It is thought that $\delta B/B \sim 10^{-1}$ in the relevant wavelength range in the interstellar medium. An estimate of the collision mean free path is then $\ell(100 \, \text{GeV}) \sim 10^{14}$ m. Now, the thickness of our galaxy's interstellar disk of gas is roughly $L \sim 3 \times 10^{18}$ m $\sim 10^4 \ell$. Therefore, an estimate of the cosmic-ray anisotropy is

$\sim \ell/L \sim 3 \times 10^{-5}$, roughly compatible with the measurements. Although this discussion is an oversimplification, it does demonstrate that the cosmic rays in the interplanetary, interstellar, and intergalactic media can be scattered efficiently by magnetosonic waves. This allows the particle transport to be impeded without much loss of energy, so that the theory of cosmic-ray acceleration (Ex. 23.12) and scattering (this section) together can account for the particle fluxes observed as a function of energy and direction at Earth.

A good question to ask at this point is "Where do the Alfvén waves come from?" The answer turns out to be that they are maintained as part of a turbulence spectrum and created by the cosmic rays themselves through the growth of plasma instabilities. To proceed further and give a more quantitative description of this interaction, we must go beyond a purely fluid description and explore the motions of individual particles. This is where we shall turn in the next few chapters, culminating with a return to the interaction of cosmic rays with Alfvén waves in Sec. 23.4.1.

Bibliographic Note

For intuitive insight into magnetohydrodynamics, we recommend Shercliff (1965).

For textbook introductions to magnetohydrodynamics, we recommend the relevant chapters of Landau, Pitaevskii, and Lifshitz (1979), Schmidt (1979), Boyd and Sanderson (2003), and Bellan (2006). For far greater detail, we recommend a textbook that deals solely with magnetohydrodynamics: Goedbloed and Poedts (2004) and its advanced supplement Goedbloed, Keppens, and Poedts (2010). For a very readable treatment from the viewpoint of an engineer, with applications to engineering and metallurgy, see Davidson (2001).

For the theory of MHD instabilities and applications to magnetic confinement, see the above references, and also Bateman (1978) and the collection of early papers edited by Jeffrey and Taniuti (1966). For applications to astrophysics and space physics, see Parker (1979), Parks (2004), and Kulsrud (2005).

PLASMA PHYSICS

A *plasma* is a gas that is significantly ionized (through heating or photoionization) and thus is composed of electrons and ions, and that has a low enough density to behave classically (i.e., to obey Maxwell-Boltzmann statistics rather than Fermi-Dirac or Bose-Einstein). Plasma physics originated in the nineteenth century, with the study of gas discharges (Crookes, 1879). In 1902, Heaviside and Kennelly realized that plasma is also the key to understanding the propagation of radio waves across the Atlantic. The subject received a further boost in the early 1950s, with the start of the controlled (and the uncontrolled) thermonuclear fusion program. The various confinement devices described in the preceding chapter are intended to hold plasma at temperatures as high as $\sim 10^8$ K, high enough for fusion to begin; the difficulty of this task has turned out to be an issue of plasma physics as much as of MHD. The next new venue for plasma research was extraterrestrial. Although it was already understood that Earth was immersed in a tenuous outflow of ionized hydrogen known as the solar wind, the dawn of the space age in 1957 also initiated experimental *space plasma physics*. More recently, the interstellar and intergalactic media beyond the solar system as well as exotic astronomical objects like quasars and pulsars have allowed us to observe plasmas under quite extreme conditions, irreproducible in any laboratory experiment.

The dynamical behavior of a plasma is more complex than the dynamics of the gases and fluids we have met so far. This dynamical complexity has two main origins:

1. The dominant form of interparticle interaction in a plasma, Coulomb scattering, is so weak that the mean free paths of the electrons and ions are often larger than the plasma's macroscopic lengthscales. This allows the particles' momentum distribution functions to deviate seriously from their equilibrium Maxwellian forms and, in particular, to be highly anisotropic.

2. The electromagnetic fields in a plasma are of long range. This allows charged particles to couple to one another electromagnetically and act in concert as modes of excitation (plasma waves) that behave like single dynamical entities

(plasmons). Much of plasma physics consists of the study of the properties and interactions of these modes.

The dynamical behavior of a plasma depends markedly on frequency. At the lowest frequencies, the ions and electrons are locked together by electrostatic forces and behave like an electrically conducting fluid; this is the regime of MHD (Chap. 19). At somewhat higher frequencies, the electrons and the ions can move relative to each other, behaving like two separate, interpenetrating fluids; we study this two-fluid regime in Chap. 21. At still higher frequencies, complex dynamics are supported by momentum space anisotropies and can be analyzed using a variant of the kinetic-theory collisionless Boltzmann equation that we introduced in Chap. 3. We study such dynamics in Chap. 22. In the two-fluid and collisionless-Boltzmann analyses of Chaps. 21 and 22, we focus on phenomena that can be treated as linear perturbations of an equilibrium state. However, the complexities and long mean free paths of plasmas also produce rich nonlinear phenomena; we study some of these in Chap. 23. But first, as a foundation for the dynamical studies in Chaps. 21, 22, and 23, we develop in Chap. 20 detailed insights into the microscopic structure of a plasma.

20

The Particle Kinetics of Plasma

The study of individual particles frequently gives insight
into the behavior of an ionized gas.

LYMAN SPITZER (1962)

20.1 Overview

The preceding chapter, Chap. 19, can be regarded as a transition from fluid mechanics to plasma physics. In the context of a magnetized plasma, it describes equilibrium and low-frequency dynamical phenomena using fluid-mechanics techniques. In this chapter, we prepare for more sophisticated descriptions of a plasma by introducing a number of elementary foundational concepts peculiar to plasmas and by exploring a plasma's structure on the scale of individual particles using elementary techniques from kinetic theory.

Specifically, in Sec. 20.2, we identify the region of densities and temperatures in which matter, in statistical equilibrium, takes the form of a plasma, and we meet specific examples of plasmas that occur in Nature and in the laboratory. Then in Sec. 20.3, we study two phenomena that are important for plasmas: the collective manner in which large numbers of electrons and ions shield out the electric field of a charge in a plasma (Debye shielding), and the oscillations of a plasma's electrons relative to its ions (plasma oscillations).

In Sec. 20.4, we study the Coulomb scattering by which a plasma's electrons and ions deflect an individual charged particle from straight-line motion and exchange energy with it. We then examine the statistical properties of large numbers of such Coulomb scatterings—most importantly, the rates (inverse timescales) for the velocities of a plasma's electrons and ions to isotropize, and the rates for them to thermalize. Our calculations reveal that Coulomb scattering is so weak that, in most plasmas encountered in Nature, it is unlikely to produce isotropized or thermalized velocity distributions. In Sec. 20.5, we give a brief preview of the fact that in real plasmas the scattering of electrons and ions off collective plasma excitations (*plasmons*) often isotropizes and thermalizes their velocities far faster than would Coulomb scattering. Thus many real plasmas are far more isotropic and thermalized than our Coulomb-scattering analyses suggest. (We explore this "anomalous" behavior in Chaps. 22 and 23.) In Sec. 20.5, we also use the statistical properties of Coulomb scatterings to derive a plasma's transport coefficients—specifically, its electrical and thermal conductivities—for situations where Coulomb scattering dominates over particle-plasmon scattering.

Most plasmas are significantly magnetized. This introduces important new features into their dynamics, which we describe in Sec. 20.6: cyclotron motion (the spiraling of particles around magnetic field lines), a resulting anisotropy of the plasma's pressure (different pressures along and orthogonal to the field lines), and the split of a plasma's adiabatic index into four different adiabatic indices for four different types of compression. Finally, in Sec. 20.7, we examine the motion of an individual charged particle in a slightly inhomogeneous and slowly time-varying magnetic field, and we describe *adiabatic invariants,* which control that motion in easily understood ways.

20.2   20.2 Examples of Plasmas and Their Density-Temperature Regimes

The density-temperature regime in which matter behaves as a nonrelativistic plasma is shown in Fig. 20.1. In this figure, and in most of Part VI, we confine our attention to pure hydrogen plasma comprising protons and electrons. Many plasmas contain large fractions of other ions, which can have larger charges and do have greater masses than protons. This generalization introduces few new issues of principle so, for simplicity, we eschew it and focus on a hydrogen plasma.

The boundaries of the plasma regime in Fig. 20.1 are dictated by the following considerations.

20.2.1   20.2.1 Ionization Boundary

We are mostly concerned with fully ionized plasmas, even though partially ionized plasmas, such as the ionosphere, are often encountered in physics, astronomy, and engineering, and more complex plasmas (such as dusty plasmas) can also be important. ionization boundary The plasma regime's ionization boundary is the bottom curve in Fig. 20.1, at a temperature of a few thousand Kelvins. This boundary is dictated by chemical equilibrium for the reaction

$$H \leftrightarrow p + e, \tag{20.1}$$

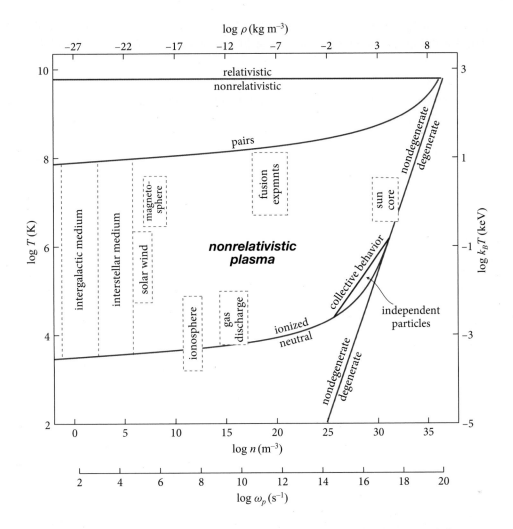

FIGURE 20.1 The density-temperature regime in which matter, made largely of hydrogen, behaves as a nonrelativistic plasma. The densities and temperatures of specific examples of plasmas are indicated by dashed lines. The number density n of electrons (and also of protons) is shown horizontally at the bottom, and the corresponding mass density ρ is shown at the top. The temperature T is shown at the left in Kelvins, and at the right $k_B T$ is shown in keV, thousands of electron volts. The bottom-most scale is the plasma frequency, discussed in Sec. 20.3.3. This figure is a more detailed version of Fig. 3.7.

as described by the Saha equation (Ex. 5.10):

$$\frac{n_e n_p}{n_H} = \frac{(2\pi m_e k_B T)^{3/2}}{h^3} e^{-I_P/(k_B T)}. \tag{20.2}$$

Here n_e, n_p, and n_H are the number densities of electrons, protons, and neutral hydrogen atoms (at the relevant temperatures hydrogen molecules have dissociated into individual atoms), respectively; T is temperature; m_e is the electron rest mass; h is Planck's constant; k_B is Boltzmann's constant; and $I_P = 13.6$ eV is the ionization energy of hydrogen (i.e., the binding energy of its ground state). Notice that the prefactor in Eq. (20.2) is $\lambda_{T\text{dB}}^{-3}$ with $\lambda_{T\text{dB}} \equiv h/\sqrt{2\pi m_e k_B T}$, the electrons' thermal de Broglie wavelength. The bottom boundary in Fig. 20.1 is that of 50% ionization [i.e., $n_e = n_p = n_H = \rho/(2m_H)$, with m_H the mass of a hydrogen atom]; but because of the exponential factor in Eq. (20.2), the line of 90% ionization is virtually indistinguishable from that of 50% ionization on the scale of the figure. Using the rough equivalence $1\,\text{eV} \cong 10^4\,\text{K}$, we might have expected that the ionization boundary would correspond to a temperature $T \sim I_P/k_B \sim 10^5\,\text{K}$. However, this is true only near the degeneracy boundary (see below). When the plasma is strongly nondegenerate, ionization occurs at a significantly lower temperature due to the vastly greater number of states available to an electron when free than when bound in a hydrogen atom. Equivalently, at low densities, once a hydrogen atom has been broken up into an electron plus a proton, the electron (or proton) must travel a large distance before encountering another proton (or electron) with which to recombine, making a new hydrogen atom; as a result, equilibrium occurs at a lowered temperature, where the ionization rate is thereby lowered to match the smaller recombination rate.

20.2.2 Degeneracy Boundary

The electrons, with their small rest masses, become degenerate more easily than the protons or hydrogen atoms. The slanting line on the right side of Fig. 20.1 is the plasma's boundary of electron degeneracy. This boundary is determined by the demand that the mean occupation number of the electrons' single-particle quantum states not be $\ll 1$. In other words, the volume of phase space per electron—the product of the volumes of real space $\sim n_e^{-1}$ and of momentum space $\sim (m_e k_B T)^{3/2}$ occupied by each electron—should be comparable with the elementary quantum mechanical phase-space volume given by the uncertainty principle, h^3. Inserting the appropriate factors of order unity [cf. Eq. (3.41)], this relation becomes the boundary equation:

$$n_e \simeq 2\frac{(2\pi m_e k_B T)^{3/2}}{h^3}. \tag{20.3}$$

When the electrons become degenerate (rightward of the degeneracy line in Fig. 20.1), as they do in a metal or a white dwarf, the electron de Broglie wavelength becomes large compared with the mean interparticle spacing, and quantum mechanical considerations are of paramount importance.

20.2.3 Relativistic Boundary

Another important limit arises when the electron thermal speeds become relativistic. This occurs when

$$T \sim m_e c^2/k_B \sim 6 \times 10^9\,\text{K} \tag{20.4}$$

(top horizontal line in Fig. 20.1). Although we shall not consider them much further, the properties of relativistic plasmas (above this line) are mostly analogous to those of nonrelativistic plasmas (below it).

20.2.4 Pair-Production Boundary

Finally, for plasmas in statistical equilibrium, electron-positron pairs are created in profusion at high enough temperatures. In Ex. 5.9 we showed that, for $k_B T \ll m_e c^2$ but T high enough that pairs begin to form, the density of positrons divided by that of protons is

$$\frac{n_+}{n_p} = \frac{1}{2y[y + (1 + y^2)^{1/2}]}, \quad \text{where } y \equiv \frac{1}{4} n_e \left(\frac{h}{\sqrt{2\pi m_e k_B T}} \right)^3 e^{m_e c^2/(k_B T)}. \quad (20.5)$$

Setting this expression to unity gives the pair-production boundary. This boundary curve, labeled "pairs" in Fig. 20.1, is similar in shape to the ionization boundary but shifted in temperature by $\sim 2 \times 10^4 \sim \alpha_F^{-2}$, where α_F is the fine structure constant. This is because we are now effectively "ionizing the vacuum" rather than a hydrogen atom, and the "ionization potential of the vacuum" is $\sim 2 m_e c^2 = 4 I_P/\alpha_F^2$.

pair-production boundary

We shall encounter a plasma above the pair-production boundary, and thus with a profusion of electron-positron pairs, in our discussion of the early universe in Secs. 28.4.1 and 28.4.2.

20.2.5 Examples of Natural and Human-Made Plasmas

Figure 20.1 and Table 20.1 show the temperature-density regions for the following plasmas:

- *Laboratory gas discharge.* The plasmas created in the laboratory by electric currents flowing through hot gas (e.g., in vacuum tubes, spark gaps, welding arcs, and neon and fluorescent lights).

- *Controlled thermonuclear fusion experiments.* The plasmas in which experiments for controlled thermonuclear fusion are carried out (e.g., in tokamaks).

- *Ionosphere.* The part of Earth's upper atmosphere (at heights of ~ 50–300 km) that is partially photoionized by solar ultraviolet radiation.

- *Magnetosphere.* The plasma of high-speed electrons and ions that are locked on Earth's dipolar magnetic field and slide around on its field lines at several Earth radii.

- *Sun's core.* The plasma at the center of the Sun, where fusion of hydrogen to form helium generates the Sun's heat.

- *Solar wind.* The wind of plasma that blows off the Sun and outward through the region between the planets.

- *Interstellar medium.* The plasma, in our galaxy, that fills the region between the stars. This plasma exhibits a fairly wide range of density and temperature as a result of such processes as heating by photons from stars, heating and compression by shock waves from supernovae, and cooling by thermal emission of radiation.

TABLE 20.1: Representative electron number densities n, temperatures T, and magnetic field strengths B, together with derived plasma parameters in a variety of environments

Plasma	n (m^{-3})	T (K)	B (T)	λ_D (m)	N_D	ω_p (s^{-1})	ν_{ee} (s^{-1})	ω_c (s^{-1})	r_L (m)
Gas discharge	10^{16}	10^4	—	10^{-4}	10^4	10^{10}	10^5	—	—
Fusion experiments	10^{20}	10^8	10	10^{-4}	10^8	10^{12}	10^4	10^{12}	10^{-5}
Ionosphere	10^{12}	10^3	10^{-5}	10^{-3}	10^5	10^8	10^3	10^6	10^{-1}
Magnetosphere	10^7	10^7	10^{-8}	10^2	10^{10}	10^5	10^{-8}	10^3	10^4
Sun's core	10^{32}	10^7	—	10^{-11}	1	10^{18}	10^{16}	—	—
Solar wind	10^6	10^5	10^{-9}	10	10^{11}	10^5	10^{-6}	10^2	10^4
Interstellar medium	10^5	10^4	10^{-10}	10	10^{10}	10^4	10^{-5}	10	10^4
Intergalactic medium	10^{-1}	10^6	—	10^5	10^{15}	10^2	10^{-12}	—	—

Notes: The derived parameters, discussed later in this chapter, are: λ_D, Debye length; N_D, Debye number; ω_p, plasma frequency; ν_{ee}, equilibration rate for electron energy; ω_c, electron cyclotron frequency; and r_L, electron Larmor radius. Values are given to order of magnitude, as all of these environments are quite inhomogeneous. — indicates this quantity is irrelevant.

- *Intergalactic medium.* The plasma that fills the space outside galaxies and clusters of galaxies (the locale of almost all the universe's plasma). We shall meet the properties and evolution of this intergalactic plasma in our study of cosmology, Sec. 28.3.1.

Characteristic plasma properties in these various environments are collected in Table 20.1. In the next three chapters we study applications from all these environments.

EXERCISES

Exercise 20.1 *Derivation: Boundary of Degeneracy*
Show that the condition $n_e \ll (2\pi m_e k_B T)^{3/2}/h^3$ [cf. Eq. (20.3)] that electrons be nondegenerate is equivalent to the following statements.

(a) The mean separation between electrons, $l \equiv n_e^{-1/3}$, is large compared to the electrons' thermal de Broglie wavelength, $\lambda_{T\,dB} = h/\sqrt{2\pi m_e k_B T}$.

(b) The uncertainty in the location of an electron drawn at random from the thermal distribution is small compared to the average inter-electron spacing.

(c) The quantum mechanical zero-point energy associated with squeezing each electron into a region of size $l = n_e^{-1/3}$ is small compared to the electron's mean thermal energy $k_B T$.

In this section, we introduce two key ideas that are associated with most of the collective effects in plasma dynamics: Debye shielding (also called Debye screening) and plasma oscillations.

20.3.1 Debye Shielding

Any charged particle inside a plasma attracts other particles with opposite charge and repels those with the same charge, thereby creating a net cloud of opposite charges around itself. This cloud shields the particle's own charge from external view (i.e., it causes the particle's Coulomb field to fall off exponentially at large radii, rather than falling off as $1/r^2$).[1]

This Debye shielding (or screening) of a particle's charge can be demonstrated and quantified as follows. Consider a single fixed test charge Q surrounded by a plasma of protons and electrons. Denote the average densities of protons and electrons—considered as smooth functions of radius r from the test charge—by $n_p(r)$ and $n_e(r)$, and let the mean densities of electrons and protons (averaged over a large volume) be $\bar{n}$. (The mean electron and proton densities must be equal because of overall charge neutrality, which is enforced by the electrostatic restoring force that we explore in Sec. 20.3.3.) Then the electrostatic potential $\Phi(r)$ outside the particle satisfies Poisson's equation, which we write in SI units:[2]

$$\nabla^2 \Phi = -\frac{(n_p - n_e)e}{\epsilon_0} - \frac{Q}{\epsilon_0}\delta(\mathbf{r}).$$

(20.6)

(We denote the positive charge of a proton by $+e$ and the negative charge of an electron by $-e$.)

A proton at radius r from the particle has an electrostatic potential energy $e\Phi(r)$. Correspondingly, the number density of protons at radius r is altered from $\bar{n}$ by the Boltzmann factor $\exp[-e\Phi/(k_B T)]$; and, similarly, the density of electrons is altered by $\exp[+e\Phi/(k_B T)]$:

$$n_p = \bar{n}\exp[-e\Phi/(k_B T)] \simeq \bar{n}[1 - e\Phi/(k_B T)],$$

$$n_e = \bar{n}\exp[+e\Phi/(k_B T)] \simeq \bar{n}[1 + e\Phi/(k_B T)],$$

(20.7)

where we have made a Taylor expansion of the Boltzmann factor valid for $|e\Phi| \ll k_B T$. By inserting the linearized versions of Eqs. (20.7) into Eq. (20.6), we obtain

$$\nabla^2 \Phi = \frac{2e^2\bar{n}}{\epsilon_0 k_B T}\Phi - \frac{Q}{\epsilon_0}\delta(\mathbf{r}).$$

(20.8)

---

1. Analogous effects are encountered in condensed-matter physics and quantum electrodynamics.
2. For those who prefer Gaussian units, the translation is most easily effected by the transformations $4\pi\epsilon_0 \to 1$ and $\mu_0/(4\pi) \to 1$, and inserting factors of c by inspection using dimensional analysis. It is also useful to recall that $1\,\text{T} \equiv 10^4$ Gauss and that the charge on an electron is $-1.602 \times 10^{-19}\,\text{C} \equiv -4.803 \times 10^{-10}$ esu.

The spherically symmetric solution to this equation,

$$\Phi = \frac{Q}{4\pi\epsilon_0 r}e^{-\sqrt{2}\,r/\lambda_D},$$

(20.9)

has the form of a Coulomb field with an exponential cutoff. The characteristic length-scale of the exponential cutoff,

Debye length

$$\lambda_D \equiv \left(\frac{\epsilon_0 k_B T}{\bar{n}e^2}\right)^{1/2} = 69\left(\frac{T/1\,\mathrm{K}}{\bar{n}/1\,\mathrm{m}^{-3}}\right)^{1/2}\mathrm{m},$$

(20.10)

is called the *Debye length*. It is a rough measure of the size of the Debye shielding cloud that the charged particle carries with itself.

The charged particle could be some foreign charged object (not a plasma electron or proton), or equally well, it could be one of the plasma's own electrons or protons. Thus, we can think of each electron in the plasma as carrying with itself a positively charged Debye shielding cloud of size λ_D, and each proton as carrying a negatively charged cloud. Each electron and proton not only carries its own cloud; it also plays a role as one of the contributors to the clouds around other electrons and protons.

20.3.2

20.3.2 Collective Behavior

A charged particle's Debye cloud is almost always made of a huge number of electrons and very nearly the same number of protons. It is only a tiny, time-averaged excess of electrons over protons (or protons over electrons) that produces the cloud's net charge and the resulting exponential decay of the electrostatic potential. Ignoring this tiny excess, the mean number of electrons in the cloud and the mean number of protons are roughly

Debye number

$$N_D \equiv \bar{n}\frac{4\pi}{3}\lambda_D^3 = 1.4\times10^6\frac{(T/1\,\mathrm{K})^{3/2}}{(\bar{n}/1\,\mathrm{m}^{-3})^{1/2}}.$$

(20.11)

This *Debye number* is large compared to unity throughout the density-temperature regime of plasmas, except for the tiny lower right-hand corner of Fig. 20.1. The boundary of that region (labeled "collective behavior"/"independent particles") is given by $N_D = 1$. The upper left-hand side of that boundary has $N_D \gg 1$ and is called the regime of collective behavior, because a huge number of particles are collectively responsible for the Debye cloud; this leads to a variety of collective dynamical phenomena in the plasma. The lower right-hand side has $N_D < 1$ and is called the regime of independent particles, because in it collective phenomena are of small importance. (In this regime Φ on the right-hand side of Eq. (20.8) is replaced by $(kT/e)\sinh[e\Phi/(k_B T)]$.) In this book we restrict ourselves to the extensive regime of collective behavior and ignore the tiny regime of independent particles.

Characteristic values for the Debye length and Debye number in a variety of environments are listed in Table 20.1.

Of all the dynamical phenomena that can occur in a plasma, perhaps the most important is a relative oscillation of the plasma's electrons and protons. A simple, idealized version of this *plasma oscillation* is depicted in Fig. 20.2. Suppose for the moment that the protons are all fixed, and displace the electrons rightward (in the x direction) with respect to the protons by an amount ξ, thereby producing a net negative charge per unit area $-e\bar{n}\xi$ at the right end of the plasma, a net positive charge per unit area $+e\bar{n}\xi$ at the left end, and a corresponding electric field $E = e\bar{n}\xi/\epsilon_0$ in the x direction throughout the plasma. The electric field pulls on the plasma's electrons and protons, giving the electrons an acceleration $d^2\xi/dt^2 = -eE/m_e$ and the protons an acceleration smaller by $m_e/m_p = 1/1{,}836$, which we neglect. The result is an equation of motion for the electrons' collective displacement:

$$\frac{d^2\xi}{dt^2} = -\frac{e}{m_e}E = -\frac{e^2\bar{n}}{\epsilon_0 m_e}\xi. \tag{20.12}$$

Since Eq. (20.12) is a harmonic-oscillator equation, the electrons oscillate sinusoidally, $\xi = \xi_o\,\cos(\omega_p t)$, at the *plasma* (angular) *frequency*

$$\boxed{\omega_p \equiv \left(\frac{\bar{n}e^2}{\epsilon_0 m_e}\right)^{1/2} = 56.3\left(\frac{\bar{n}}{1\,\mathrm{m}^{-3}}\right)^{1/2}\,\mathrm{s}^{-1}.} \tag{20.13}$$

plasma frequency, for plasma oscillations

Notice that this frequency of plasma oscillations depends only on the plasma density $\bar{n}$ and not on its temperature. Note that, if we define the electron thermal speed to be $v_e \equiv (k_B T_e/m_e)^{1/2}$, then $\omega_p \equiv v_e/\lambda_D$. In other words, a thermal electron travels roughly a Debye length in a plasma oscillation period. We can think of the Debye length as the electrostatic correlation length, and the plasma period as the electrostatic correlation time.

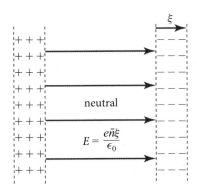

FIGURE 20.2 Idealized depiction of the displacement of electrons relative to protons, which occurs during plasma oscillations.

Characteristic values for the plasma frequency in a variety of environments are listed in Table 20.1 and depicted on the bottom-most scale of Fig. 20.1.

20.4 Coulomb Collisions

20.4

In this section and the next, we study transport coefficients (electrical and thermal conductivities) and the establishment of local thermodynamic equilibrium in a plasma under the hypothesis that Coulomb collisions are the dominant source of scattering for both electrons and protons. In fact, as we shall see later, Coulomb scattering is usually a less-effective scattering mechanism than *collisionless* processes mediated by fluctuating electromagnetic fields (plasmons).

20.4.1 Collision Frequency

20.4.1

Consider first, as we did in our discussion of Debye shielding, a single test particle—let it be an electron—interacting with background field particles—let these be protons for the moment. The test electron moves with speed v_e. The field protons will move much more slowly if the electron and protons are near thermodynamic equilibrium (as the proton masses are much greater than that of the electron), so the protons can be treated, for the moment, as at rest. When the electron flies by a single proton, we can characterize the encounter using an *impact parameter b*, which is what the distance of closest approach would have been if the electron were not deflected; see Fig. 20.3. The electron will be scattered by the Coulomb field of the proton, a process sometimes called *Rutherford scattering*. If the deflection angle is small, $\theta_D \ll 1$, we can approximate its value by computing the perpendicular impulse exerted by the proton's Coulomb field, integrating along the electron's unperturbed straight-line trajectory:

Rutherford scattering

$$m_e v_e \theta_D = \int_{-\infty}^{+\infty} \frac{e^2 b}{4\pi\epsilon_o (b^2 + v_e^2 t^2)^{3/2}} dt = \frac{e^2}{2\pi\epsilon_o v_e b}, \tag{20.14a}$$

which implies that

$$\theta_D = b_o/b \text{ for } b \gg b_o, \tag{20.14b}$$

where

$$b_o \equiv \frac{e^2}{2\pi\epsilon_0 m_e v_e^2}. \tag{20.14c}$$

When $b \lesssim b_o$, this approximation breaks down, and the deflection angle is of order 1 radian.[3]

Below we shall need to know how much energy the electron loses, for the large-impact-parameter case. That energy loss, $-\Delta E$, is equal to the energy gained by the

---

3. A more careful calculation gives $2\tan(\theta_D/2) = b_o/b$; see, e.g., Bellan (2006, Assignment 1 in Sec. 1.13).

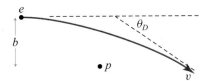

FIGURE 20.3 The geometry of a Coulomb collision.

proton. Since the proton is initially at rest, and since momentum conservation implies it gains a momentum $\Delta p = m_e v_e \theta_D$, ΔE must be

$$\Delta E = -\frac{(\Delta p)^2}{2m_p} = -\frac{m_e}{m_p}\left(\frac{b_o}{b}\right)^2 E, \quad \text{for } b \gg b_o. \tag{20.15}$$

Here $E = \frac{1}{2}m_e v_e^2$ is the electron's initial energy.

We turn from an individual Coulomb collision to the net, statistically averaged effect of many collisions. The first thing we compute is the mean time t_D required for the orbit of the test electron to be deflected by an angle of order 1 radian from its initial direction, and the inverse of t_D, which we call the "deflection rate" or "deflection frequency" and denote $\nu_D = 1/t_D$. If the dominant source of this deflection were a single large-angle scattering event, then the relevant cross section would be $\sigma = \pi b_o^2$ (since all impact parameters $\lesssim b_o$ produce large-angle scatterings), and the mean deflection time and frequency would be

$$\nu_D \equiv \frac{1}{t_D} = n\sigma v_e = n\pi b_o^2 v_e. \tag{20.16}$$

time and frequency for
a single large-angle
deflection of an electron
by a proton

Here n is the proton number density, which is the same as the electron number density in our hydrogen plasma.

The cumulative, random-walk effects of many small-angle scatterings off field protons actually produce a net deflection of order 1 radian in a time shorter than this. As the directions of the individual scatterings are random, the mean deflection angle after many scatterings vanishes. However, the mean-square deflection angle, $\langle \Theta^2 \rangle = \sum_{\text{all encounters}} \theta_D{}^2$, will not vanish. That mean-square deflection angle, during a time t, accumulates to

$$\langle \Theta^2 \rangle = \int_{b_{\min}}^{b_{\max}} \left(\frac{b_o}{b}\right)^2 n v_e t\, 2\pi b\, db = n2\pi b_o^2 v_e t \, \ln\left(\frac{b_{\max}}{b_{\min}}\right). \tag{20.17}$$

mean-square deflection
from small-angle
scatterings

Here the factor $(b_o/b)^2$ in the integrand is the squared deflection angle $\theta_D{}^2$ for impact parameter b, and the remaining factor $n v_e t\, 2\pi b\, db$ is the number of encounters that occur with impact parameters between b and $b + db$ during time t. The integral diverges logarithmically at both its lower limit $b_{\min}$ and its upper limit $b_{\max}$. In Sec. 20.4.2 we discuss the physical origins of and values of the cutoffs $b_{\min}$ and $b_{\max}$. The value of t that makes the mean-square deflection angle $\langle \Theta^2 \rangle$ equal to unity is, to within factors of order unity, the deflection time t_D (and inverse deflection frequency v_D^{-1}):

time and frequency for large net deflection of an electron by protons

$$\boxed{v_D^{ep} = \frac{1}{t_D^{ep}} = n2\pi b_o^2 v_e \ln\Lambda = \frac{ne^4 \ln\Lambda}{2\pi \epsilon_0^2 m_e^2 v_e^3}, \quad \text{where } \Lambda = b_{\max}/b_{\min}.} \tag{20.18a}$$

Here the superscript ep indicates that the test particle is an electron and the field particles are protons. Notice that this deflection frequency is larger, by a factor $2\ln\Lambda$, than the frequency (20.16) for a single large-angle scattering.

We must also consider the repulsive collisions of our test electron with field electrons. Although we are no longer justified in treating the field electrons as being at rest, the impact parameter for a large angle deflection is still $\sim b_0$, so Eq. (20.18a) is also appropriate to this case, in order of magnitude:

time and frequency for large net deflection of an electron by electrons

$$\boxed{v_D^{ee} = \frac{1}{t_D^{ee}} \sim v_D^{ep} = n2\pi b_0^2 v_e \ln\Lambda = \frac{ne^4 \ln\Lambda}{2\pi \epsilon_0^2 m_e^2 v_e^3}.} \tag{20.18b}$$

Finally, and in the same spirit, we can compute the collision frequency for the protons. Because electrons are so much lighter than protons, proton-proton collisions will be more effective in deflecting protons than are proton-electron collisions. Therefore, the proton collision frequency is given by Eq. (20.18b) with the electron subscripts replaced by proton subscripts:

time and frequency for large net deflection of a proton by protons

$$\boxed{v_D^{pp} = \frac{1}{t_D^{pp}} \sim \frac{ne^4 \ln\Lambda}{2\pi \epsilon_0^2 m_p^2 v_p^3}.} \tag{20.18c}$$

20.4.2 The Coulomb Logarithm

20.4.2

The maximum impact parameter $b_{\max}$, which appears in $\Lambda \equiv b_{\max}/b_{\min}$, is the Debye length λ_D, since for impact parameters $b \gg \lambda_D$ the Debye shielding screens out a field particle's Coulomb field, while for $b \ll \lambda_D$ Debye shielding is unimportant.

The minimum impact parameter $b_{\min}$ has different values, depending on whether quantum mechanical wave-packet spreading is important for the test particle during the collision. Because of wave-packet spreading, the nearest the test particle can come to a field particle is the test particle's de Broglie wavelength: $b_{\min} = \hbar/mv$. However, if the de Broglie wavelength is smaller than b_0, then the effective value of $b_{\min}$ will be simply b_0. Therefore, in $\ln\Lambda = \ln(b_{\max}/b_{\min})$ (the *Coulomb logarithm*), we have

Coulomb logarithm

$$b_{\min} = \max[b_o = e^2/(2\pi\epsilon_0 m_e v_e^2), \ \hbar/(m_e v_e)], \quad \text{and} \quad b_{\max} = \lambda_D \quad \text{for test electrons;}$$
$$b_{\min} = \max[b_o = e^2/(2\pi\epsilon_0 m_p v_p^2), \ \hbar/(m_p v_p)], \quad \text{and} \quad b_{\max} = \lambda_D \quad \text{for test protons.}$$

$$\tag{20.19}$$

Over most of the accessible range of density and temperature for a plasma, we have $3 \lesssim \ln\Lambda \lesssim 30$. Therefore, if we set

$$\ln \Lambda \simeq 10, \qquad (20.20)$$

our estimate is good to a factor ~ 3. For numerical values of $\ln \Lambda$ in a thermalized plasma, see Spitzer (1962, Table 5.1).

Logarithmic terms, closely related to the Coulomb logarithm, arise in a variety of other situations where one is dealing with a field whose potential varies as $1/r$ and force as $1/r^2$—and for the same reason: one encounters integrals of the form $\int (1/r) dr$ that diverge logarithmically. Specific examples are (i) the *Gaunt factors*, which arise in the theory of bremsstrahlung radiation (electromagnetic waves from Coulomb scattering of charged particles), (ii) *Coulomb wave functions* (solutions to the Schrödinger equation for the quantum mechanical theory of Coulomb scattering), and (iii) the Shapiro time delay for electromagnetic waves that travel past a general relativistic gravitating body, such as the Sun (Sec. 27.2.4).

Exercise 20.2 *Problem: The Coulomb Logarithm*

(a) Express the Coulomb logarithm in terms of the Debye number, N_D, in the classical regime, where $b_{min} \sim b_0$.

(b) What range of electron temperatures corresponds to the classical regime, and what range to the quantum regime?

(c) Use the representative parameters from Table 20.1 to evaluate Coulomb logarithms for the Sun's core, a tokamak, and the interstellar medium, and verify that they lie in the range $3 \lesssim \ln \Lambda \lesssim 30$.

Exercise 20.3 *Example: Parameters for Various Plasmas*

Estimate the Debye length λ_D, the Debye number N_D, the plasma frequency $f_p \equiv \omega_p/2\pi$, and the electron deflection timescale $t_D^{ee} \sim t_D^{ep}$, for the following plasmas.

(a) An atomic bomb explosion in Earth's atmosphere 1 ms after the explosion. (Use the Sedov-Taylor similarity solution for conditions behind the bomb's shock wave; Sec. 17.6.)

(b) The ionized gas that enveloped the Space Shuttle (Box 17.4) as it reentered Earth's atmosphere.

(c) The expanding universe during its early evolution, just before it became cool enough for electrons and protons to combine to form neutral hydrogen (i.e., just before ionization "turned off"). (As we shall discuss in Chap. 28, the universe today is filled with blackbody radiation, produced in the big bang, that has a temperature $T = 2.7$ K, and the universe today has a mean density of hydrogen $\rho \sim 4 \times 10^{-28}$ kg m^{-3}. Extrapolate backward in time to infer the density and temperature at the epoch just before ionization turned off.)

20.4.3 Thermal Equilibration Rates in a Plasma

Suppose that a hydrogen plasma is heated in some violent way (e.g., by a shock wave). Such heating will typically give the plasma's electrons and protons a non-Maxwellian velocity distribution. Coulomb collisions then, as time passes (and in the absence of more violent disruptions), will force the particles to exchange energy in random ways, gradually driving them into thermal equilibrium. As we shall see, thermal equilibration is achieved at different rates for the electrons and the protons. Correspondingly, the following three timescales are all different:

$$t_{ee}^{\text{eq}} \equiv \left(\begin{array}{c} \text{time required for electrons to equilibrate with one another,} \\ \text{achieving a near-Maxwellian velocity distribution} \end{array} \right),$$

$$t_{pp}^{\text{eq}} \equiv \left(\text{time for protons to equilibrate with one another} \right),$$

$$t_{ep}^{\text{eq}} \equiv \left(\text{time for electrons to equilibrate with protons} \right).$$

$$(20.21)$$

In this section we compute these three equilibration times.

ELECTRON-ELECTRON EQUILIBRATION

To evaluate t_{ee}^{eq}, we assume that the electrons begin with typical individual energies of order $k_B T_e$, where T_e is the temperature to which they are going to equilibrate, but their initial velocity distribution is rather non-Maxwellian. Then we can choose a typical electron as the "test particle." We have argued that Coulomb interactions with electrons and protons are comparably effective in deflecting test electrons. However, they are not comparably effective in transferring energy. When the electron collides with a stationary proton, the energy transfer is

$$\frac{\Delta E}{E} \simeq -\frac{m_e}{m_p} \theta_D{}^2 \qquad (20.22)$$

[Eq. (20.15)]. This is smaller than the typical energy transfer in an electron-electron collision by the ratio m_e/m_p. Therefore, the collisions between electrons are responsible for establishing an electron Maxwellian distribution function.

The alert reader may spot a problem at this point. According to Eq. (20.22), electrons always lose energy to protons and never gain it. This would cause the electron temperature to continue to fall below the proton temperature, in clear violation of the second law of thermodynamics. Actually what happens in practice is that, if we allow for nonzero proton velocities, then the electrons can gain energy from some electron-proton collisions. This is also the case for the electron-electron collisions of immediate concern.

The most accurate formalism for dealing with this situation is the Fokker-Planck formalism, discussed in Sec. 6.9 and made explicit for Coulomb scattering in Ex. 20.8. Fokker-Planck is appropriate because, as we have shown, many weak scatterings dominate the few strong scatterings. If we use the Fokker-Planck approach to compute an energy equilibration time for a nearly Maxwellian distribution of electrons with temperature T_e, then it turns out that a simple estimate, based on combining the deflection time (20.16) with the typical energy transfer (20.22) (with $m_p \to m_e$), and on assuming a typical electron velocity $v_e = (3k_B T_e/m_e)^{1/2}$, gives an answer good to a factor of 2. It is actually convenient to express this electron energy equilibration timescale using its reciprocal, the *electron-electron equilibration rate*, v_{ee}. This facilitates comparison with the other frequencies characterizing the plasma. The true Fokker-Planck result for electrons near equilibrium is then

$$\boxed{\begin{aligned} v_{ee} = \frac{1}{t_{ee}} &= \frac{n\sigma_T c \ln \Lambda}{2\pi^{1/2}} \left(\frac{k_B T_e}{m_e c^2}\right)^{-3/2} \\ &= 2.6 \times 10^{-5} \left(\frac{n}{1\,\text{m}^{-3}}\right) \left(\frac{T_e}{1\,\text{K}}\right)^{-3/2} \left(\frac{\ln \Lambda}{10}\right)\,\text{s}^{-1} \end{aligned}}$$

(20.23a)   electron-electron equilibration rate

where we have used the Thomson cross section

$$\sigma_T = \frac{8\pi}{3}\left(\frac{e^2}{4\pi\epsilon_0 m_e c^2}\right)^2 = 6.65 \times 10^{-29}\,\text{m}^2. \qquad (20.23\text{b})$$

PROTON-PROTON EQUILIBRATION

As for proton deflections [Eq. (20.18c)], so also for proton energy equilibration: the light electrons are far less effective at influencing the protons than are other protons. Therefore, the protons achieve a thermal distribution by equilibrating with one another, and their *proton-proton equilibration rate* can be written down immediately from Eq. (20.23a) by replacing the electron masses and temperatures with the proton values:

$$\boxed{\begin{aligned} v_{pp} = \frac{1}{t_{pp}} &= \frac{n\sigma_T c \ln \Lambda}{2\pi^{1/2}} \left(\frac{m_e}{m_p}\right)^{1/2}\left(\frac{k_B T_p}{m_e c^2}\right)^{-3/2} \\ &= 6.0 \times 10^{-7} \left(\frac{n}{1\,\text{m}^{-3}}\right) \left(\frac{T_p}{1\,\text{K}}\right)^{-3/2} \left(\frac{\ln \Lambda}{10}\right)\,\text{s}^{-1} \end{aligned}}$$

(20.23c)   proton-proton equilibration rate

ELECTRON-PROTON EQUILIBRATION

Finally, if the electrons and protons have different temperatures, we should compute the timescale for the two species to equilibrate with each other. This again is easy to estimate using the energy-transfer equation (20.22): $t_{ep} \simeq (m_p/m_e)t_{ee}$. The more accurate Fokker-Planck result for the *electron-proton equilibration rate* is again close to this value and is given by

$$\boxed{\begin{aligned} v_{ep} = \frac{1}{t_{ep}} &= \frac{2n\sigma_T c \ln \Lambda}{\pi^{1/2}} \left(\frac{m_e}{m_p}\right)\left(\frac{k_B T_e}{m_e c^2}\right)^{-3/2} \\ &= 5.6 \times 10^{-8} \left(\frac{n}{1\,\text{m}^{-3}}\right) \left(\frac{T_e}{1\,\text{K}}\right)^{-3/2} \left(\frac{\ln \Lambda}{10}\right)\,\text{s}^{-1} \end{aligned}}$$

(20.23d)   electron-proton equilibration rate

Thus, at the same density and temperature, protons require $\sim(m_p/m_e)^{1/2} = 43$ times longer to reach thermal equilibrium among themselves than do the electrons, and proton-electron equilibration takes a time $\sim (m_p/m_e) = 1{,}836$ longer than electron-electron equilibration.

20.4.4 Discussion

Table 20.1 lists the electron-electron equilibration rates for a variety of plasma environments. Generally, they are very small compared with the plasma frequencies. For example, if we take parameters appropriate to fusion experiments (e.g., a tokamak), we find that $\nu_{ee} \sim 10^{-8}\omega_p$ and $\nu_{ep} \sim 10^{-11}\omega_p$. In fact, the equilibration time is comparable, to order of magnitude, with the total plasma confinement time ~ 0.1 s (cf. Sec. 19.3). The disparity between ν_{ee} and ω_p is even greater in the interstellar medium. For this reason most plasmas are well described as collisionless, and we must anticipate that the particle distribution functions will depart significantly from Maxwellian.

EXERCISES

Exercise 20.4 *Derivation: Electron-Electron Equilibration Rate*
Using the non-Fokker-Planck arguments outlined in the text, compute an estimate of the electron-electron equilibration rate, and show that it agrees with the Fokker-Planck result, Eq. (20.23a), to within a factor of 2.

Exercise 20.5 *Problem: Dependence of Thermal Equilibration on Charge and Mass*
Compute the ion equilibration rate for a pure He^3 plasma with electron density 10^{20} m^{-3} and temperature 10^8 K.

Exercise 20.6 *Example: Stopping of α-Particles*
A 100-MeV α-particle is incident on a plastic object. Estimate the distance that it will travel before coming to rest. This is known as the particle's *range*. [Hints: (i) Debye shielding does not occur in a plastic. (Why?) (ii) The α-particle loses far more energy to Coulomb scattering off electrons than off atomic nuclei. (Why?) Estimate the electron density as $n_e = 2 \times 10^{29}$ m^{-3}. (iii) Ignore relativistic corrections and refinements, such as the so-called *density effect*. (iv) Consider the appropriate values of b_{max} and b_{min}.]

Exercise 20.7 *Problem: Equilibration Time for a Globular Star Cluster*
Collections of stars have many similarities to a plasma of electrons and ions. These similarities arise from the fact that in both cases the interaction between the individual particles (stars, or ions and electrons) is a radial, $1/r^2$ force. The principal difference is that the force between stars is always attractive, so there is no analog of Debye shielding. One consequence of this difference is that a plasma can be spatially homogeneous and static, when one averages over lengthscales large compared to the interparticle separation; but a collection of stars cannot be—the stars congregate into clusters that are held together by the stars' mutual gravity.

A globular star cluster is an example. A typical globular cluster is a nearly spherical swarm of stars with the following parameters: cluster radius $\equiv R = 10$ light-years;

total number of stars in the cluster $\equiv N = 10^6$; and mass of a typical star $\equiv m = 0.4$ solar masses $= 8 \times 10^{32}$ g. Each star moves on an orbit of the average, "smeared out" gravitational field of the entire cluster. Since that smeared-out gravitational field is independent of time, each star conserves its total energy (kinetic plus gravitational) as it moves. Actually, the total energy is only approximately conserved. Just as in a plasma, gravitational "Coulomb collisions" of the star with other stars produce changes in the star's energy.

(a) What is the mean time t_E for a typical star in a globular cluster to change its energy substantially? Express your answer, accurate to within a factor of $\sim$3, in terms of N, R, and m. Evaluate it numerically and compare it with the age of the universe.

(b) The cluster evolves substantially on the timescale t_E. What type of evolution would you expect to occur? What type of stellar energy distribution would you expect to result from this evolution?[4]

Exercise 20.8 **Example and Challenge: Fokker-Planck Formalism* *for Coulomb Collisions* **T2**

Consider two families of particles that interact via Coulomb collisions (e.g., electrons and protons, or electrons and electrons). Regard one family as test particles, with masses m and charges q and a kinetic-theory distribution function $f(\mathbf{v}) = dN/dV_x dV_v$ that is homogeneous in space and thus only a function of the test particles' velocities $\mathbf{v}$. (For discussion of this distribution function, see Sec. 3.2.3, and in greater detail, Sec. 22.2.1.) Regard the other family as field particles, with masses m' and charges q' and a spatially homogeneous distribution function $f_F(\mathbf{v}')$.

The Fokker-Planck equation (Sec. 6.9.3) can be used to give a rather accurate description of how the field particles' distribution function $f(\mathbf{v}, t)$ evolves due to Coulomb collisions with the test particles. In this exercise, you will work out explicit versions of this Fokker-Planck equation. These explicit versions are concrete foundations for deducing Eqs. (20.23) for the electron and proton equilibration rates, Eqs. (20.26) for the thermal and electrical conductivities of a plasma, and other evolutionary effects of Coulomb collisions.

(a) Explain why, for a spatially homogeneous plasma with the only electromagnetic fields present being the Coulomb fields of the plasma particles, the collisionless Boltzmann transport equation (3.65) (with $\mathcal{N}$ replaced by f and p_j by v_j) becomes simply $\partial f/\partial t = 0$. Then go on to explain why the collision terms in the collisional Boltzmann transport equation (3.66) are of Fokker-Planck form, Eq. (6.106a):

$$\frac{\partial f}{\partial t} = -\frac{\partial}{\partial v_j}\left[A_j(\mathbf{v}) f(\mathbf{v}, t)\right] + \frac{1}{2}\frac{\partial^2}{\partial v_j \partial v_k}\left[B_{jk}(\mathbf{v}) f(\mathbf{v}, t)\right], \qquad \text{(20.24a)}$$

---

4. For a detailed discussion, see, e.g., Binney and Tremaine (2003).

where A_j and B_{jk} are, respectively, the slowdown rate and velocity diffusion rate of the test particles due to Coulomb scatterings off the field particles:

$$A_j = \lim_{\Delta t \to 0} \left(\frac{\overline{\Delta v_j}}{\Delta t} \right), \quad B_{jk} = \lim_{\Delta t \to 0} \left(\frac{\overline{\Delta v_j \Delta v_k}}{\Delta t} \right)$$

[Eqs. (6.106b) and (6.106c)].

(b) Analyze, in the center-of-mass reference frame, the small-angle Coulomb scattering of a test particle off a field particle (with some chosen impact parameter b and relative velocity $\mathbf{v}_{\mathrm{rel}} = \mathbf{v} - \mathbf{v}' =$ difference between lab-frame velocities), and then transform to the laboratory frame where the field particles on average are at rest. Then add up the scattering influences of all field particles on the chosen test particle (i.e., integrate over b and over $\mathbf{v}'$) to show that the test particle's rate of change of velocity is

$$A_j = \Gamma \frac{\partial}{\partial v_j} h(\mathbf{v}). \tag{20.24b}$$

Here the constant Γ and the *first Rosenbluth potential h* are

$$\Gamma = \frac{q^2 q'^2 \ln \Lambda}{4\pi \epsilon_0^2 m^2}, \quad h(\mathbf{v}) = \frac{m}{\mu} \int \frac{f_F(\mathbf{v}')}{|\mathbf{v} - \mathbf{v}'|} d\mathcal{V}_{v'}. \tag{20.24c}$$

In addition, $\mu = mm'/(m + m')$ is the reduced mass, which appears in the center-of-mass-frame analysis of the Coulomb collision; $\Lambda = b_{\max}/b_{\min}$ is the ratio of the maximum to the minimum possible impact parameters as in Sec. 20.4.2; and $d\mathcal{V}_{v'} = dv'_x dv'_y dv'_z$ is the volume element in velocity space. Although $b_{\min}$ depends on the particles' velocities, without significant loss of accuracy we can evaluate it for some relevant mean velocity and pull it out from under the integral, as we have done. This is because $\Lambda = b_{\max}/b_{\min}$ appears only in the logarithm.

(c) By the same method as in part (b), show that the velocity diffusion rate is

$$B_{jk}(\mathbf{v}) = \Gamma \frac{\partial^2}{\partial v_j \partial v_k} g(\mathbf{v}), \tag{20.24d}$$

where the *second Rosenbluth potential* is

$$g(\mathbf{v}) = \int |\mathbf{v} - \mathbf{v}'| f_F(\mathbf{v}') d\mathcal{V}_{v'}. \tag{20.24e}$$

(d) Show that, with A_j and B_{jk} expressed in terms of the Rosenbluth potentials [Eqs. (20.24b) and (20.24d)], the Fokker-Planck equation (20.24a) can be brought into the following alternative form:

$$\frac{\partial f}{\partial t} = \frac{\Gamma m}{2} \frac{\partial}{\partial v_j} \int \frac{\partial^2 |\mathbf{v} - \mathbf{v}'|}{\partial v_j \partial v_k} \left[\frac{f_F(\mathbf{v}')}{m} \frac{\partial f(\mathbf{v})}{\partial v_k} - \frac{f(\mathbf{v})}{m'} \frac{\partial f_F(\mathbf{v}')}{\partial v'_k} \right] d\mathcal{V}_{v'}. \tag{20.25}$$

[Hint: Manipulate the partial derivatives in a wise way and perform integrations by parts.]

(e) Using Eq. (20.25), show that, if the two species—test and field—are thermalized at the same temperature, then the test-particle distribution function will remain unchanged: $\partial f/\partial t = 0$.

For details of the solution to this exercise see, for example, the original paper by Rosenbluth, MacDonald, and Judd (1957), or the pedagogical treatments in standard plasma physics textbooks (e.g., Boyd and Sanderson, 2003, Sec. 8.4; Bellan, 2006, Sec. 13.2).

The explicit form (20.24) of the Fokker-Planck equation has been solved approximately for idealized situations. It has also been integrated numerically for more realistic situations, so as to evolve the distribution functions of interacting particle species and, for example, watch them thermalize. (See, e.g., Bellan, 2006, Sec. 13.2; Boyd and Sanderson, 2003, Secs. 8.5, 8.6; and Shkarofsky, Johnston, and Bachynski, 1966, Secs. 7.5–7.11.) An extension to a plasma with an applied electric field and/or temperature gradient (using techniques developed in Sec. 3.7.3) has been used to deduce the electrical and thermal conductivities discussed in the next section (see, e.g., Shkarofsky, Johnston, and Bachynski, 1966, Chap. 8).

20.5 Transport Coefficients

Because electrons have far lower masses than ions, they have far higher typical speeds at fixed temperature and are much more easily accelerated (i.e., they are much more mobile). As a result, it is the motion of the electrons, not the ions, that is responsible for the transport of heat and charge through a plasma. In the spirit of the discussion above, we can compute transport properties, such as the electric conductivity and thermal conductivity, on the presumption that it is Coulomb collisions that determine the electron mean free paths and that magnetic fields are unimportant. (Later we will see that collisionless effects—scattering off plasmons—usually provide a more serious impediment to charge and heat flow than Coulomb collisions and thus dominate the conductivities.)

20.5.1 Coulomb Collisions

First consider an electron exposed to a constant, accelerating electric field E. The electron's typical field-induced velocity is along the direction of the electric field and has magnitude $-eE/(m_e v_D)$, where v_D is the deflection frequency (rate) evaluated in Eqs. (20.18a) and (20.18b). We call this a *drift velocity*, because it is superposed on the electrons' collision-induced isotropic velocity distribution. The associated current density is $j \sim ne^2 E/(m_e v_D)$, and the electrical conductivity therefore is $\kappa_e \sim ne^2/(m_e v_D)$. (Note that electron-electron collisions conserve momentum and thus do not impede the flow of current, so electron-proton collisions, which happen about as frequently, produce all the electrical resistance and are thus responsible for this κ_e.)

The thermal conductivity can likewise be estimated by noting that a typical electron travels a mean distance $\ell \sim v_e/\nu_D$ between large net deflections, from an initial location where the average temperature is different from the final location's temperature by an amount $\Delta T \sim \ell|\nabla T|$. The heat flux transported by the electrons is therefore $\sim n v_e k_B \Delta T$, which should be equated to $-\kappa \nabla T$. We thereby obtain the electron contribution to the thermal conductivity as $\kappa \sim n k_B^2 T/(m_e \nu_D)$.

Computations based on the Fokker-Planck approach (Spitzer and Harm, 1953; see also Ex. 20.8) produce equations for the electrical and thermal conductivities that agree with the above estimates within factors of order ten:

electrical and thermal
conductivities when
Coulomb collisions are
responsible for the
resistivity

$$\kappa_e = 4.9 \left(\frac{e^2}{\sigma_T c \ln \Lambda m_e} \right) \left(\frac{k_B T_e}{m_e c^2} \right)^{3/2} = 1.5 \times 10^{-3} \left(\frac{T_e}{1\,\mathrm{K}} \right)^{3/2} \left(\frac{\ln \Lambda}{10} \right)^{-1} \Omega^{-1}\,\mathrm{m}^{-1},$$

(20.26a)

$$\kappa = 19.1 \left(\frac{k_B c}{\sigma_T \ln \Lambda} \right) \left(\frac{k_B T_e}{m_e c^2} \right)^{5/2} = 4.4 \times 10^{-11} \left(\frac{T_e}{1\,\mathrm{K}} \right)^{5/2} \left(\frac{\ln \Lambda}{10} \right)^{-1} \mathrm{W}\,\mathrm{m}^{-1}\,\mathrm{K}^{-1}.$$

(20.26b)

Here σ_T is the Thomson cross section, Eq. (20.23b). Note that neither transport coefficient depends explicitly on the density; increasing the number of charge or heat carriers is compensated by the reduction in their mean free paths.

20.5.2 Anomalous Resistivity and Anomalous Equilibration

We have demonstrated that the theoretical Coulomb interaction between charged particles gives very long mean free paths. Correspondingly, the electrical and thermal conductivities (20.26) are very large in practical, geophysical, and astrophysical applications. Is this the way that real plasmas behave? The answer is almost always "no."

mechanism for
anomalous resistivity
and equilibration

As we shall show in the next three chapters, a plasma can support a variety of modes of collective excitation (plasmons), in which large numbers of electrons and/or ions move in collective, correlated fashions that are mediated by electromagnetic fields that they create. When the modes of a plasma are sufficiently excited (which is common), the electromagnetic fields carried by the excitations can be much more effective than Coulomb scattering at deflecting the orbits of individual electrons and ions and at feeding energy into or removing it from the electrons and ions. Correspondingly, the electrical and thermal conductivities will be reduced. The reduced transport coefficients are termed *anomalous*, as is the scattering by plasmons that controls them. Providing quantitative calculations of the anomalous scattering and these anomalous transport coefficients is one of the principal tasks of nonlinear plasma physics—as we start to discuss in Chap. 22.

Exercise 20.9 *Example: Bremsstrahlung*

Bremsstrahlung (or *free-free* radiation) arises when electrons are accelerated by ions and the changing electric dipole moment that they create leads to the emission of photons. This process can only be fully described using quantum mechanics, though the low-frequency spectrum can be calculated accurately using classical electromagnetism. Here we make a "back of the envelope" estimate of the absorption coefficient and convert it into an emission coefficient.

(a) Consider an electron with average speed v oscillating in an electromagnetic wave of electric field amplitude E and (slow) angular frequency ω. Show that its energy of oscillation is $E_{\rm osc} \sim e^2 E^2 m^{-1} \omega^{-2}$.

(b) Let the electron encounter an ion and undergo a rapid deflection. Explain why the electron will, on average, gain energy of $\sim E_{\rm osc}$ as a result of the encounter. Hence estimate the average rate of electron heating. (You may ignore the cumulative effects of distant encounters.)

(c) Recognizing that classical absorption at low frequency is the difference between quantum absorption and stimulated emission (e.g., Sec. 10.2.1), convert the result from part (b) into a spontaneous emission rate, assuming that the electron velocity is typical for a gas in thermal equilibrium at temperature T.

(d) Hence show that the free-free cooling rate (power radiated per unit volume) of an electron-proton plasma with electron density n_e and temperature T is $\sim n_e^2 \alpha_F \sigma_T c^2 \left(k_B T m_e\right)^{1/2}$, where $\alpha_F = e^2/(4\pi\epsilon_0 \hbar c) = 7.3 \times 10^{-3}$ is the fine structure constant, and $\sigma_T = 6.65 \times 10^{-29}$ m^2 is the Thomson cross section. In a more detailed analysis, the long-range $(1/r^2)$ character of the electric force gives rise to a multiplicative logarithmic factor, called the Gaunt factor, analogous to the Coulomb logarithm of Sec. 20.4.2.

This approach is useful when the plasma is so dense that the emission rate is significantly different from the rate in a vacuum (Boyd and Sanderson, 2003, Chap. 9).

Exercise 20.10 *Challenge and Example: Thermoelectric Transport Coefficients*

(a) Consider a plasma in which the magnetic field is so weak that it presents little impediment to the flow of heat and electric current. Suppose that the plasma has a gradient ∇T_e of its electron temperature and also has an electric field **E**. It is a familiar fact that the temperature gradient will cause heat to flow and the electric field will create an electric current. Not so familiar, but somewhat obvious if one stops to think about it, is that the temperature gradient also creates an electric current and the electric field also causes heat flow. Explain in physical terms why this is so.

(b) So long as the mean free path of an electron between substantial deflections by electrons and protons, $\ell_{D,e} = (3k_B T_e/m_e)^{1/2} t_{D,e}$, is short compared to the lengthscale for substantial temperature change, $T_e/|\nabla T_e|$, and short compared to the lengthscale for the electrons to be accelerated to near the speed of light by the electric field, $m_e c^2 / (eE)$, the fluxes of heat $\mathbf{q}$ and of electric charge $\mathbf{j}$ will be governed by nonrelativistic electron diffusion and will be linear in ∇T and $\mathbf{E}$:

$$\mathbf{q} = -\kappa \nabla T - \beta \mathbf{E}, \quad \mathbf{j} = \kappa_e \mathbf{E} + \alpha \nabla T. \tag{20.27}$$

The coefficients κ (heat conductivity), κ_e (electrical conductivity), β, and α are called *thermoelectric transport coefficients*. Use kinetic theory (Chap. 3), in a situation where $\nabla T = 0$, to derive the conductivity equations $\mathbf{j} = \kappa_e \mathbf{E}$ and $\mathbf{q} = -\beta \mathbf{E}$, and the following approximate formulas for the transport coefficients:

$$\kappa_e \sim \frac{n e^2 t_{D,e}}{m_e}, \quad \beta \sim \frac{k_B T}{e} \kappa_e. \tag{20.28a}$$

Show that, aside from a coefficient of order unity, this κ_e, when expressed in terms of the plasma's temperature and density, reduces to the Fokker-Planck result Eq. (20.26a).

(c) Use kinetic theory, in a situation where $\mathbf{E} = 0$ and the plasma is near thermal equilibrium at temperature T, to derive the conductivity equations $\mathbf{q} = -\kappa \nabla T$ and $\mathbf{j} = \alpha \nabla T$, and the approximate formulas

$$\kappa \sim k_B n \frac{k_B T}{m_e} t_{D,e}, \quad \alpha \sim \frac{e}{k_B T} \kappa. \tag{20.28b}$$

Show that, aside from a coefficient of order unity, this κ reduces to the Fokker-Planck result Eq. (20.26b).

(d) It can be shown (Spitzer and Harm, 1953) that for a hydrogen plasma

$$\frac{\alpha \beta}{\kappa_e \kappa} = 0.581. \tag{20.28c}$$

By studying the entropy-governed probability distributions for fluctuations away from statistical equilibrium, one can derive another relation among the thermo-electric transport coefficients, the *Onsager relation* (Kittel, 2004, Secs. 33, 34; Reif, 2008, Sec. 15.8):

$$\beta = \alpha T + \frac{5}{2} \frac{k_B T_e}{e} \kappa_e; \tag{20.28d}$$

Eqs. (20.28c) and (20.28d) determine α and β in terms of κ_e and κ. Show that your approximate values of the transport coefficients, Eqs. (20.28a) and (20.28b), are in rough accord with Eqs. (20.28c) and (20.28d).

(e) If a temperature gradient persists for sufficiently long, it will give rise to sufficient charge separation in the plasma to build up an electric field (called a "secondary field") that prevents further charge flow. Show that this cessation of charge flow suppresses the heat flow. The total heat flux is then $\mathbf{q} = -\kappa_{T\,\text{effective}} \nabla T$, where

$$\kappa_{T\,\text{effective}} = \left(1 - \frac{\alpha \beta}{\kappa_e \kappa}\right) \kappa = 0.419\kappa. \tag{20.29}$$

20.6 Magnetic Field

20.6.1 Cyclotron Frequency and Larmor Radius

Many of the plasmas that we will encounter are endowed with a strong magnetic field. This causes the charged particles to travel along helical orbits about the field direction rather than move rectilinearly between collisions.

If we denote the magnetic field by **B**, then the equation of motion for a nonrelativistic electron becomes

$$m_e \frac{d\mathbf{v}}{dt} = -e\, \mathbf{v} \times \mathbf{B}, \tag{20.30}$$

which gives rise to a constant speed $v_{\parallel}$ parallel to the magnetic field and a circular motion perpendicular to the field with angular velocity sometimes denoted ω_{ce} and sometimes ω_c:

$$\boxed{\omega_{ce} = \omega_c = \frac{eB}{m_e} = 1.76 \times 10^{11} \left(\frac{B}{1\,\mathrm{T}} \right)\,\mathrm{s}^{-1}.} \tag{20.31}$$

(electron) cyclotron frequency

This angular velocity is called the *electron cyclotron frequency,* or simply the *cyclotron frequency;* and it is sometimes called the *gyro frequency.* Notice that this cyclotron frequency depends only on the magnetic field strength B and not on the plasma's density n or the electron velocity (i.e., the plasma temperature T). Nor does it depend on the angle between **v** and **B** (the *pitch angle, α*).

The radius of the electron's gyrating (spiraling) orbit, projected perpendicular to the direction of the magnetic field, is called the *Larmor radius* and is given by

$$\boxed{r_L = \frac{v_{\perp}}{\omega_{ce}} = \frac{v \sin \alpha}{\omega_{ce}} = 5.7 \times 10^{-9} \left(\frac{v_{\perp}}{1\,\mathrm{km\,s^{-1}}} \right) \left(\frac{B}{1\,\mathrm{T}} \right)^{-1}\,\mathrm{m},} \tag{20.32}$$

Larmor radius

where $v_{\perp}$ is the electron's velocity projected perpendicular to the field. Protons (and other ions) in a plasma also undergo cyclotron motion. Because the proton mass is larger by $m_p/m_e = 1{,}836$ than the electron mass, its angular velocity

$$\boxed{\omega_{cp} = \frac{eB}{m_p} = 0.96 \times 10^{8}\,\mathrm{s}^{-1} \left(\frac{B}{1\,\mathrm{T}} \right)} \tag{20.33}$$

proton cyclotron frequency

is 1,836 times lower. The quantity ω_{cp} is called the *proton cyclotron frequency* or *ion cyclotron frequency.* The sense of gyration is, of course, opposite to that of the electrons. If the protons have similar temperatures to the electrons, their speeds are typically $\sim \sqrt{m_p/m_e} = 43$ times smaller than those of the electrons, and their typical Larmor radii are $\sim$43 times larger than those of the electrons.

We demonstrated in Sec. 20.3.3 that all the electrons in a plasma can oscillate in phase at the plasma frequency. The electrons' cyclotron motions can also be coherent. Such coherent motions are called *cyclotron resonances* or *cyclotron oscillations,* and we shall study them in Chap. 21. Ion cyclotron resonances can also occur. Characteristic electron cyclotron frequencies and Larmor radii are tabulated in Table 20.1. As shown

cyclotron resonances (oscillations)

there, the cyclotron frequency, like the plasma frequency, is typically far larger than the rates for Coulomb-mediated energy equilibration.

20.6.2 Validity of the Fluid Approximation

20.6.2

MAGNETOHYDRODYNAMICS

In Chap. 19, we developed the magnetohydrodynamic (MHD) description of a magnetized plasma. We described the plasma by its density and temperature (or equivalently, its pressure). Under what circumstances, for a plasma, is this description accurate? The answer to this question turns out to be quite complex, and a full discussion would go well beyond the scope of this book. Some aspects, however, are easy to describe. A fluid description ought to be acceptable when (i) the timescales τ that characterize the macroscopic flow are long compared with the time required to establish Maxwellian equilibrium (i.e., $\tau \gg \nu_{ep}^{-1}$), and (ii) the excitation level of collective wave modes is so small that the modes do not interfere seriously with the influence of Coulomb collisions. Unfortunately, these conditions rarely apply. (One type of plasma where they might be a quite good approximation is that in the interior of the Sun.)

Magnetohydrodynamics can still provide a moderately accurate description of a plasma, even if the electrons and ions are not fully equilibrated, when the electrical conductivity can be treated as very large and the thermal conductivity as very small. Then we can treat the magnetic Reynolds number [Eq. (19.9c)] as effectively infinite and the plasma as an adiabatic perfect fluid (as we assumed in much of Chap. 19). It is not so essential that the actual particle distribution functions be Maxwellian, merely that they have second moments that can be associated with a (roughly defined) common temperature.

Quite often in plasma physics almost all of the dissipation is localized—for example, to the vicinity of a shock front or a site of magnetic-field-line reconnection—and the remainder of the flow can be treated using MHD. The MHD description then provides a boundary condition for a fuller plasma physical analysis of the dissipative region. This approach simplifies the analysis of such situations.

GENERALIZATIONS OF MHD

The great advantage of fluid descriptions, and the reason physicists abandon them with such reluctance, is that they are much simpler than other descriptions of a plasma. One only has to cope with the fluid pressure, density, and velocity and does not have to deal with an elaborate statistical description of the positions and velocities of all the particles. Generalizations of the simple fluid approximation have therefore been devised that can extend the domain of validity of simple MHD ideas.

One extension, which we develop in the following chapter, is to treat the protons and the electrons as two separate fluids and derive dynamical equations that describe their (coupled) evolution. Another extension, which we describe now, is to acknowledge that, in most plasmas:

conditions for validity of MHD approximation

two-fluid approximation

1. the cyclotron period is very short compared with the Coulomb collision time (and with the anomalous scattering time), and

2. the timescale on which energy is transferred back and forth among the electrons, the protons, and the electromagnetic field is intermediate between ω_c^{-1} and ν_{ee}^{-1}.

Intuitively, these assumptions allow the electron and proton velocity distributions to become axisymmetric with respect to the magnetic field direction, though not fully isotropic. In other words, we can characterize the plasma using a density and two separate components of pressure, one associated with motion along the direction of the magnetic field, and the other with gyration around the field lines. **anisotropic fluid in presence of magnetic field**

For simplicity, let us just consider the electrons and their stress tensor, which we can write as

$$T_e^{jk} = \int \mathcal{N}_e \, p^j p^k \frac{d\mathcal{V}_p}{m} \tag{20.34}$$

[Eq. (3.32d)], where $\mathcal{N}_e$ is the electron number density in phase space and $d\mathcal{V}_p = dp_x dp_y dp_z$ is the volume element in phase space. If we orient Cartesian axes so that the direction of $\mathbf{e}_z$ is parallel to the local magnetic field, then

$$||T_e^{jk}|| = \begin{bmatrix} P_{e\perp} & 0 & 0 \\ 0 & P_{e\perp} & 0 \\ 0 & 0 & P_{e||} \end{bmatrix}, \tag{20.35}$$

anisotropic pressure

where $P_{e\perp}$ is the electron pressure perpendicular to $\mathbf{B}$, and $P_{e||}$ is the electron pressure parallel to $\mathbf{B}$. Now suppose that there is a compression or expansion on a timescale that is long compared with the cyclotron period but short compared with the Coulomb collision and anomalous scattering timescales. Then we should not expect $P_{e\perp}$ to be equal to $P_{e||}$, and we anticipate that they will evolve with density according to different laws.

The adiabatic indices governing $P_\perp$ and $P_{||}$ in such a situation are easily derived from kinetic-theory arguments (Ex. 20.11). For compression perpendicular to $\mathbf{B}$ and no change of length along $\mathbf{B}$, we have

$$\gamma_\perp \equiv \left(\frac{\partial \ln P_\perp}{\partial \ln \rho} \right)_s = 2, \quad \gamma_{||} \equiv \left(\frac{\partial \ln P_{||}}{\partial \ln \rho} \right)_s = 1; \tag{20.36a}$$

collisionless, anistotropic adiabatic indices

for compression parallel to $\mathbf{B}$ and no change of transverse area, we have

$$\gamma_\perp \equiv \left(\frac{\partial \ln P_\perp}{\partial \ln \rho} \right)_s = 1, \quad \gamma_{||} \equiv \left(\frac{\partial \ln P_{||}}{\partial \ln \rho} \right)_s = 3. \tag{20.36b}$$

By contrast, if the expansion is sufficiently slow that Coulomb collisions are effective (though not so slow that heat conduction can operate), then we expect the velocity distribution to maintain isotropy and both components of pressure to evolve according to the law appropriate to a monatomic gas:

collisional, isotropic adiabatic index

$$\gamma = \left(\frac{\partial \ln P_\perp}{\partial \ln \rho}\right)_s = \left(\frac{\partial \ln P_\|}{\partial \ln \rho}\right)_s = \frac{5}{3}.$$

(20.37)

20.6.3 Conductivity Tensor

As is evident from the foregoing remarks, if we are in a regime where Coulomb scattering really does determine the particle mean free path, then an extremely small magnetic field strength suffices to ensure that individual particles complete gyrational orbits before they collide. Specifically, for electrons, the deflection time t_D, given by Eq. (20.18a,b), exceeds ω_c^{-1} if

$$B \gtrsim 10^{-16} \left(\frac{n}{1\,\mathrm{m}^{-3}}\right)\left(\frac{T_e}{1\,\mathrm{K}}\right)^{-3/2} \mathrm{T}.$$

(20.38)

This is almost always the case. It is also almost always true for the ions.

When inequality (20.38) is satisfied and also anomalous scattering is negligible on the timescale ω_c^{-1}, the transport coefficients must be generalized to form tensors. Let us compute the electrical conductivity tensor for a plasma in which a steady electric field $\mathbf{E}$ is applied. Once again orienting our coordinate system so that the magnetic field is parallel to $\mathbf{e}_z$, we can write down an equation of motion for the electrons by balancing the electromagnetic acceleration with the average rate of loss of momentum due to collisions or anomalous scattering:

electron equation of motion

$$-e(\mathbf{E} + \mathbf{v} \times \mathbf{B}) - m_e \nu_D \mathbf{v} = 0.$$

(20.39)

Solving for the velocity, we obtain

$$\begin{pmatrix} v_x \\ v_y \\ v_z \end{pmatrix} = -\frac{e}{m_e \nu_D(1 + \omega_c^2/\nu_D^2)} \begin{pmatrix} 1 & \omega_c/\nu_D & 0 \\ -\omega_c/\nu_D & 1 & 0 \\ 0 & 0 & 1 + \omega_c^2/\nu_D^2 \end{pmatrix} \begin{pmatrix} E_x \\ E_y \\ E_z \end{pmatrix}.$$

(20.40)

As the current density is $\mathbf{j}_e = -ne\mathbf{v} = \kappa_e \mathbf{E}$, the electrical conductivity tensor is given by

anisotropic electric conductivity

$$\kappa_e = \frac{ne^2}{m_e \nu_D(1 + \omega_c^2/\nu_D^2)} \begin{pmatrix} 1 & \omega_c/\nu_D & 0 \\ -\omega_c/\nu_D & 1 & 0 \\ 0 & 0 & 1 + \omega_c^2/\nu_D^2 \end{pmatrix}.$$

(20.41)

It is apparent from the form of this conductivity tensor that, when $\omega_c \gg \nu_D$ (as is almost always the case), the conductivity perpendicular to the magnetic field is greatly inhibited, whereas that along the magnetic field is unaffected. Similar remarks apply to the flow of heat. It is therefore often assumed that only transport parallel to the field is effective. However, as is made clear in the next section, if the plasma is inhomogeneous, then cross-field transport can be quite rapid in practice.

Exercise 20.11 *Example and Derivation: Adiabatic Indices for Compression of a Magnetized Plasma*

Consider a plasma in which, in the local mean rest frame of the electrons, the electron stress tensor has the form (20.35) with $\mathbf{e}_z$ the direction of the magnetic field. The following analysis for the electrons can be carried out independently for the ions, resulting in the same formulas.

(a) Show that

$$P_{e\|} = nm_e\langle v_{\|}^2\rangle, \quad P_{e\perp} = \frac{1}{2}nm_e\langle |\mathbf{v}_\perp|^2\rangle, \tag{20.42}$$

where $\langle v_{\|}^2\rangle$ is the mean-square electron velocity parallel to $\mathbf{B}$, and $\langle |\mathbf{v}_\perp|^2\rangle$ is the mean-square velocity orthogonal to $\mathbf{B}$. (The velocity distributions are not assumed to be Maxwellian.)

(b) Consider a fluid element with length l along the magnetic field and cross sectional area A orthogonal to the field. Let $\bar{\mathbf{v}}$ be the mean velocity of the electrons ($\bar{\mathbf{v}} = 0$ in the mean electron rest frame), and let θ and σ_{jk} be the expansion and shear of the mean electron motion as computed from $\bar{\mathbf{v}}$ (Sec. 13.7.1). Show that

$$\frac{dl/dt}{l} = \frac{1}{3}\theta + \sigma^{jk}b_jb_k, \quad \frac{dA/dt}{A} = \frac{2}{3}\theta - \sigma^{jk}b_jb_k, \tag{20.43}$$

where $\mathbf{b} = \mathbf{B}/|\mathbf{B}| = \mathbf{e}_z$ is a unit vector in the direction of the magnetic field.

(c) Assume that the timescales for compression and shearing are short compared with those for Coulomb scattering and anomalous scattering: $\tau \ll t_{D,e}$. Show, using the laws of energy and particle conservation, that

$$\frac{1}{\langle v_{\|}^2\rangle}\frac{d\langle v_{\|}^2\rangle}{d} = -\frac{2}{l}\frac{dl}{dt},$$

$$\frac{1}{\langle v_\perp^2\rangle}\frac{d\langle v_\perp^2\rangle}{dt} = -\frac{1}{A}\frac{dA}{dt}, \tag{20.44}$$

$$\frac{1}{n}\frac{dn}{dt} = -\frac{1}{l}\frac{dl}{dt} - \frac{1}{A}\frac{dA}{dt}.$$

(d) Show that

$$\frac{1}{P_{e\|}}\frac{dP_{e\|}}{dt} = -3\frac{dl/dt}{l} - \frac{dA/dt}{A} = -\frac{5}{3}\theta - 2\sigma^{jk}b_jb_k,$$

$$\frac{1}{P_{e\perp}}\frac{dP_{e\perp}}{dt} = -\frac{dl/dt}{l} - 2\frac{dA/dt}{A} = -\frac{5}{3}\theta + \sigma^{jk}b_jb_k. \tag{20.45}$$

(e) Show that, when the fluid is expanding or compressing entirely perpendicular to **B**, with no expansion or compression along **B**, the pressures change in accord with the adiabatic indices of Eq. (20.36a). Show, similarly, that when the fluid expands or compresses along **B**, with no expansion or compression in the perpendicular direction, the pressures change in accord with the adiabatic indices of Eq. (20.36b).

(f) Hence derive the so-called *double adiabatic* or *CGL* (Chew, Goldberger, and Low, 1956) equations of state:

$$P_{\parallel} \propto n^3/B^2, \qquad P_{\perp} \propto nB, \tag{20.46}$$

valid for changes on timescales long compared with the cyclotron period but short compared with all Coulomb collision and anomalous scattering times.

Exercise 20.12 *Problem: Relativistic Electron Motion*
Use the relativistic equation of motion to show that the relativistic electron cyclotron frequency is $\omega_c = eB/(\gamma m_e)$, where $\gamma = 1/\sqrt{1 - (v/c)^2}$ is the electron Lorentz factor. What is the relativistic electron Larmor radius? What is the relativistic proton cyclotron frequency and Larmor radius?

Exercise 20.13 *Example: Ultra-High-Energy Cosmic Rays*
The most energetic (ultra-high-energy) cosmic rays are probably created with energies up to $\sim 1\,\mathrm{ZeV} = 10^{21}\,\mathrm{eV}$ in sources roughly 100 million light-years away.

(a) Show that they start with the "energy of a well-hit baseball and the momentum of a snail." By the time they arrive at Earth, their energies are likely to have been reduced by a factor of ~ 3.

(b) Assuming that the intergalactic magnetic field is $\sim 10^{-13}\,\mathrm{T} \sim 10^{-9}\,\mathrm{G}$, compute the cosmic rays' Larmor radii, assuming that they are protons and iron nuclei.

(c) Do you expect their arrival directions to point back to their sources?

(d) Suppose that an ultra-high-energy cosmic ray collides with a stationary nitrogen nucleus in our atmosphere. How much energy becomes available in the center-of-mass frame? Compare this with the energy at the Large Hadron Collider, where protons, each with energy $\sim 7\,\mathrm{TeV}$, collide head on.

20.7

20.7 Particle Motion and Adiabatic Invariants

In the next three chapters we shall meet a variety of plasma phenomena that can be understood in terms of the orbital motions of individual electrons and ions. These phenomena typically entail motions in an electromagnetic field that is nearly but not quite spatially homogeneous on the scale of the Larmor radius r_L and nearly but not quite constant during a cyclotron period $2\pi/\omega_c$. In this section, in preparation for the next three chapters, we review charged-particle motion in such nearly homogeneous, nearly time-independent fields.

Since the motions of electrons are usually of greater interest than those of ions, we presume throughout this section that the charged particle is an electron. We denote its charge by $-e$ and its mass by m_e.

20.7.1 Homogeneous, Time-Independent Magnetic Field and No Electric Field

From the nonrelativistic version of the Lorentz force equation, $d\mathbf{v}/dt = -(e/m_e)\mathbf{v} \times \mathbf{B}$, one readily deduces that an electron in a homogeneous, time-independent magnetic field $\mathbf{B}$ moves with uniform velocity $\mathbf{v}_{||}$ parallel to the field, and it moves perpendicular to the field in a circular orbit with the cyclotron frequency $\omega_c = eB/m_e$ and Larmor radius $r_L = m_e v_\perp/(eB)$. Here $v_\perp$ is the electron's time-independent transverse speed (speed perpendicular to $\mathbf{B}$).

20.7.2 Homogeneous, Time-Independent Electric and Magnetic Fields

Suppose that the homogeneous magnetic field $\mathbf{B}$ is augmented by a homogeneous electric field $\mathbf{E}$, and assume initially that $|\mathbf{E} \times \mathbf{B}| < B^2 c$. Then examine the electric and magnetic fields in a new reference frame, one that moves with the velocity

$$\mathbf{v}_D = \frac{\mathbf{E} \times \mathbf{B}}{B^2}, \tag{20.47}$$

relative to the original frame. Note that the moving frame's velocity $\mathbf{v}_D$ is perpendicular to both the magnetic field and the electric field. From the Lorentz transformation law for the electric field, $\mathbf{E}' = \gamma(\mathbf{E} + \mathbf{v}_D \times \mathbf{B})$, we infer that in the moving frame the electric field and the magnetic field are parallel to each other. As a result, in the moving frame the electron's motion perpendicular to the magnetic field is purely circular; and, correspondingly, in the original frame its perpendicular motion consists of a *drift* with velocity $\mathbf{v}_D$, and superposed on that drift, a circular motion (Fig. 20.4). In other words, the electron moves in a circle whose center (the electron's *guiding center*) drifts with velocity $\mathbf{v}_D$. Notice that the drift velocity (20.47) is independent of the electron's charge and mass, and thus is the same for ions as for electrons. This drift is called the "$\mathbf{E} \times \mathbf{B}$ drift."

guiding-center approximation; $\mathbf{E} \times \mathbf{B}$ drift

When the component of the electric field orthogonal to $\mathbf{B}$ is so large that the drift velocity computed from Eq. (20.47) exceeds the speed of light, the electron's guiding center, of course, cannot move with that velocity. Instead, the electric field drives the electron up to higher and higher velocities as time passes, but in a sinusoidally modulated manner. Ultimately, the electron velocity becomes arbitrarily close to the speed of light.

breakdown of guiding-center approximation when B is large

When a uniform, time-independent gravitational field $\mathbf{g}$ accompanies a uniform, time-independent magnetic field $\mathbf{B}$, its effect on an electron will be the same as that

guiding-center approximation in gravitational field

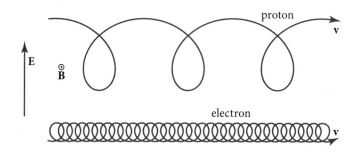

FIGURE 20.4 The proton motion (upper trajectory) and electron motion (lower trajectory) orthogonal to the magnetic field, when there are constant electric and magnetic fields with $|\mathbf{E} \times \mathbf{B}| < B^2 c$. Each electron and proton moves in a circle with a superposed drift velocity $\mathbf{v}_D$ given by Eq. (20.47).

of an electric field $\mathbf{E}_{\text{equivalent}} = -(m_e/e)\mathbf{g}$: The electron's guiding center will acquire a drift velocity

$$\mathbf{v}_D = -\frac{m_e}{e}\frac{\mathbf{g} \times \mathbf{B}}{B^2}, \tag{20.48}$$

and similarly for a proton. This *gravitational drift* velocity is typically very small.

20.7.3 ### 20.7.3 Inhomogeneous, Time-Independent Magnetic Field

When the electric field vanishes, but the magnetic field is spatially inhomogeneous and time-independent, and the inhomogeneity scale is large compared to the Larmor radius r_L of the electron's orbit, the electron motion again is nicely described in terms of a guiding center.

First consider the effects of a curvature of the field lines (Fig. 20.5a). Suppose that the speed of the electron along the field lines is $v_{||}$. We can think of this as a longitudinal guiding-center motion. As the field lines bend in, say, the direction of the unit vector $\mathbf{n}$ with radius of curvature R, this longitudinal guiding-center motion experiences the **curvature drift** acceleration $\mathbf{a} = v_{||}^2\mathbf{n}/R$. That acceleration is equivalent to the effect of an electric field $\mathbf{E}_{\text{effective}} = (-m_e/e)v_{||}^2\mathbf{n}/R$, and it therefore produces a transverse drift of the guiding center with $\mathbf{v}_D = (\mathbf{E}_{\text{effective}} \times \mathbf{B})/B^2$. Since the curvature R of the field line and the direction $\mathbf{n}$ of its bend are given by $B^{-2}(\mathbf{B} \cdot \boldsymbol{\nabla})\mathbf{B} = \mathbf{n}/R$, this *curvature drift* velocity is

$$\mathbf{v}_D = -\frac{m_e v_{||}^2}{e}\mathbf{B} \times \frac{(\mathbf{B} \cdot \boldsymbol{\nabla})\mathbf{B}}{B^4}. \tag{20.49}$$

Notice that the magnitude of this drift is

$$v_D = \frac{r_L}{R}\frac{v_{||}}{v_\perp}v_{||}. \tag{20.50}$$

This particle drift and others discussed below also show up as fluid drifts in a magnetized plasma; see Ex. 21.1.

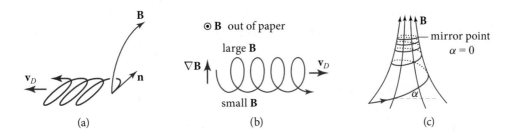

FIGURE 20.5 An electron's motion in a time-independent, inhomogeneous magnetic field. (a) The drift induced by the curvature of the field lines. (b) The drift induced by a transverse gradient of the magnitude of the magnetic field. (c) The change in electron pitch angle induced by a longitudinal gradient of the magnitude of the magnetic field.

A second kind of inhomogeneity is a transverse spatial gradient ∇B of the magnitude of **B**. As is shown in Fig. 20.5b, such a gradient causes the electron's circular motion to be tighter (smaller radius of curvature of the circle) in the region of larger B than in the region of smaller B; this difference in radii of curvature clearly induces a drift. It is straightforward to show that the resulting *gradient drift* velocity is

$$\mathbf{v}_D = \frac{-m_e v_\perp^2}{2e} \frac{\mathbf{B} \times \nabla B}{B^3}.$$ (20.51)

The third and final kind of inhomogeneity is a longitudinal gradient of the magnitude of **B** (Fig. 20.5c). Such a gradient results from the magnetic field lines converging toward (or diverging away from) one another. The effect of this convergence (or divergence) is most easily inferred in a frame that moves longitudinally with the electron. In such a frame the magnetic field changes with time, $\partial \mathbf{B}'/\partial t \neq 0$, and correspondingly, the resultant electric field satisfies $\nabla \times \mathbf{E}' = -\partial \mathbf{B}'/\partial t$. The kinetic energy of the electron as measured in this longitudinally moving frame is the same as the transverse energy $\frac{1}{2} m_e v_\perp^2$ in the original frame. This kinetic energy is forced to change by the electric field $\mathbf{E}'$. The change in energy during one circuit around the magnetic field is

$$\Delta \left(\frac{1}{2} m_e v_\perp^2 \right) = -e \oint \mathbf{E}' \cdot d\mathbf{l} = e \int \frac{\partial \mathbf{B}'}{\partial t} \cdot d\mathbf{A} = e \left(\frac{\omega_c}{2\pi} \Delta B \right) \pi r_L^2 = \frac{m_e v_\perp^2}{2} \frac{\Delta B}{B}.$$ (20.52)

Here the second expression in the equation involves a line integral once around the electron's circular orbit. The third expression involves a surface integral over the interior of the orbit and has $\partial \mathbf{B}'/\partial t$ parallel to $d\mathbf{A}$. In the fourth the time derivative of the magnetic field has been expressed as $(\omega_c/(2\pi))\Delta B$, where ΔB is the change in magnetic field strength along the electron's guiding center during one circular orbit.

Equation (20.52) can be rewritten as a conservation law along the world line of the electron's guiding center:

$$\frac{m_e v_\perp^2}{2B} = \text{const.}$$ (20.53)

conservation of enclosed
flux (i.e., of magnetic
moment)

Notice that the conserved quantity $m_e v_\perp^2/(2B)$ is proportional to the total magnetic flux threading the electron's circular orbit, $\pi r_L^2 B$. Thus the electron moves along the field lines in such a manner as to keep constant the magnetic flux enclosed in its orbit; see Fig. 20.5c. A second interpretation of Eq. (20.53) is in terms of the magnetic moment created by the electron's circulatory motion. That moment is $\boldsymbol{\mu} = -m_e v_\perp^2/(2B^2)\mathbf{B}$, and its magnitude is the conserved quantity

$$\mu = \frac{m_e v_\perp^2}{2B} = \text{const.} \tag{20.54}$$

An important consequence of the conservation law (20.54) is a gradual change in the electron's pitch angle,

$$\alpha \equiv \tan^{-1}(v_\parallel/v_\perp), \tag{20.55}$$

as it spirals along the converging field lines. Because there is no electric field in the original frame, the electron's total kinetic energy is conserved in that frame:

$$E_{\text{kin}} = \frac{1}{2} m_e (v_\parallel^2 + v_\perp^2) = \text{const.} \tag{20.56}$$

This expression, together with the constancy of $\mu = m_e v_\perp^2/(2B)$ and the definition of the electron pitch angle [Eq. (20.55)], implies that the pitch angle varies with magnetic field strength as

$$\tan^2 \alpha = \frac{E_{\text{kin}}}{\mu B} - 1. \tag{20.57}$$

Notice that as the field lines converge, B increases in magnitude, and α decreases. Ultimately, when B reaches a critical value $B_{\text{crit}} = E_{\text{kin}}/\mu$, the pitch angle α goes to zero. The electron then "reflects" off the strong-field region and starts moving back toward weak fields, with increasing pitch angle. The location at which the electron reflects is called the electron's *mirror point*.

Figure 20.6 shows two examples of this mirroring. The first example is a "magnetic bottle" (Ex. 20.14). Electrons whose pitch angles at the center of the bottle are sufficiently small have mirror points in the bottle and thus cannot leak out. The second example is the van Allen belts of Earth. Electrons (and also ions) travel up and down the magnetic field lines of the van Allen belts, reflecting at mirror points.

It is not hard to show that the gradient of $\mathbf{B}$ can be split up into the three pieces we have studied: a curvature with no change of $B = |\mathbf{B}|$ (Fig. 20.5a), a change of B orthogonal to the magnetic field (Fig. 20.5b), and a change of B along the magnetic field (Fig. 20.5c). When (as we have assumed) the lengthscales of these changes are far greater than the electron's Larmor radius, their effects on the electron's motion superpose linearly.

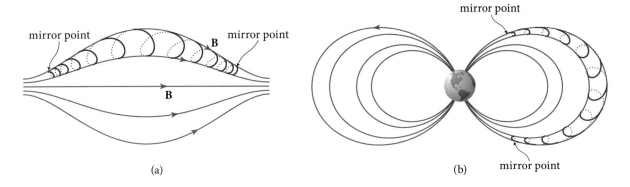

FIGURE 20.6 Two examples of the mirroring of particles in an inhomogeneous magnetic field. (a) A magnetic bottle. (b) Earth's van Allen belts.

20.7.4 A Slowly Time-Varying Magnetic Field

20.7.4

When the magnetic field changes on timescales long compared to the cyclotron period $2\pi/\omega_c$, its changes induce alterations of the electron's orbit that can be deduced with the aid of *adiabatic invariants*—quantities that are nearly constant when the field changes adiabatically, in other words, slowly (see, e.g., Lifshitz and Pitaevskii, 1981; Northrop, 1963). The conserved magnetic moment $\mu = m_e v_\perp^2/(2B)$ associated with an electron's transverse, circular motion is an example of an adiabatic invariant. We proved its invariance in Eqs. (20.52) and (20.53) (in Sec. 20.7.3, where we were working in a reference frame in which the magnetic field changed slowly, and associated with that change was a weak electric field). This adiabatic invariant can be shown to be, aside from a constant multiplicative factor $2\pi m_e/e$, the action associated with the electron's circular motion: $J_\phi = \oint p_\phi d\phi$. Here ϕ is the angle around the circular orbit, and $p_\phi = (m_e v_\perp - eA_\phi)r_L$ is the ϕ component of the electron's canonical momentum. The action J_ϕ is a well-known adiabatic invariant.

 When a slightly inhomogeneous magnetic field varies slowly in time, not only is $\mu = m_e v_\perp^2/(2B)$ adiabatically invariant (conserved), so also are two other actions. One is the action associated with motion from one mirror point of the magnetic field to another and back:

$$J_{||} = \oint \mathbf{p}_{||} \cdot d\,\mathbf{l}. \tag{20.58}$$

Here $\mathbf{p}_{||} = m_e \mathbf{v}_{||} - e\mathbf{A}_{||} = m_e \mathbf{v}_{||}$ is the generalized (canonical) momentum along the field line, and $d\,\mathbf{l}$ is distance along the field line. Thus the adiabatic invariant is the spatial average $\langle v_{||}\rangle$ of the longitudinal speed of the electron, multiplied by m_e and twice the distance Δl between mirror points: $J_{||} = 2m_e\langle v_{||}\rangle\Delta l$.

 The other (third) adiabatic invariant is the action associated with the drift of the guiding center: an electron mirroring back and forth along the field lines drifts sideways, and by its drift it traces out a 2-dimensional surface to which the magnetic

adiabatic invariants

magnetic moment as circular-motion action

longitudinal, mirror-point action

guiding-center-drift action

field is tangent (e.g., the surface of the center of the magnetic bottle in Fig. 20.6a, rotated around the horizontal axis). The action of the electron's drift around this magnetic surface turns out to be proportional to the total magnetic flux enclosed by the surface. Thus, if the field geometry changes slowly, the magnetic flux enclosed by the magnetic surface on which the electron's guiding center moves is adiabatically conserved.

accuracy of adiabatic invariants

How nearly constant are the adiabatic invariants? The general theory of adiabatic invariants shows that, so long as the temporal changes of the magnetic field structure are smooth enough to be described by analytic functions of time, the fractional failures of the adiabatic invariants to be conserved are of order $e^{-\tau/P}$. Here τ is the timescale on which the field changes, and P is the period of the motion associated with the adiabatic invariant. (This period P is $2\pi/\omega_c$ for the invariant μ; it is the mirroring period for the longitudinal action, and it is the drift period for the magnetic flux enclosed in the electron's magnetic surface.) Because the exponential $e^{-\tau/P}$ dies out so quickly with increasing timescale τ, the adiabatic invariants are conserved to very high accuracy when $\tau \gg P$.

20.7.5

20.7.5 Failure of Adiabatic Invariants; Chaotic Orbits

When the magnetic field changes as fast as or more rapidly than a cyclotron orbit (in space or in time), then the adiabatic invariants fail, and the charged-particle orbits may even be chaotic in some cases.

failure of adiabatic invariants and chaotic motion near a magnetic neutral line

An example is charged-particle motion near a neutral line (also called an X-line) of a magnetic field. Near an X-line, the field has the hyperbolic shape $\mathbf{B} = B_0(y\,\mathbf{e}_x + \gamma\,x\,\mathbf{e}_y)$ with constants B_0 and γ; see Fig. 19.13 for an example that occurs in magnetic-field-line reconnection. The X-line is the z-axis, $(x, y) = (0, 0)$, on which the field vanishes. Near there the Larmor radius becomes arbitrarily large, far larger than the scale on which the field changes. Correspondingly, no adiabatic invariants exist near the X-line, and it turns out that some charged-particle orbits near there are chaotic in the sense of extreme sensitivity to initial conditions (Sec. 15.6.4). An electric field in the $\mathbf{e}_z$ direction (orthogonal to $\mathbf{B}$ and parallel to the X-line) enhances the chaos (for details, see, e.g., Martin, 1986).

EXERCISES

Exercise 20.14 *Example: Mirror Machine*
One method for confining hot plasma is to arrange electric coils so as to make a mirror machine in which the magnetic field has the geometry sketched in Fig. 20.6a. Suppose that the magnetic field in the center is 1 T and the field strength at the two necks is 10 T, and that plasma is introduced with an isotropic velocity distribution near the center of the bottle.

(a) What fraction of the plasma particles will escape?

(b) Sketch the pitch-angle distribution function for the particles that remain.

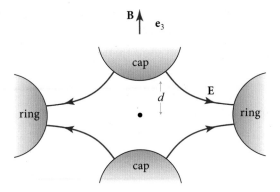

FIGURE 20.7 Penning trap for localizing individual charged particles. The magnetic field is uniform and parallel to the vertical axis of symmetry $\mathbf{e}_z$. The electric field is maintained between a pair of hyperboloidal caps and a hyperboloidal ring.

(c) Suppose that Coulomb collisions cause particles to diffuse in pitch angle α with a diffusion coefficient

$$D_{\alpha\alpha} \equiv \left\langle \frac{\Delta\alpha^2}{\Delta t} \right\rangle = t_D^{-1}. \tag{20.59}$$

Estimate how long it will take most of the plasma to escape the mirror machine.

(d) What do you suspect will happen in practice?

Exercise 20.15 *Challenge: Penning Trap*

A clever technique for studying the behavior of individual electrons or ions is to entrap them using a combination of electric and magnetic fields. One of the simplest and most useful devices is the *Penning trap* (see, e.g., Brown and Gabrielse, 1986). Basically this comprises a uniform magnetic field B combined with a hyperboloidal electrostatic field that is maintained between hyperboloidal electrodes as shown in Fig. 20.7. The electrostatic potential has the form $\Phi(\mathbf{x}) = \Phi_0(z^2 - x^2/2 - y^2/2)/(2d^2)$, where Φ_0 is the potential difference maintained across the electrodes, and d is the minimum axial distance from the origin to the hyperboloidal cap (as well as being $1/\sqrt{2}$ times the minimum radius of the ring electrode).

(a) Show that the potential satisfies Laplace's equation, as it must.

(b) Now consider an individual charged particle in the trap. Show that three separate oscillations can be excited:

    (i)   cyclotron orbits in the magnetic field with angular frequency ω_c,

    (ii) "magnetron" orbits produced by $\mathbf{E} \times \mathbf{B}$ drift around the axis of symmetry with angular frequency ω_m (which you should compute), and

(iii) axial oscillations parallel to the magnetic field with angular frequency ω_z (which you should also compute).

Assume that $\omega_m \ll \omega_z \ll \omega_c$, and show that $\omega_z^2 \simeq 2\omega_m\omega_c$.

(c) Typically, the potential difference across the electrodes is ~ 10 V, the magnetic field strength is $B \sim 6$ T, and the radius of the ring and the height of the caps above the center of the traps are ~ 3 mm. Estimate the three independent angular frequencies for electrons and ions, verifying the ordering $\omega_m \ll \omega_z \ll \omega_c$. Also estimate the maximum velocities associated with each of these oscillations if the particle is to be retained in the trap.

(d) Solve the classical equation of motion exactly, and demonstrate that the magnetron motion is formally unstable.

Penning traps have been used to perform measurements of the electron-proton mass ratio and the magnetic moment of the electron with unprecedented precision.

Bibliographic Note

For a thorough treatment of the particle kinetics of plasmas, see Shkarofsky, Johnston, and Bachynski (1966). For less detailed treatments, we recommend the relevant portions of Spitzer (1962), Bittencourt (2004), and Bellan (2006).

For applications of the Fokker-Planck equation to Coulomb scattering in plasmas, we like Shkarofsky, Johnston, and Bachynski (1966, Chaps. 7 and 8), Boyd and Sanderson (2003, Chap. 8), Kulsrud (2005, Chap. 8), and Bellan (2006, Chap. 13).

For charged-particle motion in inhomogeneous and time-varying magnetic fields, we recommend Northrop (1963); Jackson (1999, Chap. 12), which is formulated using special relativity; Bittencourt (2004, Chaps. 2–4); and Bellan (2006, Chap. 3).

21

Waves in Cold Plasmas: Two-Fluid Formalism

Waves from moving sources: Adagio. Andante. Allegro moderato.

OLIVER HEAVISIDE (1912)

21.1 Overview

The growth of plasma physics came about, in the early twentieth century, through studies of oscillations in electric discharges and the contemporaneous development of means to broadcast radio waves over great distances by reflecting them off Earth's ionosphere. It is therefore not surprising that most early plasma-physics research was devoted to describing the various modes of wave propagation. Even in the simplest, linear approximation for a plasma sufficiently cold that thermal effects are unimportant, we will see that the variety of possible modes is immense.

In the previous chapter, we introduced several length- and timescales, most importantly the Larmor (gyro) radius, the Debye length, the plasma period, the cyclotron (gyro) period, the collision time (inverse collision frequency), and the equilibration times (inverse collision rates). To these we must now add the wavelength and period of the wave under study. The wave's characteristics are controlled by the relative sizes of these parameters, and in view of the large number of parameters, there is a bewildering number of possibilities. If we further recognize that plasmas are collisionless, so there is no guarantee that the particle distribution functions can be characterized by a single temperature, then the possibilities multiply further.

Fortunately, the techniques needed to describe the propagation of linear wave perturbations in a particular equilibrium configuration of a plasma are straightforward and can be amply illustrated by studying a few simple cases. In this chapter, we follow this course by restricting our attention to one class of modes, those where we can either ignore completely the thermal motions of the ions and electrons that compose the plasma (i.e., treat these species as *cold*) or include them using just a velocity dispersion or temperature. We can then apply our knowledge of fluid dynamics by treating the ions and electrons separately as fluids acted on by electromagnetic forces. This is called the *two-fluid formalism* for plasmas. In the next chapter, we explore when and how waves are sensitive to the actual distribution of particle speeds by developing the more sophisticated *kinetic-theory formalism* and using it to study waves in *warm plasmas.*

We begin our two-fluid study of plasma waves in Sec. 21.2 by deriving a very general wave equation, which governs weak waves in a homogeneous plasma that may or may not have a magnetic field and also governs electromagnetic waves in any other dielectric medium. That wave equation and the associated dispersion relation for the wave modes depend on a dielectric tensor, which must be derived from an examination of the motions of the electrons and protons (or other charge carriers) inside the wave. Those motions are described, in our two-fluid (electron-fluid and proton-fluid) model by equations that we write down in Sec. 21.3.

In Sec. 21.4, we use those two-fluid equations to derive the dielectric tensor, and then we combine with our general wave equation from Sec. 21.2 to obtain the dispersion relation for wave modes in a uniform, unmagnetized plasma. The modes fall into two classes: (i) *Transverse or electromagnetic waves,* with the electric field **E** perpendicular to the wave's propagation direction. These are modified versions of electromagnetic waves in vacuum. As we shall see, they can propagate only at frequencies above the plasma frequency; at lower frequencies they become evanescent. (ii) *Longitudinal waves,* with **E** parallel to the propagation direction, which come in two species: *Langmuir waves* and *ion-acoustic waves.* Longitudinal waves are a melded combination of sound waves in a fluid and electrostatic plasma oscillations; their restoring force is a mixture of thermal pressure and electrostatic forces.

In Sec. 21.5, we explore how a uniform magnetic field changes the character of these waves. The **B** field makes the plasma anisotropic but axially symmetric. As a result, the dielectric tensor, dispersion relation, and wave modes have much in common with those in an anisotropic but axially symmetric dielectric crystal, which we studied in the context of nonlinear optics in Chap. 10. A plasma, however, has a much richer set of characteristic frequencies than does a crystal (electron plasma frequency, electron cyclotron frequency, ion cyclotron frequency, etc.). As a result, even in the regime of weak linear waves and a cold plasma (no thermal pressure), the plasma has a far greater richness of modes than does a crystal.

In Sec. 21.5.1, we derive the general dispersion relation that encompasses all these cold-magnetized-plasma modes, and in Secs. 21.5.2 and 21.5.3, we explore the special cases of modes that propagate parallel to and perpendicular to the magnetic field. Then in Sec. 21.5.4, we examine a practical example: the propagation of radio waves in Earth's ionosphere, where it is a good approximation to ignore the ion motion and work with a one-fluid (i.e., electron-fluid) theory. Having gained insight into simple cases (parallel modes, perpendicular modes, and one-fluid modes), we return in Sec. 21.5.5 to the full class of linear modes in a cold, magnetized, two-fluid plasma and briefly describe some tools by which one can make sense of them all.

Finally, in Sec. 21.6, we turn to the question of plasma stability. In Sec. 14.6.1 and Chap. 15, we saw that fluid flows that have sufficient shear are unstable; perturbations can feed off the relative kinetic energy of adjacent regions of the fluid and use that energy to power an exponential growth. In plasmas, with their long mean free paths, there can similarly exist kinetic energies of relative, ordered motion in velocity space; and perturbations, feeding off those energies, can grow exponentially. To study this phenomenon in full requires the kinetic-theory description of a plasma, which we develop in Chap. 22; but in Sec. 21.6 we gain insight into a prominent example of such a velocity-space instability by analyzing two cold plasma streams moving through each other. We illustrate the resulting *two-stream instability* by a short discussion of particle beams that are created in disturbances on the surface of the Sun and propagate out through the solar wind.

21.2 Dielectric Tensor, Wave Equation, and General Dispersion Relation

We begin our study of waves in plasmas by deriving a general wave equation that applies equally well to electromagnetic waves in unmagnetized plasmas; in magnetized plasmas; and in any other kind of dielectric medium, such as an anisotropic crystal. This wave equation is the same one we used in our study of nonlinear optics in Chap. 10 [Eqs. (10.50) and (10.51a)], and the derivation is essentially a linearized variant of the one we gave in Chap. 10 [Eqs. (10.16a)–(10.22b)].

When a wave propagates through a plasma (or other dielectric), it entails a relative motion of electrons and protons (or other charge carriers). Assuming the wave has small enough amplitude to be linear, those charge motions can be embodied in an oscillating polarization (electric dipole moment per unit volume) $\mathbf{P}(\mathbf{x}, t)$, which is related to the plasma's (or dielectric's) varying charge density ρ_e and current density $\mathbf{j}$ in the usual way:

polarization vector

$$\rho_e = -\boldsymbol{\nabla} \cdot \mathbf{P}, \qquad \mathbf{j} = \frac{\partial \mathbf{P}}{\partial t}. \qquad (21.1)$$

(These relations enforce charge conservation, $\partial \rho_e / \partial t + \nabla \cdot \mathbf{j} = 0$.) When these ρ_e and $\mathbf{j}$ are inserted into the standard Maxwell equations for $\mathbf{E}$ and $\mathbf{B}$, one obtains

Maxwell equations

$$\nabla \cdot \mathbf{E} = -\frac{\nabla \cdot \mathbf{P}}{\epsilon_0}, \quad \nabla \cdot \mathbf{B} = 0, \quad \nabla \times \mathbf{E} = -\frac{\partial \mathbf{B}}{\partial t}, \quad \nabla \times \mathbf{B} = \mu_0 \frac{\partial \mathbf{P}}{\partial t} + \frac{1}{c^2} \frac{\partial \mathbf{E}}{\partial t}.$$

(21.2)

If the plasma is endowed with a uniform magnetic field $\mathbf{B}_o$, that field can be left out of these equations, as its divergence and curl are guaranteed to vanish. Thus we can regard $\mathbf{E}$, $\mathbf{B}$, and $\mathbf{P}$ in these Maxwell equations as the perturbed quantities associated with the waves.

From a detailed analysis of the response of the charge carriers to the oscillating $\mathbf{E}$ and $\mathbf{B}$ fields, one can deduce a linear (frequency-dependent and wave-vector-dependent) relationship between the waves' electric field $\mathbf{E}$ and the polarization $\mathbf{P}$:

$$P_j = \epsilon_0 \chi_{jk} E_k.$$

(21.3)

tensorial electric susceptibility

Here ϵ_0 is the vacuum permittivity, and χ_{jk} is a dimensionless, tensorial electric susceptibility [cf. Eq. (10.21)]. A different, but equivalent, viewpoint on the relationship between $\mathbf{P}$ and $\mathbf{E}$ can be deduced by taking the time derivative of Eq. (21.3); setting $\partial \mathbf{P} / \partial t = \mathbf{j}$; assuming a sinusoidal time variation $e^{-i\omega t}$ so that $\partial \mathbf{E} / \partial t = -i\omega \mathbf{E}$; and then reinterpreting the result as Ohm's law with a tensorial electric conductivity κ_{ejk}:

tensorial electric conductivity

$$j_j = \kappa_{ejk} E_k, \qquad \kappa_{ejk} = -i\omega \epsilon_0 \chi_{jk}.$$

(21.4)

Evidently, for sinusoidal waves the electric susceptibility χ_{jk} and the electric conductivity κ_{ejk} embody the same information about the wave-particle interactions.

dielectric tensor

That information is also embodied in a third object: the dimensionless dielectric tensor ϵ_{jk}, which relates the electric displacement $\mathbf{D}$ to the electric field $\mathbf{E}$:

$$D_j \equiv \epsilon_0 E_j + P_j = \epsilon_0 \epsilon_{jk} E_k, \qquad \epsilon_{jk} = \delta_{jk} + \chi_{jk} = \delta_{jk} + \frac{i}{\epsilon_0 \omega} \kappa_{ejk}.$$

(21.5)

In the next section, we derive the value of the dielectric tensor ϵ_{jk} for waves in an unmagnetized plasma, and in Sec. 21.4.1, we derive it for a magnetized plasma.

Using the definition $\mathbf{D} = \epsilon_0 \mathbf{E} + \mathbf{P}$, we can eliminate $\mathbf{P}$ from Eqs. (21.2), thereby obtaining the familiar form of Maxwell's equations for dielectric media with no nonpolarization-based charges or currents:

$$\nabla \cdot \mathbf{D} = 0, \qquad \nabla \cdot \mathbf{B} = 0, \qquad \nabla \times \mathbf{E} = -\frac{\partial \mathbf{B}}{\partial t}, \qquad \nabla \times \mathbf{B} = \mu_0 \frac{\partial \mathbf{D}}{\partial t}.$$

(21.6)

general wave equation in a dielectric

By taking the curl of the third of these equations and combining with the fourth and with $D_j = \epsilon_0 \epsilon_{jk} E_k$, we obtain the wave equation that governs the perturbations:

$$\nabla^2 \mathbf{E} - \boldsymbol{\nabla}(\boldsymbol{\nabla} \cdot \mathbf{E}) - \boldsymbol{\epsilon} \cdot \frac{1}{c^2} \frac{\partial^2 \mathbf{E}}{\partial t^2} = 0, \tag{21.7}$$

where $\boldsymbol{\epsilon}$ is our index-free notation for ϵ_{jk}. [This is the same as the linearized approximation Eq. (10.50) to our nonlinear-optics wave equation (10.22a).] Specializing to a plane-wave mode with wave vector $\mathbf{k}$ and angular frequency ω, so $\mathbf{E} \propto e^{i\mathbf{k}\mathbf{x}} e^{-i\omega t}$, we convert this wave equation into a homogeneous, algebraic equation for the Cartesian components of the electric vector E_j (cf. Box 12.2):

$$L_{ij} E_j = 0, \tag{21.8}$$

algebratized wave equation

where

$$L_{ij} = k_i k_j - k^2 \delta_{ij} + \frac{\omega^2}{c^2} \epsilon_{ij}. \tag{21.9}$$

algebratized wave operator

We call Eq. (21.8) the *algebratized wave equation*, and L_{ij} the *algebratized wave operator*.

The algebratized wave equation (21.8) can have a solution only if the determinant of the 3-dimensional matrix L_{ij} vanishes:

$$\det ||L_{ij}|| \equiv \det \left\| k_i k_j - k^2 \delta_{ij} + \frac{\omega^2}{c^2} \epsilon_{ij} \right\|. \tag{21.10}$$

general dispersion relation for electromagnetic waves in a dielectric medium

This is a polynomial equation for the angular frequency as a function of the wave vector (with ω and $\mathbf{k}$ appearing not only explicitly in L_{ij} but also implicitly in the functional form of ϵ_{jk}). Each solution $\omega(\mathbf{k})$ of this equation is the dispersion relation for a particular wave mode. Therefore, we can regard Eq. (21.10) as the general dispersion relation for plasma waves—and for linear electromagnetic waves in any other kind of dielectric medium.

To obtain an explicit form of the dispersion relation (21.10), we must give a prescription for calculating the dielectric tensor $\epsilon_{ij}(\omega, \mathbf{k})$, or equivalently [cf. Eq. (21.5)] the conductivity tensor $\kappa_{e\,ij}$ or the susceptibility tensor χ_{ij}. The simplest prescription involves treating the electrons and ions as independent fluids; so we digress, briefly, from our discussion of waves to present the two-fluid formalism for plasmas.

21.3 Two-Fluid Formalism

21.3

A plasma necessarily contains rapidly moving electrons and ions, and their individual responses to an applied electromagnetic field depend on their velocities. In the simplest model of these responses, we average over all the particles in a species (electrons or protons in this case) and treat them collectively as a fluid. Now, the fact that all the electrons are treated as one fluid does not mean that they have to collide with one

two-fluid formalism

another. In fact, as we have already emphasized in Chap. 20, electron-electron collisions are quite rare, and we can usually ignore them. Nevertheless, we can still define a mean fluid velocity for both the electrons and the protons by averaging over their total velocity distribution functions just as we would for a gas:

mean fluid velocity

$$\mathbf{u}_s = \langle \mathbf{v} \rangle_s; \quad s = p, e,$$

(21.11)

where the subscripts p and e refer to protons and electrons. Similarly, for each fluid we define a pressure tensor using the fluid's dispersion of particle velocities:

pressure tensor

$$\mathbf{P}_s = n_s m_s \langle (\mathbf{v} - \mathbf{u}_s) \otimes (\mathbf{v} - \mathbf{u}_s) \rangle$$

(21.12)

[cf. Eqs. (20.34) and (20.35)].

The density n_s and mean velocity $\mathbf{u}_s$ of each species s must satisfy the equation of continuity (particle conservation):

particle conservation

$$\frac{\partial n_s}{\partial t} + \mathbf{\nabla} \cdot (n_s \mathbf{u}_s) = 0.$$

(21.13a)

They must also satisfy an equation of motion: the law of momentum conservation (i.e., the Euler equation with the Lorentz force added to the right-hand side):

momentum conservation

$$n_s m_s \left(\frac{\partial \mathbf{u}_s}{\partial t} + (\mathbf{u}_s \cdot \mathbf{\nabla}) \mathbf{u}_s \right) = -\mathbf{\nabla} \cdot \mathbf{P}_s + n_s q_s (\mathbf{E} + \mathbf{u}_s \times \mathbf{B}).$$

(21.13b)

Here we have neglected the tiny influence of collisions between the two species. In these equations and below, $q_s = \pm e$ is the particles' charge (positive for protons and negative for electrons). Note that, as collisions are ineffectual, we cannot assume that the pressure tensors are isotropic.

Although the particle- and momentum-conservation equations (21.13) for each species (electron or proton) are formally decoupled from the equations for the other species, there is actually a strong physical coupling induced by the electromagnetic field. The two species together produce $\mathbf{E}$ and $\mathbf{B}$ through their joint charge density and current density:

charge and current density

$$\rho_e = \sum_s q_s n_s, \qquad \mathbf{j} = \sum_s q_s n_s \mathbf{u}_s,$$

(21.14)

and those $\mathbf{E}$ and $\mathbf{B}$ fields strongly influence the electron and proton fluids' dynamics via their equations of motion (21.13b).

EXERCISES

Exercise 21.1 *Problem: Fluid Drifts in a Magnetized Plasma*
We developed a one-fluid (MHD) description of plasma in Chap. 19, and in Chap. 20 we showed how to describe the orbits of individual charged particles in a magnetic field that varies slowly compared with the particles' orbital periods and radii. In this

chapter, we describe the plasma as two or more cold fluids. We relate these three approaches in this exercise.

(a) Generalize Eq. (21.13b) to a single fluid as:

$$\rho \mathbf{a} = \rho \mathbf{g} - \nabla \cdot \mathbf{P} + \rho_e \mathbf{E} + \mathbf{j} \times \mathbf{B},$$ (21.15)

where $\mathbf{a}$ is the fluid acceleration, and $\mathbf{g}$ is the acceleration of gravity, and write the pressure tensor as $\mathbf{P} = P_\perp \mathbf{g} + (P_{||} - P_\perp)\mathbf{B} \otimes \mathbf{B}/B^2$, where we suppress the subscript s.[1] Show that the component of current density perpendicular to the local magnetic field is

$$\mathbf{j}_\perp = \frac{\mathbf{B} \times \nabla \cdot \mathbf{P}}{B^2} + \rho_e \frac{\mathbf{E} \times \mathbf{B}}{B^2} + \rho \frac{(\mathbf{g} - \mathbf{a}) \times \mathbf{B}}{B^2}.$$ (21.16a)

(b) Use vector identities to rewrite Eq. (21.16a) in the form:

$$\mathbf{j}_\perp = P_{||} \frac{\mathbf{B} \times (\mathbf{B} \cdot \nabla)\mathbf{B}}{B^4} + P_\perp \frac{\mathbf{B} \times \nabla B}{B^3} - (\nabla \times \mathbf{M})_\perp + \rho_e \frac{\mathbf{E} \times \mathbf{B}}{B^2} + \rho \frac{(\mathbf{g} - \mathbf{a}) \times \mathbf{B}}{B^2},$$ (21.16b)

where

$$\mathbf{M} = P_\perp \frac{\mathbf{B}}{B^3}$$ (21.16c)

is the *magnetization.*

(c) Identify the first term of Eq. (21.16b) with the curvature drift (20.49) and the second term with the gradient drift (20.51).

(d) Using a diagram, explain how the magnetization (20.54)—the magnetic moment per unit volume—can contribute to the current density. In particular, consider what might happen at the walls of a cavity containing plasma.[2] Argue that there should also be a local magnetization current parallel to the magnetic field, even when there is no net drift of the particles.

(e) Associate the final two terms of Eq. (21.16b) with the "$\mathbf{E} \times \mathbf{B}$" drift (20.47) and the gravitational drift (20.48). Explain the presence of the acceleration in the gravitational drift.

(f) Discuss how to combine these contributions to rederive the standard formulation of MHD, and specify some circumstances under which MHD might be a poor approximation.

1. Note that when writing the pressure tensor this way, we are implicitly assuming that it has only two components, along and perpendicular to the local magnetic field. Physically, this is reasonable if the particles that make up the fluid are *magnetized,* i.e., the particles' orbital periods and radii are short compared to the scales on which the magnetic field changes.
2. This is an illustration of a general theorem in statistical mechanics due to Niels Bohr in 1911, which essentially shows that magnetization cannot arise in classical physics. This theorem may have had a role in the early development of quantum mechanics.

Note that this single-fluid formalism does not describe the component of the current parallel to the magnetic field, $\mathbf{j}_{||}$. For this, we need to introduce an effective collision frequency for the electrons and ions either with waves (Chap. 23) or with other particles (Chap. 20). If we assume that the electrical conductivity is perfect, then $\mathbf{E} \cdot \mathbf{B} = 0$, and $\mathbf{j}_{||}$ is essentially fixed by the boundary conditions.

21.4

21.4 Wave Modes in an Unmagnetized Plasma

We now specialize to waves in a homogeneous, unmagnetized electron-proton plasma.

unperturbed plasma

First consider an unperturbed plasma in the absence of a wave, and work in a frame in which the proton fluid velocity $\mathbf{u}_p$ vanishes. By assumption, the equilibrium is spatially uniform. If there were an electric field, then charges would quickly flow to neutralize it; so there can be no electric field, and hence (since $\mathbf{\nabla} \cdot \mathbf{E} = \rho_e / \epsilon_0$) no net charge density. Therefore, the electron density must equal the proton density. Furthermore, there can be no net current, as this would lead to a magnetic field; so since the proton current $e\, n_p \mathbf{u}_p$ vanishes, the electron current $= -e\, n_e \mathbf{u}_e$ must also vanish. Hence the electron fluid velocity $\mathbf{u}_e$ must vanish in our chosen frame. Thus in an equilibrated homogeneous, unmagnetized plasma, $\mathbf{u}_e$, $\mathbf{u}_p$, $\mathbf{E}$, and $\mathbf{B}$ all vanish in the protons' mean rest frame.

Now apply an electromagnetic perturbation. This will induce a small, oscillating fluid velocity $\mathbf{u}_s$ in both the proton and electron fluids. It should not worry us that the fluid velocity is small compared with the random speeds of the constituent particles; the same is true in any subsonic gas dynamical flow, but the fluid description remains good there and also here.

21.4.1

21.4.1 Dielectric Tensor and Dispersion Relation for a Cold, Unmagnetized Plasma

Continuing to keep the plasma unmagnetized, let us further simplify matters (until Sec. 21.4.3) by restricting ourselves to a cold plasma, so the tensorial pressures vanish: $\mathbf{P}_s = 0$. As we are only interested in linear wave modes, we rewrite the two-fluid equations (21.13), just retaining terms that are first order in perturbed quantities [i.e., dropping the $(\mathbf{u}_s \cdot \mathbf{\nabla})\mathbf{u}_s$ and $\mathbf{u}_s \times \mathbf{B}$ terms]. Then, focusing on a plane-wave mode, $\propto \exp[i(\mathbf{k} \cdot \mathbf{x} - \omega t)]$, we bring the equation of motion (21.13b) into the form

linearized perturbative equation of motion

$$\boxed{-i\omega n_s m_s \mathbf{u}_s = q_s n_s \mathbf{E}} \tag{21.17}$$

for each species, $s = p, e$. From this, we can immediately deduce the linearized current density:

$$\mathbf{j} = \sum_s n_s q_s \mathbf{u}_s = \sum_s \frac{i n_s q_s^2}{m_s \omega} \mathbf{E}, \tag{21.18}$$

from which we infer that the conductivity tensor κ_e has Cartesian components

$$\kappa_{e\,ij} = \sum_s \frac{in_s q_s^2}{m_s \omega} \delta_{ij}, \tag{21.19}$$

where δ_{ij} is the Kronecker delta. Note that the conductivity is purely imaginary, which means that the current oscillates out of phase with the applied electric field, which in turn implies that there is no time-averaged Ohmic energy dissipation: $\langle \mathbf{j} \cdot \mathbf{E} \rangle = 0$. Inserting the conductivity tensor (21.19) into the general equation (21.5) for the dielectric tensor, we obtain

$$\epsilon_{ij} = \delta_{ij} + \frac{i}{\epsilon_0 \omega} \kappa_{e\,ij} = \left(1 - \frac{\omega_p^2}{\omega^2}\right) \delta_{ij}. \tag{21.20}$$

Here and throughout this chapter, the plasma frequency ω_p is slightly different from that used in Chap. 20: it includes a tiny $(1/1{,}836)$ correction due to the motion of the protons, which we neglected in our analysis of plasma oscillations in Sec. 20.3.3:

$$\boxed{\omega_p^2 = \sum_s \frac{n_s q_s^2}{m_s \epsilon_0} = \frac{ne^2}{m_e \epsilon_0} \left(1 + \frac{m_e}{m_p}\right) = \frac{ne^2}{\mu \epsilon_0},} \tag{21.21}$$

plasma frequency

where μ is the reduced mass $\mu = m_e m_p/(m_e + m_p)$. Note that because there is no physical source of a preferred direction in the plasma, the dielectric tensor (21.20) is isotropic.

Now, without loss of generality, let the waves propagate in the z direction, so $\mathbf{k} = k\mathbf{e}_z$. Then the algebratized wave operator (21.9), with ϵ given by Eq. (21.20), takes the following form:

$$\boxed{L_{ij} = \frac{\omega^2}{c^2} \begin{pmatrix} 1 - \dfrac{c^2 k^2}{\omega^2} - \dfrac{\omega_p^2}{\omega^2} & 0 & 0 \\ 0 & 1 - \dfrac{c^2 k^2}{\omega^2} - \dfrac{\omega_p^2}{\omega^2} & 0 \\ 0 & 0 & 1 - \dfrac{\omega_p^2}{\omega^2} \end{pmatrix}.} \tag{21.22}$$

algebratized wave operator and general dispersion relation for waves in a cold, unmagnetized plasma

The corresponding dispersion relation $\det||L_{jk}|| = 0$ [Eq. (21.10)] becomes

$$\left(1 - \frac{c^2 k^2}{\omega^2} - \frac{\omega_p^2}{\omega^2}\right)^2 \left(1 - \frac{\omega_p^2}{\omega^2}\right) = 0. \tag{21.23}$$

This is a polynomial equation of order 6 for ω as a function of k, so formally there are six solutions corresponding to three pairs of modes propagating in opposite directions.

Two of the pairs of modes are degenerate with frequency

$$\boxed{\omega = \sqrt{\omega_p^2 + c^2 k^2}.} \tag{21.24}$$

These are called *plasma electromagnetic modes,* and we study them in the next subsection. The remaining pair of modes exist at a single frequency:

$$\boxed{\omega = \omega_p.} \tag{21.25}$$

These must be the electrostatic plasma oscillations that we studied in Sec. 20.3.3 (though now with an arbitrary wave number k, while in Sec. 20.3.3 the wave number was assumed to be zero). In Sec. 21.4.3, we show that this is so and explore how these plasma oscillations are modified by finite-temperature effects.

21.4.2 Plasma Electromagnetic Modes

To learn the physical nature of the modes with dispersion relation $\omega = \sqrt{\omega_p^2 + c^2 k^2}$

[Eq. (21.24)], we must examine the details of their electric-field oscillations, magnetic-field oscillations, and electron and proton motions. A key to this is the algebratized wave equation $L_{ij} E_j = 0$, with L_{ij} specialized to the dispersion relation (21.24): $||L_{ij}|| = \mathrm{diag}[0, 0, (\omega^2 - \omega_p^2)/c^2]$. In this case, the general solution to $L_{ij} E_j = 0$ is an electric field that lies in the x-y plane (transverse plane) and that therefore

is orthogonal to the waves' propagation vector $\mathbf{k} = k \mathbf{e}_z$. The third of the Maxwell equations (21.2) implies that the magnetic field is

$$\mathbf{B} = (\mathbf{k}/\omega) \times \mathbf{E}, \tag{21.26}$$

which also lies in the transverse plane and is orthogonal to $\mathbf{E}$. Evidently, these modes are close analogs of electromagnetic waves in vacuum; correspondingly, they are known as the plasma's *electromagnetic modes.* The electron and proton motions in these modes, as given by Eq. (21.17), are oscillatory displacements in the direction of $\mathbf{E}$ but are out of phase with $\mathbf{E}$. The amplitudes of the fluid motions, at fixed electric-field amplitude, vary as $1/\omega$; when ω decreases, the fluid amplitudes grow.

The dispersion relation for these modes, Eq. (21.24), implies that they can only propagate (i.e., have real angular frequency when the wave vector is real) if ω exceeds the plasma frequency. As ω is decreased toward ω_p, k approaches zero, so these modes become electrostatic plasma oscillations with arbitrarily long wavelength orthogonal to the oscillation direction (i.e., they become a spatially homogeneous variant of the plasma oscillations studied in Sec. 20.3.3). At $\omega < \omega_p$ these modes become evanescent.

In their regime of propagation, $\omega > \omega_p$, these cold-plasma electromagnetic waves have a phase velocity given by

$$\mathbf{V}_{\mathrm{ph}} = \frac{\omega}{k} \hat{\mathbf{k}} = c \left(1 - \frac{\omega_p^2}{\omega^2} \right)^{-1/2} \hat{\mathbf{k}}, \tag{21.27a}$$

where $\hat{\mathbf{k}} \equiv \mathbf{k}/k$ is a unit vector in the propagation direction. Although this phase velocity exceeds the speed of light, causality is not violated, because information (and energy) propagate at the group velocity, not the phase velocity. The group velocity is readily shown to be

$$\mathbf{V_g} = \nabla_\mathbf{k} \, \omega = \frac{c^2 \mathbf{k}}{\omega} = c \left(1 - \frac{\omega_p^2}{\omega^2} \right)^{1/2} \hat{\mathbf{k}}, \qquad (21.27\text{b}) \qquad \text{group velocity}$$

which is less than c as it must be.

These cold-plasma electromagnetic modes transport energy and momentum just like wave modes in a fluid. There are three contributions to the waves' mean (time-averaged) energy density: the electric, the magnetic, and the kinetic-energy densities. (If we had retained the pressure, then an additional contribution would come from the internal energy.) To compute these mean energy densities, we must form the time average of products of physical quantities. Now, we have used the complex representation to denote each of our oscillating quantities (e.g., E_x), so we must be careful to remember that $A = ae^{i(\mathbf{k}\cdot\mathbf{x}-\omega t)}$ is an abbreviation for the real part of this quantity—which is the physical A. It is easy to show (Ex. 21.3) that the time-averaged value of the physical A times the physical B (which we shall denote by $\langle AB \rangle$) is given in terms of their complex amplitudes by

$$\langle AB \rangle = \frac{AB^* + A^*B}{4}, \qquad (21.28) \qquad \begin{array}{l}\text{time-averaged product}\\\text{in terms of complex}\\\text{amplitudes}\end{array}$$

where * denotes a complex conjugate.

Using Eqs. (21.26) and (21.27a), we can write the magnetic energy density in the form $\langle B^2 \rangle/(2\mu_0) = (1 - \omega_p^2/\omega^2)\epsilon_0\langle E^2 \rangle/2$. Using Eq. (21.17), the particle kinetic energy is $\sum_s n_s m_s \langle u_s^2 \rangle/2 = (\omega_{pe}^2/\omega^2)\epsilon_0\langle E^2 \rangle/2$. Summing these contributions and using Eq. (21.28), we obtain

$$U = \frac{\epsilon_0 E E^*}{4} + \frac{B B^*}{4\mu_0} + \sum_s \frac{n_s m_s u_s u_s^*}{4} \qquad \text{energy density}$$

$$= \frac{\epsilon_0 E E^*}{2}. \qquad (21.29\text{a})$$

The mean energy flux in the wave is carried (to quadratic order) by the electromagnetic field and is given by the Poynting flux. (The kinetic energy flux vanishes to this order.) A straightforward calculation gives

$$\mathbf{F}_{\text{EM}} = \langle \mathbf{E} \times \mathbf{B} \rangle = \frac{\mathbf{E} \times \mathbf{B}^* + \mathbf{E}^* \times \mathbf{B}}{4} = \frac{E E^* \mathbf{k}}{2\mu_0 \omega} \qquad \text{energy flux}$$

$$= U\mathbf{V_g}, \qquad (21.29\text{b})$$

where we have used $\mu_0 = c^{-2}\epsilon_0^{-1}$. We therefore find that the energy flux is the product of the energy density and the group velocity, as is true quite generally (cf. Sec. 6.3). (If it were not true, then a localized wave packet, which propagates at the group velocity, would move along a different trajectory from its energy, and we would wind up with energy in regions with vanishing amplitude!)

Exercise 21.2 *Derivation: Phase and Group Velocities for Electromagnetic Modes*
Derive Eqs. (21.27) for the phase and group velocities of electromagnetic modes in a plasma.

Exercise 21.3 *Derivation: Time-Averaging Formula*
Verify Eq. (21.28).

Exercise 21.4 *Problem: Collisional Damping in an Electromagnetic Wave Mode*
Consider a transverse electromagnetic wave mode propagating in an unmagnetized, partially ionized gas in which the electron-neutral collision frequency is ν_e. Include the effects of collisions in the electron equation of motion (21.17), by introducing a term $-n_e m_e \nu_e \mathbf{u}_e$ on the right-hand side. Ignore ion motion and electron-ion and electron-electron collisions.

Derive the dispersion relation when $\omega \gg \nu_e$, and show by explicit calculation that the rate of loss of energy per unit volume ($-\nabla \cdot \mathbf{F}_{\mathrm{EM}}$, where $\mathbf{F}_{\mathrm{EM}}$ is the Poynting flux) is balanced by the Ohmic heating of the plasma. [Hint: It may be easiest to regard ω as real and $\mathbf{k}$ as complex.]

21.4.3

21.4.3 Langmuir Waves and Ion-Acoustic Waves in Warm Plasmas

longitudinal electrostatic plasma oscillations

For our case of a cold, unmagnetized plasma, the third pair of modes embodied in the dispersion relation (21.23) only exists at a single frequency, the plasma frequency: $\omega = \omega_p$. These modes' wave equation $L_{ij} E_j = 0$ with $||L_{ij}|| = \mathrm{diag}(-k^2, -k^2, 0)$ [Eq. (21.22) with $\omega^2 = \omega_p^2$] implies that $\mathbf{E}$ points in the z direction (i.e., along $\mathbf{k}$, which is to say the longitudinal direction). Maxwell's equations then imply $\mathbf{B} = 0$, and the equation of motion (21.17) implies that the fluid displacements are also in the direction of $\mathbf{E}$—the longitudinal direction. Clearly, these modes, like electromagnetic modes in the limit $k = 0$ and $\omega = \omega_p$, are electrostatic plasma oscillations. However, in this case, where the spatial variations of $\mathbf{E}$ and $\mathbf{u}_s$ are along the direction of oscillation instead of perpendicular to it, k is not constrained to vanish; instead, all wave numbers are allowed. This means that the plasma can undergo plane-parallel oscillations at $\omega = \omega_p$ with displacements in some Cartesian z direction and with any arbitrary z-dependent amplitude that one might wish. But these oscillations cannot transport energy: because ω is independent of $\mathbf{k}$, their group velocity, $\mathbf{V}_g = \nabla_\mathbf{k}\,\omega$, vanishes. Their phase velocity, $\mathbf{V}_{\mathrm{ph}} = (\omega_p/k)\hat{\mathbf{k}}$, by contrast, is finite.

finite temperature converts longitudinal plasma oscillations into Langmuir waves

So far, we have confined ourselves to wave modes in cold plasmas. When thermal motions are turned on, the resulting thermal pressure gradients convert longitudinal plasma oscillations, at finite wave number k, into propagating, energy-transporting, longitudinal modes called *Langmuir waves*.[3] As we have already intimated, because

---

3. The chemist Irving Langmuir observed these waves and introduced the name "plasma" for ionized gas in 1927.

the plasma is collisionless, we must turn to kinetic theory (Chap. 22) to understand the thermal effects fully. However, using the present chapter's two-fluid formalism and with the guidance of physical arguments, we can deduce the leading-order effects of finite temperature.

In our physical arguments, we assume that the electrons are thermalized with one another at a temperature T_e, the protons are thermalized at temperature T_p, and T_e and T_p may differ (because the timescale for electrons and protons to exchange energy is so much longer than the timescales for electrons to exchange energy among themselves and for protons to exchange energy among themselves; see Sec. 20.4.3).

Physically, the key to the Langmuir waves' propagation is the warm electrons' thermal pressure. (The proton pressure is unimportant here, because the protons oscillate electrostatically with an amplitude that is tiny compared to the electron amplitude; nevertheless, as we shall see below, the proton pressure is important in other ways.)

In an adiabatic sound wave *in a fluid* (where the particle mean free paths are small compared to the wavelength), we relate the pressure perturbation to the density perturbation by assuming that the entropy is held constant in each fluid element. In other words, we write $\nabla P = C^2 m \nabla n$, where $C = [\gamma P/(nm)]^{1/2}$ is the adiabatic sound speed, n is the particle density, m is the particle mass, and γ is the adiabatic index [which is equal to the specific heat ratio $\gamma = c_P/c_V$ (Ex. 5.4)], whose value is 5/3 for a monatomic gas.

However, the electron gas *in the plasma we are considering* is collisionless on the short timescale of a perturbation period, and we are only interested in the tensorial pressure gradient parallel to $\mathbf{k}$ (which we take to point in the z direction), $\delta P_{e\,zz,z}$. We can therefore ignore all electron motion perpendicular to the wave vector, as it is not coupled to the parallel motion. The electron motion is then effectively 1-dimensional, since there is only one (translational) degree of freedom. The relevant specific heat at constant volume is therefore just $k_B/2$ per electron, while that at constant pressure is $3k_B/2$, giving $\gamma = 3$.[4] The effective sound speed for the electron gas is then $C = (3k_B T_e/m_e)^{1/2}$, and correspondingly, the perturbations of longitudinal electron pressure and electron density are related by

$$\frac{\delta P_{e\,zz}}{m_e \delta n_e} = C^2 = \frac{3k_B T_e}{m_e}. \tag{21.30a}$$

This is one of the equations governing Langmuir waves. The others are the linearized equation of continuity (21.13a), which relates the electrons' density perturbation to the longitudinal component of their fluid velocity perturbation:

$$\delta n_e = n_e \frac{k}{\omega} u_{e\,z}, \tag{21.30b}$$

---

4. We derived this longitudinal adiabatic index $\gamma = 3$ by a different method in Ex. 20.11e [Eq. (20.36b)] in the context of a plasma with a magnetic field along the longitudinal (or z) direction. It is valid also in our unmagnetized case, because the magnetic field has no influence on longitudinal electron motions.

the linearized equation of motion (21.13b), which relates $u_{e\,z}$ and $\delta P_{e\,zz}$ to the longitudinal component of the oscillating electric field:

$$-i\omega n_e m_e u_{e\,z} = -ik\delta P_{e\,zz} - n_e e E_z,\qquad(21.30c)$$

and the linearized form of Poisson's equation $\nabla \cdot \mathbf{E} = \rho_e/\epsilon_0$, which relates E_z to δn_e:

$$ikE_z = -\frac{\delta n_e e}{\epsilon_0}.\qquad(21.30d)$$

Equations (21.30) are four equations for three ratios of the perturbed quantities. By combining these equations, we obtain a condition that must be satisfied for them to have a solution:

Bohm-Gross dispersion relation

$$\boxed{\omega^2 = \omega_{pe}^2 + \frac{3k_B T_e}{m_e}k^2 = \omega_{pe}^2(1 + 3k^2\lambda_D^2).}\qquad(21.31)$$

Here $\lambda_D = \sqrt{\epsilon_0 k_B T_e/(n_e e^2)}$ is the Debye length [Eq. (20.10)]. Equation (21.31) is the *Bohm-Gross* dispersion relation for Langmuir waves.

From this dispersion relation, we deduce the phase speed of a Langmuir wave:

phase speed

$$V_{\rm ph} = \frac{\omega}{k} = \left(\frac{k_B T_e}{m_e}\right)^{1/2}\left(3 + \frac{1}{k^2\lambda_D^2}\right)^{1/2}.\qquad(21.32)$$

Evidently, when the reduced wavelength $\lambda/(2\pi) = 1/k$ is less than or of order the Debye length ($k\lambda_D \gtrsim 1$), the phase speed becomes comparable with the electron thermal speed. It is then possible for individual electrons to transfer energy between adjacent compressions and rarefactions in the wave. As we shall see in Sec. 22.3, when we recover Eq. (21.31) from a kinetic treatment, the resulting energy transfers **Landau damping** damp the wave; this is called *Landau damping*. Therefore, the Bohm-Gross dispersion relation is only valid for reduced wavelengths much longer than the Debye length (i.e., $k\lambda_D \ll 1$; Fig. 21.1).

In our analysis of Langmuir waves we have ignored the proton motion. This is justified as long as the proton thermal speeds are small compared to the electron thermal speeds (i.e., $T_p \ll m_p T_e/m_e$), which will almost always be the case. Proton motion is, **properties of ion-acoustic waves:** however, not ignorable in a second type of plasma wave that owes its existence to finite temperature: *ion-acoustic waves* (also called ion-sound waves). These are waves **low frequencies; electrons locked to protons** that propagate with frequencies far below the electron plasma frequency—frequencies so low that the electrons remain locked electrostatically to the protons, keeping the plasma charge neutral and preventing electromagnetic fields from participating in the oscillations. As for Langmuir waves, we can derive the ion-acoustic dispersion relation using fluid theory combined with physical arguments.

Using kinetic theory in the next chapter, we shall see that ion-acoustic waves can propagate only when the proton temperature is small compared with the electron temperature: $T_p \ll T_e$; otherwise they are damped. (Such a temperature disparity

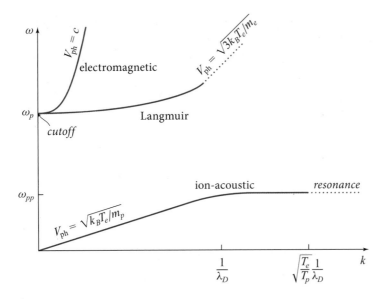

FIGURE 21.1 Dispersion relations for electromagnetic waves, Langmuir waves, and ion-acoustic waves in an unmagnetized plasma, whose electrons are thermalized with one another at temperature T_e, and whose protons are thermalized at a temperature T_p that might not be equal to T_e. In the dotted regions the waves are strongly damped, according to kinetic-theory analyses (Chap. 22). Ion-acoustic waves are wiped out by that damping at all k unless $T_p \ll T_e$ (as is assumed on the horizontal axis), in which case they survive on the non-dotted part of their curve.

can be produced, e.g., when a plasma passes through a shock wave, and it can be maintained for a long time because Coulomb collisions are so ineffective at restoring $T_p \sim T_e$; cf. Sec. 20.4.3.) Because $T_p \ll T_e$, the proton pressure can be ignored, and the waves' restoring force is provided by electron pressure. Now, in an ion-acoustic wave—by contrast with a Langmuir wave—the individual thermal electrons can travel over many wavelengths during a single wave period, so the electrons remain isothermal as their mean motion oscillates in lock-step with the protons' mean motion. Correspondingly, the electrons' effective (1-dimensional) specific-heat ratio is $\gamma_{\text{eff}} = 1$.

require low proton temperature; restoring force is electron pressure and inertia is proton mass

Although the electrons provide the ion-acoustic waves' restoring force, the inertia of the electrostatically locked electrons and protons is almost entirely that of the heavy protons. Correspondingly, the waves' phase velocity is

$$\mathbf{V}_{ia} = \left(\frac{\gamma_{\text{eff}} P_e}{n_p m_p} \right)^{1/2} \hat{k} = \left(\frac{k_B T_e}{m_p} \right)^{1/2} \hat{k} \tag{21.33}$$

phase velocity

(cf. Ex. 21.5), and the dispersion relation is $\omega = V_{\text{ph}} k = (k_B T_e / m_p)^{1/2} k$.

From this phase velocity and our physical description of these ion-acoustic waves, it should be evident that they are the magnetosonic waves of MHD theory (Sec. 19.7.2), in the limit that the plasma's magnetic field is turned off.

In Ex. 21.5, we show that the character of these waves is modified when their wavelength becomes of order the Debye length (i.e., when $k\lambda_D \sim 1$). The dispersion relation then becomes

dispersion relation

$$\omega = \left(\frac{k_B T_e/m_p}{1 + \lambda_D^2 k^2}\right)^{1/2} k, \tag{21.34}$$

which means that for $k\lambda_D \gg 1$, the waves' frequency approaches the *proton plasma frequency* $\omega_{pp} \equiv \sqrt{ne^2/(\epsilon_0 m_p)} \simeq \sqrt{m_e/m_p}\,\omega_p$. A kinetic-theory treatment reveals that these waves are strongly damped when $k\lambda_D \gtrsim \sqrt{T_e/T_p}$. These features of the ion-acoustic dispersion relation are shown in Fig. 21.1.

proton plasma frequency

The regime in which ion-acoustic (magnetosonic) waves can propagate, $\omega < \omega_{pp}$, is quite generally the regime in which the electrons and ions are locked to each other, making magnetohydrodynamics a good appoximation to the plasma's dynamics.

MHD regime

EXERCISES

Exercise 21.5 *Derivation: Ion-Acoustic Waves*

Ion-acoustic waves can propagate in an unmagnetized plasma when the electron temperature T_e greatly exceeds the ion temperature T_p. In this limit, the electron density n_e can be approximated by $n_e = n_0 \exp[e\Phi/(k_B T_e)]$, where n_0 is the mean electron density, and Φ is the electrostatic potential.

(a) Show that for ion-acoustic waves that propagate in the z direction, the nonlinear equations of continuity, the motion for the ion (proton) fluid, and Poisson's equation for the potential take the form

$$\frac{\partial n}{\partial t} + \frac{\partial(nu)}{\partial z} = 0,$$

$$\frac{\partial u}{\partial t} + u\frac{\partial u}{\partial z} = -\frac{e}{m_p}\frac{\partial \Phi}{\partial z},$$

$$\frac{\partial^2 \Phi}{\partial z^2} = -\frac{e}{\epsilon_0}\left(n - n_0 e^{e\Phi/(k_B T_e)}\right). \tag{21.35}$$

Here n is the proton density, and u is the proton fluid velocity (which points in the z direction).

(b) Linearize Eqs. (21.35), and show that the dispersion relation for small-amplitude ion-acoustic modes is

$$\omega = \omega_{pp}\left(1 + \frac{1}{\lambda_D^2 k^2}\right)^{-1/2} = \left(\frac{k_B T_e/m_p}{1 + \lambda_D^2 k^2}\right)^{1/2} k, \tag{21.36}$$

where λ_D is the Debye length.

Exercise 21.6 *Challenge: Ion-Acoustic Solitons*

In this exercise we explore nonlinear effects in ion-acoustic waves (Ex. 21.5), and show that they give rise to solitons that obey the same KdV equation as governs solitonic water waves (Sec. 16.3). This version of the solitons is only mildly nonlinear. In Sec. 23.6 we will generalize to strong nonlinearity.

(a) Introduce a bookkeeping expansion parameter ε whose numerical value is unity,[5] and expand the ion density, ion velocity, and potential in the forms

$$n = n_0(1 + \varepsilon n_1 + \varepsilon^2 n_2 + \ldots),$$

$$u = (k_B T_e/m_p)^{1/2}(\varepsilon u_1 + \varepsilon^2 u_2 + \ldots),$$

$$\Phi = (k_B T_e/e)(\varepsilon \Phi_1 + \varepsilon^2 \Phi_2 + \ldots). \tag{21.37}$$

Here n_1, u_1, and Φ_1 are small compared to unity, and the factors of ε tell us that, as the wave amplitude is decreased, these quantities scale proportionally to one another, while n_2, u_2, and Φ_2 scale proportionally to the squares of n_1, u_1, and Φ_1, respectively. Change independent variables from (t, z) to (τ, η), where

$$\eta = \sqrt{2}\varepsilon^{1/2}\lambda_D^{-1}[z - (k_B T_e/m_p)^{1/2}t],$$

$$\tau = \sqrt{2}\varepsilon^{3/2}\omega_{pp}t. \tag{21.38}$$

Explain, now or at the end, the chosen powers $\varepsilon^{1/2}$ and $\varepsilon^{3/2}$. By substituting Eqs. (21.37) and (21.38) into the nonlinear equations (21.35), equating terms of the same order in ε, and then setting $\varepsilon = 1$ (bookkeeping parameter!), show that n_1, u_1, and Φ_1 *each* satisfy the KdV equation (16.32):

$$\frac{\partial \zeta}{\partial \tau} + \zeta\frac{\partial \zeta}{\partial \eta} + \frac{\partial^3 \zeta}{\partial \eta^3} = 0. \tag{21.39}$$

(b) In Sec. 16.3 we discussed the exact, single-soliton solution (16.33) to this KdV equation. Show that for an ion-acoustic soliton, this solution propagates with the physical speed $(1 + n_{1o})(k_B T_e/m_p)^{1/2}$ (where n_{1o} is the value of n_1 at the peak of the soliton), which is greater the larger is the wave's amplitude.

21.4.4 Cutoffs and Resonances

Electromagnetic waves, Langmuir waves, and ion-acoustic waves in an unmagnetized plasma provide examples of *cutoffs* and *resonances*.

A *cutoff* is a frequency at which a wave mode ceases to propagate because its wave number k there becomes zero. Langmuir and electromagnetic waves at $\omega \to \omega_p$ are examples; see their dispersion relations in Fig. 21.1. For concreteness, consider a monochromatic radio-frequency electromagnetic wave propagating upward into Earth's ionosphere at some nonzero angle to the vertical (left side of Fig. 21.2), and neglect the effects of Earth's magnetic field. As the wave moves deeper (higher) into the ionosphere, it encounters a rising electron density n and correspondingly, a rising plasma frequency ω_p. The wave's wavelength will typically be small compared to the

cutoff, where k goes to zero

---

5. See Box 7.2 for a discussion of such bookkeeping parameters in a different context.

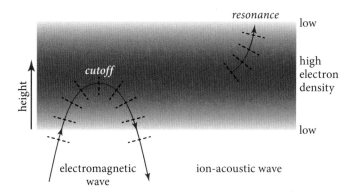

FIGURE 21.2 Cutoff and resonance illustrated by wave propagation in Earth's ionosphere. The thick, arrowed curves are rays, and the thin, dashed lines are phase fronts. The electron density is proportional to the darkness of the shading.

inhomogeneity scale for ω_p, so the wave propagation can be analyzed using geometric optics (Sec. 7.3). Across a phase front, the portion of the wave that is higher in the ionosphere will have a smaller k and thus a larger wavelength and phase speed; it thus has a greater distance between phase fronts (dashed lines). Therefore, the rays, which are orthogonal to the phase fronts, will bend away from the vertical (left side of Fig. 21.2); that is, the wave will be reflected away from the cutoff, at which $\omega_p \rightarrow \omega$ and $k \rightarrow 0$. Clearly, this behavior is quite general. Wave modes are generally reflected from regions in which slowly changing plasma conditions give rise to cutoffs.

A *resonance* is a frequency at which a wave mode ceases to propagate because its wave number k there becomes infinite (i.e., its wavelength goes to zero). Ion-acoustic waves provide an example; see their dispersion relation in Fig. 21.1. For concreteness, consider an ion-acoustic wave deep in the ionosphere, where ω_{pp} is larger than the wave's frequency ω (right side of Fig. 21.2). As the wave propagates toward the upper edge of the ionosphere, at some nonzero angle to the vertical, the portion of a phase front that is higher sees a smaller electron density and thus a smaller ω_{pp}; hence it has a larger k and shorter wavelength and thus a shorter distance between phase fronts (dashed lines). This causes the rays to bend toward the vertical (right side of Fig. 21.2). The wave is "attracted" to the region of the resonance, $\omega \rightarrow \omega_{pp}$, $k \rightarrow \infty$, where it is "Landau damped" (Chap. 22) and dies. This behavior is quite general. Wave modes are generally attracted to regions in which slowly changing plasma conditions give rise to resonances, and on reaching a resonance, they die.

We study wave propagation in the ionosphere in greater detail in Sec. 21.5.4.

waves are reflected by cutoff regions

resonance, where k becomes infinite

waves are attracted to resonance regions, and damped

21.5   21.5 Wave Modes in a Cold, Magnetized Plasma

21.5.1   21.5.1 Dielectric Tensor and Dispersion Relation

We now complicate matters somewhat by introducing a uniform magnetic field $\mathbf{B}_0$ into the unperturbed plasma. To avoid additional complications, we make the plasma cold (i.e., we omit thermal effects). The linearized equation of motion (21.13b) for each species s then becomes

$$-i\omega\mathbf{u}_s = \frac{q_s \mathbf{E}}{m_s} + \frac{q_s}{m_s}\mathbf{u}_s \times \mathbf{B}_0.$$ (21.40)

It is convenient to multiply this equation of motion by $n_s q_s/\epsilon_0$ and introduce for each species a scalar plasma frequency and scalar and vectorial cyclotron frequencies:

$$\boxed{\omega_{ps} = \left(\frac{n_s q_s^2}{\epsilon_0 m_s}\right)^{1/2}, \qquad \omega_{cs} = \frac{q_s B_0}{m_s}, \qquad \boldsymbol{\omega}_{cs} = \omega_{cs}\hat{\mathbf{B}}_0 = \frac{q_s \mathbf{B}_0}{m_s}}$$ (21.41)

electron and proton plasma frequencies, and cyclotron frequencies

[so $\omega_{pp} = \sqrt{(m_e/m_p)}\,\omega_{pe} \simeq \omega_{pe}/43$, $\omega_p = \sqrt{\omega_{pe}^2 + \omega_{pp}^2}$, $\omega_{ce} < 0$, $\omega_{cp} > 0$, and $\omega_{cp} = (m_e/m_p)|\omega_{ce}| \simeq |\omega_{ce}|/1{,}836$]. Thereby we bring the equation of motion (21.40) into the form

$$-i\omega\left(\frac{n_s q_s}{\epsilon_0}\mathbf{u}_s\right) + \boldsymbol{\omega}_{cs} \times \left(\frac{n_s q_s}{\epsilon_0}\mathbf{u}_s\right) = \omega_{ps}^2 \mathbf{E}.$$ (21.42)

By combining this equation with $\boldsymbol{\omega}_{cs}\times$(this equation) and $\boldsymbol{\omega}_{cs}\cdot$(this equation), we can solve for the fluid velocity of species s as a linear function of the electric field $\mathbf{E}$:

properties of wave modes in a cold, magnetized plasma:

$$\boxed{\frac{n_s q_s}{\epsilon_0}\mathbf{u}_s = -i\left(\frac{\omega\omega_{ps}^2}{\omega_{cs}^2 - \omega^2}\right)\mathbf{E} - \frac{\omega_{ps}^2}{(\omega_{cs}^2 - \omega^2)}\boldsymbol{\omega}_{cs}\times\mathbf{E} + \boldsymbol{\omega}_{cs}\frac{i\omega_{ps}^2}{(\omega_{cs}^2 - \omega^2)\omega}\boldsymbol{\omega}_{cs}\cdot\mathbf{E}.}$$

fluid velocities

(21.43)

(This relation is useful for deducing the physical properties of wave modes.) From this fluid velocity we can read off the current $\mathbf{j} = \sum_s n_s q_s \mathbf{u}_s$ as a linear function of $\mathbf{E}$; by comparing with Ohm's law, $\mathbf{j} = \boldsymbol{\kappa}_e \cdot \mathbf{E}$, we then obtain the tensorial conductivity $\boldsymbol{\kappa}_e$, which we insert into Eq. (21.20) to get the following expression for the dielectric tensor (in which $\mathbf{B}_0$ and hence $\boldsymbol{\omega}_{cs}$ are taken to be along the z-axis):

$$\boldsymbol{\epsilon} = \begin{pmatrix} \epsilon_1 & -i\epsilon_2 & 0 \\ i\epsilon_2 & \epsilon_1 & 0 \\ 0 & 0 & \epsilon_3 \end{pmatrix},$$ (21.44)

dielectric tensor

where

$$\boxed{\epsilon_1 = 1 - \sum_s \frac{\omega_{ps}^2}{\omega^2 - \omega_{cs}^2}, \qquad \epsilon_2 = \sum_s \frac{\omega_{ps}^2 \omega_{cs}}{\omega(\omega^2 - \omega_{cs}^2)}, \qquad \epsilon_3 = 1 - \sum_s \frac{\omega_{ps}^2}{\omega^2}.}$$

(21.45)

Let the wave propagate in the x-z plane at an angle θ to the z-axis (i.e., to the magnetic field). Then the algebratized wave operator (21.9) takes the form

$$||L_{ij}|| = \frac{\omega^2}{c^2} \begin{pmatrix} \epsilon_1 - \mathfrak{n}^2 \cos^2 \theta & -i\epsilon_2 & \mathfrak{n}^2 \sin \theta \cos \theta \\ i\epsilon_2 & \epsilon_1 - \mathfrak{n}^2 & 0 \\ \mathfrak{n}^2 \sin \theta \cos \theta & 0 & \epsilon_3 - \mathfrak{n}^2 \sin^2 \theta \end{pmatrix}, \qquad (21.46)$$

where

index of refraction

$$\boxed{\mathfrak{n} = \frac{ck}{\omega}} \qquad (21.47)$$

is the wave's index of refraction (i.e., the wave's phase velocity is $V_{\text{ph}} = \omega/k = c/\mathfrak{n}$). (Note: $\mathfrak{n}$ must not be confused with the number density of particles n.) The algebratized wave operator (21.46) will be needed when we explore the physical nature of modes, in particular, the directions of their electric fields, which satisfy $L_{ij} E_j = 0$.

From the wave operator (21.46), we deduce the waves' dispersion relation $\det||L_{ij}|| = 0$. Some algebra brings this into the form

dispersion relation

$$\boxed{\tan^2 \theta = \frac{-\epsilon_3 (\mathfrak{n}^2 - \epsilon_R)(\mathfrak{n}^2 - \epsilon_L)}{\epsilon_1 (\mathfrak{n}^2 - \epsilon_3) \left(\mathfrak{n}^2 - \epsilon_R \epsilon_L / \epsilon_1 \right)},} \qquad (21.48)$$

where

$$\boxed{\epsilon_L = \epsilon_1 - \epsilon_2 = 1 - \sum_s \frac{\omega_{ps}^2}{\omega(\omega - \omega_{cs})}, \qquad \epsilon_R = \epsilon_1 + \epsilon_2 = 1 - \sum_s \frac{\omega_{ps}^2}{\omega(\omega + \omega_{cs})}.}$$

$$(21.49)$$

21.5.2

21.5.2 Parallel Propagation

waves propagating parallel to the magnetic field

As a first step in making sense out of the general dispersion relation (21.48) for waves in a cold, magnetized plasma, let us consider wave propagation along the magnetic field, so $\theta = 0$. The dispersion relation (21.48) then factorizes to give three solutions:

dispersion relations for three modes: left, right, and plasma oscillations; Fig. 21.3

$$\mathfrak{n}^2 \equiv \frac{c^2 k^2}{\omega^2} = \epsilon_L, \quad \mathfrak{n}^2 \equiv \frac{c^2 k^2}{\omega^2} = \epsilon_R, \quad \epsilon_3 = 0. \qquad (21.50)$$

LEFT AND RIGHT MODES; PLASMA OSCILLATIONS

Consider the first solution in Eq. (21.50): $\mathfrak{n}^2 = \epsilon_L$. The algebratized wave equation $L_{ij} E_j = 0$ [with L_{ij} given by Eq. (21.46)] in this case with the wave propagating along the $\mathbf{B}(\mathbf{e}_z)$ direction, requires that the electric field direction be $\mathbf{E} \propto (\mathbf{e}_x - i\mathbf{e}_y)e^{-i\omega t}$, which we define to be a left-hand circular polarized mode. The second solution in (21.50), $\mathfrak{n}^2 = \epsilon_R$, is the corresponding right-hand circular polarized mode. From Eqs. (21.49) we see that these two modes propagate with different phase velocities (but only slightly different, if ω is far from the electron cyclotron frequency and far

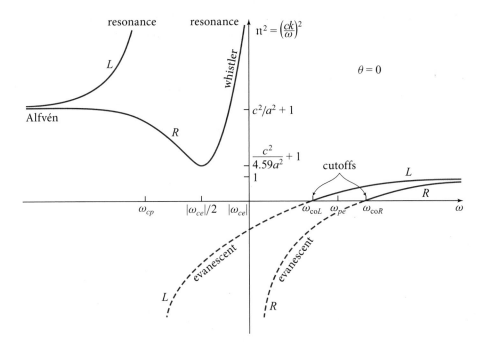

FIGURE 21.3 Square of wave refractive index for circularly polarized waves propagating along the static magnetic field in a proton-electron plasma with $\omega_{pe} > \omega_{ce}$. The horizontal, angular-frequency scale is logarithmic. L means left-hand circularly polarized ("left mode"), and R, right-hand circularly polarized ("right mode").

from the proton cyclotron frequency). The third solution in Eq. (21.50), $\epsilon_3 = 0$, is just the electrostatic plasma oscillation in which the electrons and protons oscillate parallel to and are unaffected by the static magnetic field.

As an aid to exploring the frequency dependence of the left and right modes, Fig. 21.3 shows the refractive index $\mathfrak{n} = ck/\omega$ squared as a function of ω.

HIGH-FREQUENCIES: FARADAY ROTATION

In the high-frequency limit, the refractive index for both modes is slightly less than unity and approaches the index for an unmagnetized plasma, $\mathfrak{n} = ck/\omega \simeq 1 - \frac{1}{2}\omega_p^2/\omega^2$ [cf. Eq. (21.27a)], but with a small difference between the modes given, to leading order, by

$$\mathfrak{n}_L - \mathfrak{n}_R \simeq -\frac{\omega_{pe}^2 \omega_{ce}}{\omega^3}. \tag{21.51}$$

This difference is responsible for an important effect known as *Faraday rotation*.

Suppose that a linearly polarized wave is incident on a magnetized plasma and propagates parallel to the magnetic field. We can deduce the behavior of the polarization by expanding the mode as a linear superposition of the two circularly polarized eigenmodes, left and right. These two modes propagate with slightly different phase

Faraday rotation at high frequency: difference between left and right modes velocities, so after propagating some distance through the plasma, they acquire a relative phase shift $\Delta\phi$. When one then reconstitutes the linear polarized mode from the circular eigenmodes, this phase shift is manifest in a rotation of the plane of polarization through an angle $\Delta\phi/2$ (for small $\Delta\phi$). This—together with the difference in refractive indices [Eq. (21.51)], which determines $\Delta\phi$—implies a Faraday rotation rate for the plane of polarization given by (Ex. 21.7)

rate of rotation

$$\frac{d\chi}{dz} = -\frac{\omega_{pe}^2 \omega_{ce}}{2\omega^2 c}. \tag{21.52}$$

INTERMEDIATE FREQUENCIES: CUTOFFS

As the wave frequency is reduced, the refractive index decreases to zero, first for the right circular wave, then for the left one (Fig. 21.3). Vanishing of n at a finite frequency corresponds to vanishing of k and infinite wavelength, that is, it signals a cutoff (Fig. 21.2 and associated discussion). When the frequency is lowered further, the squared refractive index becomes negative, and the wave mode becomes evanescent. Correspondingly, when a circularly polarized electromagnetic wave with constant real frequency propagates into an inhomogeneous plasma parallel to its density gradient and parallel to a magnetic field, then beyond the spatial location of the wave's cutoff, its wave number k becomes purely imaginary, and the wave dies out with distance (gradually at first, then more rapidly).

The cutoff frequencies are different for the two modes and are given by

cutoff frequencies

$$\omega_{\text{co}R,L} = \frac{1}{2}\left[\left\{(\omega_{ce} + \omega_{cp})^2 + 4(\omega_{pe}^2 + \omega_{pp}^2)\right\}^{1/2} \pm (|\omega_{ce}| - \omega_{cp})\right]$$

$$\simeq \omega_{pe} \pm \frac{1}{2}|\omega_{ce}| \quad \text{if } \omega_{pe} \gg |\omega_{ce}|, \quad \text{as is often the case.} \tag{21.53}$$

LOW FREQUENCIES: RESONANCES; WHISTLER MODES

As we lower the frequency further (Fig. 21.3), first the right mode and then the left regain the ability to propagate. When the wave frequency lies between the proton and electron gyro frequencies, $\omega_{cp} < \omega < |\omega_{ce}|$, only the right mode propagates.

whistler mode at intermediate frequencies

This mode is sometimes called a *whistler*. As the mode's frequency increases toward the electron gyro frequency $|\omega_{ce}|$ (where it first recovered the ability to propagate), its refractive index and wave vector become infinite—a signal that $\omega = |\omega_{ce}|$ is a resonance for the whistler (Fig. 21.2 and associated discussion). The physical origin

resonance with electron gyrations

of this resonance is that the wave frequency becomes resonant with the gyrational frequency of the electrons that are orbiting the magnetic field in the same sense as the wave's electric vector rotates. To quantify the strong wave absorption that occurs at this resonance, one must carry out a kinetic-theory analysis that takes account of the electrons' thermal motions (Chap. 22).

large dispersion near resonance

Another feature of the whistler is that it is highly dispersive close to resonance; its dispersion relation there is given approximately by

$$\omega \simeq \frac{|\omega_{ce}|}{1 + \omega_{pe}^2/(c^2 k^2)}. \tag{21.54}$$

Chapter 21. Waves in Cold Plasmas: Two-Fluid Formalism

The group velocity, obtained by differentiating Eq. (21.54), is given approximately by

$$\mathbf{V}_g = \nabla_\mathbf{k}\, \omega \simeq \frac{2\omega_{ce}c}{\omega_{pe}} \left(1 - \frac{\omega}{|\omega_{ce}|}\right)^{3/2} \hat{\mathbf{B}}_0. \qquad (21.55)$$

This velocity varies extremely rapidly close to resonance, so waves of slightly different frequency propagate at very different speeds.

This is the physical origin of the phenomenon by which whistlers were discovered. They were encountered in World War One by radio operators who heard, in their earphones, strange tones with rapidly changing pitch. These turned out to be whistler modes excited by lightning in the southern hemisphere; they propagated along Earth's magnetic field through the magnetosphere to the northern hemisphere. Only modes below the lowest electron gyro frequency on the waves' path (their geometric-optics ray) could propagate, and these were highly dispersed, with the lower frequencies arriving first.

There is also a resonance associated with the left circularly polarized wave, which propagates below the proton cyclotron frequency; see Fig. 21.3.

VERY LOW FREQUENCIES: ALFVÉN MODES

Finally, let us examine the very low-frequency limit of these waves (Fig. 21.3). We find that both dispersion relations $\mathrm{n}^2 = \epsilon_L$ and $\mathrm{n}^2 = \epsilon_R$ asymptote, at arbitrarily low frequencies, to

$$\omega = ak \left(1 + \frac{a^2}{c^2}\right)^{-1/2}. \qquad (21.56)$$

Here $a = B_0[\mu_0 n_e(m_p + m_e)]^{-1/2}$ is the Alfvén speed that arose in our discussion of magnetohydrodynamics [Eq. (19.75)]. In fact at very low frequencies, both modes, left and right, have become the Alfvén waves that we studied using MHD in Sec. 19.7.2. However, our two-fluid formalism reports a phase speed $\omega/k = a/\sqrt{1 + a^2/c^2}$ for these Alfvén waves that is slightly lower than the speed $\omega/k = a$ predicted by our MHD formalism. The $1/\sqrt{1 + a^2/c^2}$ correction could not be deduced using non-relativistic MHD, because that formalism neglects the displacement current. (Relativistic MHD includes the displacement current and predicts precisely this correction factor.)

Alfvén modes at very low frequencies

relativistic correction to phase speed

We can understand the physical origin of this correction by examining the particles' motions in a very-low-frequency Alfvén wave; see Fig. 21.4. Because the wave frequency is far below both the electron and the proton cyclotron frequencies, both types of particle orbit the field $\mathbf{B}_0$ many times in a wave period. When the wave's slowly changing electric field is applied, the guiding centers of both types of orbits acquire the same slowly changing drift velocity $\mathbf{v} = \mathbf{E} \times \mathbf{B}_0/B_0^2$, so the two fluid velocities also drift at this rate, and the currents associated with the proton and electron drifts cancel. However, when we consider corrections to the guiding-center response that are of higher order in ω/ω_{cp} and ω/ω_{ce}, we find that the ions drift slightly faster than the electrons, which produces a net current that modifies the magnetic field and gives rise to the $1/\sqrt{1 + a^2/c^2}$ correction to the Alfvén wave's phase speed.

interpretation as drift motion

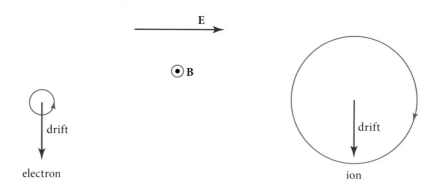

FIGURE 21.4 Gyration of electrons and ions in a very-low-frequency Alfvén wave. Although the electrons and ions gyrate with opposite senses about the magnetic field, their $\mathbf{E} \times \mathbf{B}$ drifts are similar. It is only in the next highest order of approximation that a net ion current is produced parallel to the applied electric field. Note that, if the electrons and ions have the same temperature, then the ratio of radii of the orbits is $\sim (m_e/m_i)^{1/2}$, or ~ 0.02 for protons.

interpretation as magnetic
contribution to inertia

A second way to understand this correction is by the contribution of the magnetic field to the plasma's inertial mass per unit volume (Ex. 21.8).

EXERCISES

Exercise 21.7 *Derivation: Faraday Rotation*
Derive Eq. (21.52) for Faraday rotation.

Exercise 21.8 *Example: Alfvén Waves as Plasma-Laden,*
Plucked Magnetic Field Lines
A narrow bundle of magnetic field lines with cross sectional area A, together with the plasma attached to them, can be thought of as like a stretched string. When such a string is plucked, waves travel down it with phase speed $\sqrt{T/\Lambda}$, where T is the string's tension, and Λ is its mass per unit length (Sec. 12.3.3). The plasma analog is Alfvén waves propagating parallel to the plasma-laden magnetic field.

(a) Analyzed nonrelativistically, the tension for our bundle of field lines is $T = [B^2/(2\mu_0)]A$ and the mass per unit length is $\Lambda = \rho A$, so we expect a phase velocity $\sqrt{T/\Lambda} = \sqrt{B^2/(2\mu_0\rho)}$, which is $1/\sqrt{2}$ of the correct result. Where is the error? [Hint: In addition to the restoring force on bent field lines, due to tension along the field, there is also the curvature force $(\mathbf{B} \cdot \nabla)\mathbf{B}/\mu_0$; Eq. (19.15).]

(b) In special relativity, the plasma-laden magnetic field has a tensorial inertial mass per unit volume that is discussed in Ex. 2.27. Explain why, when the field lines (which point in the z direction) are plucked so they vibrate in the x direction, the inertial mass per unit length that resists this motion is $\Lambda = (T^{00} + T^{xx})A = [\rho + B^2/(\mu_0 c^2)]A$. (In the first expression for Λ, T^{00} is the mass-energy density of plasma and magnetic field, T^{xx} is the magnetic pressure along the x direction, and the speed of light is set to unity as in Chap. 2; in the second expression, the speed of light has been restored to the equation using dimensional arguments.) Show that the magnetic contribution to this inertial mass gives the relativistic correction $1/\sqrt{1 + a^2/c^2}$ to the Alfvén waves' phase speed, Eq. (21.56).

21.5.3 Perpendicular Propagation

Turn next to waves that propagate perpendicular to the static magnetic field: $\mathbf{k} = k\mathbf{e}_x$; $\mathbf{B}_0 = B_0\mathbf{e}_z$; $\theta = \pi/2$. In this case our general dispersion relation (21.48) again has three solutions corresponding to three modes:

$$\mathfrak{n}^2 \equiv \frac{c^2k^2}{\omega^2} = \epsilon_3, \qquad \mathfrak{n}^2 \equiv \frac{c^2k^2}{\omega^2} = \frac{\epsilon_R\epsilon_L}{\epsilon_1}, \qquad \epsilon_1 = 0. \tag{21.57}$$

The first solution

$$\mathfrak{n}^2 = \epsilon_3 = 1 - \frac{\omega_p^2}{\omega^2} \tag{21.58}$$

has the same index of refraction as for an electromagnetic wave in an unmagnetized plasma [cf. Eq. (21.24)], so this is called the *ordinary mode*. In this mode the electric vector and velocity perturbation are parallel to the static magnetic field, so the field has no influence on the wave. The wave is identical to an electromagnetic wave in an unmagnetized plasma.

The second solution in Eq. (21.57),

$$\mathfrak{n}^2 = \epsilon_R\epsilon_L/\epsilon_1 = \frac{\epsilon_1^2 - \epsilon_2^2}{\epsilon_1}, \tag{21.59}$$

is known as the *extraordinary mode* and has an electric field that is orthogonal to $\mathbf{B}_0$ but not to $\mathbf{k}$. (Note that the names "ordinary" and "extraordinary" are used differently here than for waves in a nonlinear crystal in Sec. 10.6.)

The refractive indices for the ordinary and extraordinary modes are plotted as functions of frequency in Fig. 21.5. The ordinary-mode curve is dull; it is just like that in an unmagnetized plasma. The extraordinary-mode curve is more interesting. It has two cutoffs, with frequencies (ignoring the ion motions)

$$\omega_{co1,2} \simeq \left(\omega_{pe}^2 + \frac{1}{4}\omega_{ce}^2\right)^{1/2} \pm \frac{1}{2}\omega_{ce}, \tag{21.60}$$

and two resonances with strong absorption, at frequencies known as the *upper hybrid* (UH) and *lower hybrid* (LH) frequencies. These frequencies are given approximately by

$$\omega_{\mathrm{UH}} \simeq (\omega_{pe}^2 + \omega_{ce}^2)^{1/2},$$

$$\omega_{\mathrm{LH}} \simeq \left[\frac{(\omega_{pe}^2 + |\omega_{ce}|\omega_{cp})|\omega_{ce}|\omega_{cp}}{\omega_{pe}^2 + \omega_{ce}^2}\right]^{1/2}. \tag{21.61}$$

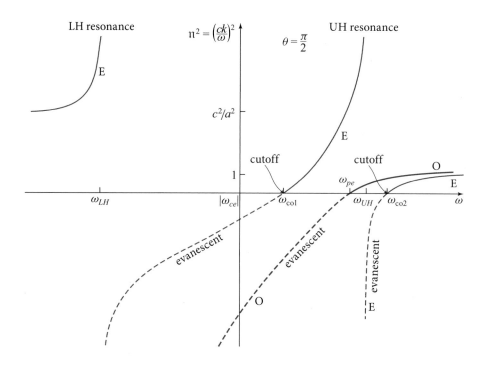

FIGURE 21.5 Square of wave refractive index $\mathfrak{n}$ as a function of frequency ω, for wave propagation perpendicular to the magnetic field in an electron ion plasma with $\omega_{pe} > \omega_{ce}$. The ordinary mode is designated by O, the extraordinary mode by E.

very low frequencies: fast magnetosonic mode and Alfvén mode

In the limit of very low frequency, the extraordinary, perpendicularly propagating mode has the same dispersion relation $\omega = ak/\sqrt{1 + a^2/c^2}$ as the parallel propagating modes [Eq. (21.56)]. In this regime the mode has become the fast magnetosonic wave, propagating perpendicularly to the static magnetic field (Sec. 19.7.2), while the parallel waves have become the Alfvén modes.

The third solution in Eq. (21.57), $\epsilon_1 = 0$, has vanishing eigenvector $\mathbf{E}$ (i.e., $L_{ij}E_j = 0$ for $\epsilon_1 = 0$ implies $E_j = 0$), so it does not represent a physical oscillation of any sort.

21.5.4 Propagation of Radio Waves in the Ionosphere; Magnetoionic Theory

The discovery that radio waves could be reflected off the ionosphere, and thereby could be transmitted over long distances,[6] revolutionized communications and stimulated intensive research on radio-wave propagation in a magnetized plasma. In this

magnetoionic theory for radio waves in ionosphere

section, we discuss radio-wave propagation in the ionosphere for waves whose prop-

---

6. This effect was predicted independently by Heaviside and Kennelly in 1902. Appleton demonstrated its existence in 1924, for which he received the Nobel prize. It was claimed by some physicists that the presence of a wave speed $> c$ violated special relativity until it was appreciated that it is the group velocity that is relevant.

agation vectors make arbitrary angles θ to the magnetic field. The approximate formalism we develop is sometimes called *magnetoionic theory*.

properties of ionosphere

The ionosphere is a dense layer of partially ionized gas between 50 and 300 km above the surface of Earth. The ionization is due to incident solar ultraviolet radiation. Although the ionization fraction increases with height, the actual density of free electrons passes through a maximum whose height rises and falls with the position of the Sun.

The electron gyro frequency varies from $\sim$0.5 to $\sim$1 MHz in the ionosphere, and the plasma frequency increases from effectively zero to a maximum that can be as high as 100 MHz; so typically, but not everywhere, $\omega_{pe} \gg |\omega_{ce}|$. We are interested in wave propagation at frequencies above the electron plasma frequency, which in turn is well in excess of the ion plasma frequency and the ion gyro frequency. It is therefore a good approximation to ignore ion motions altogether. In addition, at the altitudes of greatest interest for radio-wave propagation, the temperature is low, $T_e \sim 200$–600 K, so the cold plasma approximation is well justified. A complication that one must sometimes face in the ionosphere is the influence of collisions (see Ex. 21.4), but in this section we ignore it.

It is conventional in magnetoionic theory to introduce two dimensionless parameters:

$$X = \frac{\omega_{pe}^2}{\omega^2}, \quad Y = \frac{|\omega_{ce}|}{\omega}, \tag{21.62}$$

in terms of which (ignoring ion motions) the components (21.45) of the dielectric tensor (21.44) are

$$\epsilon_1 = 1 + \frac{X}{Y^2 - 1}, \quad \epsilon_2 = \frac{XY}{Y^2 - 1}, \quad \epsilon_3 = 1 - X. \tag{21.63}$$

In this case it is convenient to rewrite the dispersion relation $\det||L_{ij}|| = 0$ in a form different from Eq. (21.48)—a form derivable, for example, by computing explicitly the determinant of the matrix (21.46), setting

$$x = \frac{X - 1 + \mathfrak{n}^2}{1 - \mathfrak{n}^2}, \tag{21.64}$$

solving the resulting quadratic in x, and then solving for $\mathfrak{n}^2$. The result is the *Appleton-Hartree* dispersion relation:

$$\boxed{\mathfrak{n}^2 = 1 - \frac{X}{1 - \frac{Y^2 \sin^2\theta}{2(1-X)} \pm \left[\frac{Y^4 \sin^4\theta}{4(1-X)^2} + Y^2 \cos^2\theta \right]^{1/2}}.} \tag{21.65}$$

Appleton-Hartree dispersion relation

There are two commonly used approximations to this dispersion relation. The first is the *quasi-longitudinal* approximation, which is used when **k** is approximately parallel to the static magnetic field (i.e., when θ is small). In this case, just retaining

quasi-longitudinal approximation

the dominant terms in the dispersion relation, we obtain

$$\mathfrak{n}^2 \simeq 1 - \frac{X}{1 \pm Y \cos \theta}. \tag{21.66}$$

This is just the dispersion relation (21.50) for the left and right modes in strictly parallel propagation, with the substitution $B_0 \to B_0 \cos \theta$. By comparing the magnitude of the terms that we dropped from the full dispersion relation in deriving Eq. (21.66) with those that we retained, one can show that the quasi-longitudinal approximation is valid when

$$Y \sin^2 \theta \ll 2(1 - X) \cos \theta. \tag{21.67}$$

quasi-transverse approximation

The second approximation is the *quasi-transverse* approximation; it is appropriate when inequality (21.67) is reversed. In this case the two modes are generalizations of the precisely perpendicular ordinary and extraordinary modes, and their approximate dispersion relations are

$$\mathfrak{n}_O^2 \simeq 1 - X,$$

$$\mathfrak{n}_X^2 \simeq 1 - \frac{X(1 - X)}{1 - X - Y^2 \sin^2 \theta}. \tag{21.68}$$

The ordinary-mode dispersion relation (subscript O) is unchanged from the strictly perpendicular one, Eq. (21.58); the extraordinary dispersion relation (subscript X) is obtained from the strictly perpendicular one [Eq. (21.59)] by the substitution $B_0 \to B_0 \sin \theta$.

The quasi-longitudinal and quasi-transverse approximations simplify the problem of tracing rays through the ionosphere.

ionospheric reflection of AM and SW radio waves

Commercial radio stations operate in the AM (amplitude-modulated) band (0.5–1.6 MHz), the SW (short-wave) band (2.3–18 MHz), and the FM (frequency-modulated) band (88–108 MHz). Waves in the first two bands are reflected by the ionosphere and can therefore be transmitted over large surface areas (Ex. 21.12). FM waves, with their higher frequencies, are not reflected and must therefore be received as "ground waves" (waves that propagate directly, near the ground). However, they have the advantage of a larger bandwidth and consequently a higher fidelity audio output. As the altitude of the reflecting layer rises at night, short-wave communication over long distances becomes easier.

EXERCISES

Exercise 21.9 *Derivation: Appleton-Hartree Dispersion Relation*
Derive Eq. (21.65).

Exercise 21.10 *Example: Dispersion and Faraday Rotation of Pulsar Pulses*
A radio pulsar emits regular pulses at 1-s intervals, which propagate to Earth through the ionized interstellar plasma with electron density $n_e \simeq 3 \times 10^4$ m^{-3}. The pulses ob-

served at $f = 100$ MHz are believed to be emitted at the same time as those observed at much higher frequency, but they arrive with a delay of 100 ms.

(a) Explain briefly why pulses travel at the group velocity instead of the phase velocity, and show that the expected time delay of the $f = 100$-MHz pulses relative to the high-frequency pulses is given by

$$\Delta t = \frac{e^2}{8\pi^2 m_e \epsilon_0 f^2 c} \int n_e dx, \tag{21.69}$$

where the integral is along the waves' propagation path. Hence compute the distance to the pulsar.

(b) Now suppose that the pulses are linearly polarized and that their propagation is accurately described by the quasi-longitudinal approximation. Show that the plane of polarization will be Faraday rotated through an angle

$$\Delta\chi = \frac{e\Delta t}{m_e}\langle B_{\parallel}\rangle, \tag{21.70}$$

where $\langle B_{\parallel}\rangle = \int n_e \mathbf{B}\cdot d\mathbf{x}/\int n_e dx$. The plane of polarization of the pulses emitted at 100 MHz is believed to be the same as the emission plane for higher frequencies, but when the pulses arrive at Earth, the 100-MHz polarization plane is observed to be rotated through 3 radians relative to that at high frequencies. Calculate the mean parallel component of the interstellar magnetic field.

Exercise 21.11 *Challenge: Faraday Rotation Paradox*

Consider a wave mode propagating through a plasma—for example, the ionosphere—in which the direction of the background magnetic field is slowly changing. We have just demonstrated that so long as $\mathbf{B}$ is not almost perpendicular to $\mathbf{k}$, we can use the quasi-longitudinal approximation, the difference in phase velocity between the two eigenmodes is $\propto \mathbf{B}\cdot\mathbf{k}$, and the integral for the magnitude of the rotation of the plane of polarization is $\propto \int n_e \mathbf{B}\cdot d\mathbf{x}$.

Now, suppose that the parallel component of the magnetic field changes sign. It has been implicitly assumed that the faster eigenmode, which is circularly polarized in the quasi-longitudinal approximation, becomes the slower eigenmode (and vice versa) when the field is reversed, and the Faraday rotation is undone. However, if we track the modes using the full dispersion relation, we find that the faster quasi-longitudinal eigenmode remains the faster eigenmode in the quasi-perpendicular regime, and it becomes the faster eigenmode with opposite sense of circular polarization in the field-reversed quasi-longitudinal regime. Now, let there be a second field reversal where an analogous transition occurs. Following this logic, the net rotation should be $\propto \int n_e |\mathbf{B}\cdot d\mathbf{x}|$. What is going on?

Exercise 21.12 *Example: Reflection of Short Waves by the Ionosphere*

The free electron density in the night-time ionosphere increases exponentially from 10^9 m^{-3} to 10^{11} m^{-3} as the altitude increases from 100 to 200 km, and the density diminishes above this height. Use Snell's law [Eq. (7.49)] to calculate the maximum range of 10-MHz waves transmitted from Earth's surface, assuming a single ionospheric reflection.

21.5.5 CMA Diagram for Wave Modes in a Cold, Magnetized Plasma

Magnetized plasmas are anisotropic, just like many nonlinear crystals (Chap. 10). This implies that the phase speed of a propagating wave mode depends on the angle between the direction of propagation and the magnetic field. Two convenient ways are used to exhibit this anisotropy diagrammatically. The first method, due originally to Fresnel, is to construct *phase-velocity surfaces* (also called *wave-normal surfaces*), which are polar plots of the wave phase velocity $V_{ph} = \omega/k$, at fixed frequency ω, as a function of the angle θ that the wave vector $\mathbf{k}$ makes with the magnetic field; see Fig. 21.6a.

The second type of surface, used originally by Hamilton, is the *refractive-index surface*. This is a polar plot of the refractive index $\mathrm{n} = ck/\omega$ for a given frequency, again as a function of the wave vector's angle θ to $\mathbf{B}$; see Fig. 21.6b. This plot has the important property that the group velocity is perpendicular to the surface

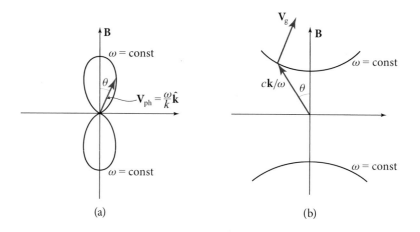

(a) (b)

FIGURE 21.6 (a) *Wave-normal surface* (i.e., phase-velocity surface) for a whistler mode propagating at an angle θ with respect to the magnetic field direction. This diagram plots the phase velocity $\mathbf{V}_{ph} = (\omega/k)\hat{\mathbf{k}}$ as a vector from the origin, with the direction of the magnetic field chosen as upward. When we fix the frequency ω of the wave, the tip of the phase velocity vector sweeps out the figure-8 curve as its angle θ to the magnetic field changes. This curve should be thought of as rotated around the vertical (magnetic-field) direction to form the figure-8 wave-normal surface. Note that there are some directions where no mode can propagate. (b) *Refractive-index surface* for the same whistler mode. Here we plot $c\mathbf{k}/\omega$ as a vector from the origin, and as its direction changes with fixed ω, this vector sweeps out the two hyperboloid-like surfaces. Since the length of the vector is $ck/\omega = \mathrm{n}$, this figure can be thought of as a polar plot of the refractive index n as a function of wave-propagation direction θ for fixed ω; hence the name "refractive-index surface." The group velocity $\mathbf{V}_g$ is orthogonal to the refractive-index surface (Ex. 21.13). Note that for this whistler mode, the energy flow (along $\mathbf{V}_g$) is focused toward the direction of the magnetic field.

(Ex. 21.13). As discussed above, the energy flow is along the direction of the group velocity and, in a magnetized plasma, this velocity can make a large angle with the wave vector.

A particularly useful application of these ideas is to a graphical representation of the various types of wave modes that can propagate in a cold, magnetized plasma (Fig. 21.7). This representation is known as the *Clemmow-Mullaly-Allis* or *CMA* diagram. The character of waves of a given frequency ω depends on the ratio of this frequency to the plasma frequency and the cyclotron frequency. This suggests defining two dimensionless numbers, $\omega_p^2/\omega^2 \equiv (\omega_{pe}^2 + \omega_{pp}^2)/\omega^2$ and $|\omega_{ce}|\omega_{cp}/\omega^2$, which are plotted on the horizontal and vertical axes of the CMA diagram. [Recall that $\omega_{pp} = \omega_{pe}\sqrt{m_e/m_p}$ and $\omega_{cp} = \omega_{ce}(m_e/m_p)$.] The CMA space defined by these two dimensionless parameters can be subdivided into sixteen regions, in each of which the propagating modes have a distinctive character. The mode properties are indicated by sketching the topological form of the *wave-normal surfaces* associated with each region.

CMA diagram, Fig. 21.7

horizontal and vertical axes

sixteen regions; topological form of wave-normal surfaces

The form of the wave-normal surface in each region can be deduced from the general dispersion relation (21.48). To deduce it, one must solve the dispersion relation for $1/n = \omega/kc = V_{ph}/c$ as a function of θ and ω, and then generate the polar plot of $V_{ph}(\theta)$.

On the CMA diagram's wave-normal curves, the characters of the parallel and perpendicular modes are indicated by labels: R and L for right and left parallel modes ($\theta = 0$), and O and X for ordinary and extraordinary perpendicular modes ($\theta = \pi/2$). As one moves across a boundary from one region to another, there is often a change of which parallel mode gets deformed continuously, with increasing θ, into which perpendicular mode. In some regions a wave-normal surface has a figure-8 shape, indicating that the waves can propagate only over a limited range of angles, $\theta < \theta_{max}$. In some regions there are two wave-normal surfaces, indicating that—at least in some directions θ—two modes can propagate; in other regions there is just one wave-normal surface, so only one mode can propagate; and in the bottom-right two regions there are no wave-normal surfaces, since no waves can propagate at these high densities and low magnetic-field strengths.

EXERCISES

Exercise 21.13 *Derivation: Refractive-Index Surface*
Verify that the group velocity of a wave mode is perpendicular to the refractive-index surface (Fig. 21.6b).

Exercise 21.14 *Problem: Exploration of Modes in the CMA Diagram*
For each of the following modes studied earlier in this chapter, identify in the CMA diagram the phase speed, as a function of frequency ω, and verify that the turning on and cutting off of the modes, and the relative speeds of the modes, are in accord with the CMA diagram's wave-normal curves.

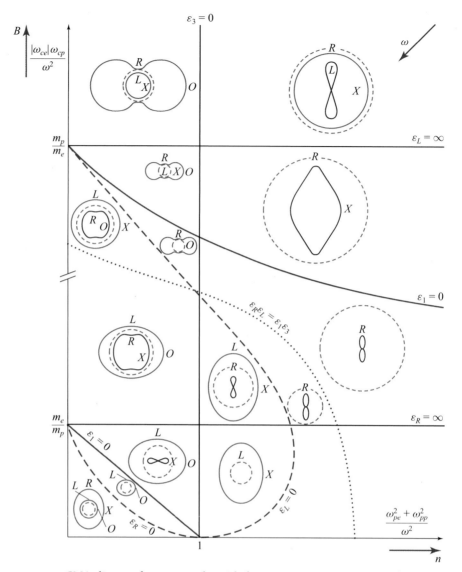

FIGURE 21.7 CMA diagram for wave modes with frequency ω propagating in a plasma with plasma frequencies ω_{pe}, ω_{pp} and gyro frequencies ω_{ce}, ω_{cp}. Plotted upward is the dimensionless quantity $|\omega_{ce}|\omega_{cp}/\omega^2$, which is proportional to B^2, so magnetic field strength also increases upward. Plotted rightward is the dimensionless quantity $(\omega_{pe}^2 + \omega_{pp}^2)/\omega^2$, which is proportional to n, so the plasma number density also increases rightward. Since both the ordinate and the abscissa scale as $1/\omega^2$, ω increases in the left-down direction. This plane is split into sixteen regions by a set of curves on which various dielectric components have special values. In each of the sixteen regions are shown two, one, or no wave-normal surfaces (phase-velocity surfaces) at fixed ω (cf. Fig. 21.6a). These surfaces depict the types of wave modes that can propagate for that region's values of frequency ω, magnetic field strength B, and electron number density n. In each wave-normal diagram the dashed circle indicates the speed of light; a point outside that circle has phase velocity $V_{\rm ph}$ greater than c; inside the circle, $V_{\rm ph} < c$. The topologies of the wave-normal surfaces and speeds relative to c are constant throughout each of the sixteen regions; they change as one moves between regions. Adapted from Boyd and Sanderson (2003, Fig. 6.12), which in turn is adapted from Allis, Buchsbaum, and Bers (1963).

(a) Electromagnetic modes in an unmagnetized plasma.

(b) Left and right modes for parallel propagation in a magnetized plasma.

(c) Ordinary and extraordinary modes for perpendicular propagation in a magnetized plasma.

21.6 Two-Stream Instability

When considered on large enough scales, plasmas behave like fluids and are subject to a wide variety of fluid dynamical instabilities. However, as we are discovering, plasmas have internal degrees of freedom associated with their velocity distributions, which offers additional opportunities for unstable wave modes to grow and for free energy to be released. A full description of velocity-space instabilities is necessarily kinetic and must await the following chapter. However, it is instructive to consider a particularly simple example, the *two-stream instability*, using cold-plasma theory, as this brings out several features of the more general theory in a particularly simple manner.

We will apply our results in a slightly unusual way, to the propagation of fast electron beams through the slowly outflowing solar wind. These electron beams are created by coronal disturbances generated on the surface of the Sun (specifically, those associated with "Type III" radio bursts). The observation of these fast electron beams was initially a puzzle, because plasma physicists knew that they should be unstable to the exponential growth of electrostatic waves. What we do in this section is demonstrate the problem. What we will not do is explain what is thought to be its resolution, since that involves nonlinear plasma-physics considerations beyond the scope of this book (see, e.g., Melrose, 1980, or Sturrock, 1994).

Consider a simple, cold (i.e., with negligible thermal motions) electron-proton plasma at rest. Ignore the protons for the moment. We can write the dispersion relation for electron plasma oscillations in the form

$$\frac{\omega_{pe}^2}{\omega^2} = 1. \tag{21.71}$$

Now allow the ions also to oscillate about their mean positions. The dispersion relation is slightly modified to

$$\frac{\omega_p^2}{\omega^2} = \frac{\omega_{pe}^2}{\omega^2} + \frac{\omega_{pp}^2}{\omega^2} = 1 \tag{21.72}$$

[cf. Eq. (21.23)]. If we were to add other components (e.g., helium ions), that would simply add extra terms to Eq. (21.72).

Next, return to Eq. (21.71) and look at it in a reference frame in which the electrons are moving with speed u. The observed wave frequency is then Doppler shifted, and so the dispersion relation becomes

$$\frac{\omega_{pe}^2}{(\omega - ku)^2} = 1, \tag{21.73}$$

dispersion relation for electron plasma oscillations in a frame where electrons move with speed u

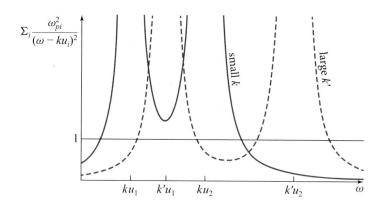

FIGURE 21.8 Left-hand side of the dispersion relation (21.74) for two cold plasma streams and two different choices of wave vector, k (small) and k' (large). For small enough k, there are only two real roots for ω.

where ω is now the angular frequency measured in this new frame. It should be evident from this how to generalize Eq. (21.72) to the case of several cold streams moving with different speeds u_i. We simply add the terms associated with each component using angular frequencies that have been appropriately Doppler shifted:

dispersion relation for electron plasma oscillations in a plasma with counterstreaming beams

$$\frac{\omega_{p1}^2}{(\omega - ku_1)^2} + \frac{\omega_{p2}^2}{(\omega - ku_2)^2} + \ldots = 1. \tag{21.74}$$

(This procedure will be justified via kinetic theory in the next chapter.)

The left-hand side of the dispersion relation (21.74) is plotted in Fig. 21.8 for the case of two cold plasma streams. Equation (21.74) in this case is a quartic in ω, and so it should have four roots. However, for small enough k only two of these roots will be real (solid curves in Fig. 21.8). The remaining two roots must be a complex-conjugate pair, and the root with the positive imaginary part corresponds to a growing mode. We have therefore shown that for small enough k the two-stream plasma will be unstable. Small electrostatic disturbances will grow exponentially to large amplitude and ultimately react back on the plasma. As we add more cold streams to the plasma, we add more modes, some of which will be unstable. This simple example demonstrates how easy it is for a plasma to tap the free energy residing in anisotropic particle-distribution functions.

two-stream instability: tapping free energy of anisotropic particle-distribution functions

Let us return to our solar-wind application and work in the rest frame of the wind ($u_1 = 0$), where the plasma frequency is $\omega_{p1} = \omega_p$. If the beam density is a small fraction α of the solar-wind density, so $\omega_{p2}^2 = \alpha \omega_p^2$, and the beam velocity (as seen in the wind's rest frame) is $u_2 = V$, then by differentiating Eq. (21.74), we find that the local minimum of the left-hand side occurs at $\omega = kV/(1 + \alpha^{1/3})$. The value of

two-stream instability for electron beams in solar wind

the left-hand side at that minimum is $\omega_p^2(1 + \alpha^{1/3})/\omega^2$. This minimum exceeds unity (thereby making two roots of the dispersion relation complex) for

$$k < \frac{\omega_p}{V}(1 + \alpha^{1/3})^{3/2}. \tag{21.75}$$

This is therefore the condition for the existence of a growing mode. The maximum value for the growth rate can be found simply by varying k. It occurs at $k = \omega_p/V$ and is

$$\omega_i = \frac{3^{1/2}\alpha^{1/3}\omega_p}{2^{4/3}}. \tag{21.76}$$

For the solar wind near Earth, we have $\omega_p \sim 10^5$ rad s^{-1}, $\alpha \sim 10^{-3}$, and $V \sim 10^4$ km s^{-1}. We therefore find that the instability should grow, thereby damping the fast electron beam, in a length of 30 km, which is much less than the distance from the Sun (1.5×10^8 km)! This describes the problem that we will not resolve here.

EXERCISES

Exercise 21.15 *Derivation: Two-Stream Instability*
Verify Eq. (21.76).

Exercise 21.16 *Problem: Relativistic Two-Stream Instability*
In a very strong magnetic field, we can consider electrons as constrained to move in 1 dimension along the direction of the magnetic field. Consider a beam of relativistic protons propagating with density n_b and speed $u_b \sim c$ through a cold electron-proton plasma along **B**. Generalize the dispersion relation (21.74) for modes with **k** ∥ **B**.

Exercise 21.17 *Example: Drift Waves*
Another type of wave mode that can be found from a fluid description of a plasma (but requires a kinetic treatment to understand completely) is a *drift wave*. Just as the two-stream instability provides a mechanism for plasmas to erase nonuniformity in velocity space, so drift waves can rapidly remove spatial irregularities.

The limiting case that we consider here is a modification of an ion-acoustic mode in a strongly magnetized plasma with a density gradient. Suppose that the magnetic field is uniform and points in the $\mathbf{e}_z$ direction. Let there be a gradient in the equilibrium density of both the electrons and the protons: $n_0 = n_0(x)$. In the spirit of our description of ion-acoustic modes in an unmagnetized, homogeneous plasma [cf. Eq. (21.33)], treat the proton fluid as cold, but allow the electrons to be warm and isothermal with temperature T_e. We seek modes of frequency ω propagating perpendicular to the density gradient [i.e., with $\mathbf{k} = (0, k_y, k_z)$].

(a) Consider the equilibrium of the warm electron fluid, and show that there must be a fluid drift velocity along the direction $\mathbf{e}_y$ of magnitude

$$V_{de} = -\frac{V_{ia}^2}{\omega_{ci}} \frac{1}{n_0} \frac{dn_0}{dx}, \tag{21.77}$$

where $V_{ia} = (k_B T_e / m_p)^{1/2}$ is the ion-acoustic speed. Explain in physical terms the origin of this drift and why we can ignore the equilibrium drift motion for the ions (protons).

(b) We limit our attention to low-frequency electrostatic modes that have phase velocities below the Alfvén speed. Under these circumstances, perturbations to the magnetic field can be ignored, and the electric field can be written as $\mathbf{E} = -\nabla \Phi$. Write down the three components of the linearized proton equation of motion in terms of the perturbation to the proton density n, the proton fluid velocity $\mathbf{u}$, and the electrostatic potential Φ.

(c) Write down the linearized equation of proton continuity, including the gradient in n_0, and combine with the equation of motion to obtain an equation for the fractional proton density perturbation at low frequencies:

$$\frac{\delta n}{n_0} = \left(\frac{(\omega_{cp}^2 k_z^2 - \omega^2 k^2) V_{ia}^2 + \omega_{cp}^2 \omega k_y V_{de}}{\omega^2 (\omega_{cp}^2 - \omega^2)} \right) \left(\frac{e\Phi}{k_B T_e} \right). \tag{21.78}$$

(d) Argue that the fractional electron-density perturbation follows a linearized Boltzmann distribution, so that

$$\frac{\delta n_e}{n_0} = \frac{e\Phi}{k_B T_e}. \tag{21.79}$$

(e) Use both the proton- and the electron-density perturbations in Poisson's equation to obtain the electrostatic drift wave dispersion relation in the low-frequency ($\omega \ll \omega_{cp}$), long-wavelength ($k\lambda_D \ll 1$) limit:

$$\omega = \frac{k_y V_{de}}{2} \pm \frac{1}{2} \left(k_y^2 V_{de}^2 + 4k_z^2 V_{ia}^2 \right)^{1/2}. \tag{21.80}$$

Describe the physical character of the mode in the additional limit $k_z \to 0$. A proper justification of this procedure requires a kinetic treatment, which also shows that, under some circumstances, drift waves can be unstable and grow exponentially.

Bibliographic Note

The definitive monograph on waves in plasmas is Stix (1992); also very good, and with a controlled-fusion orientation, is Swanson (2003).

For an elementary and lucid textbook treatment, which makes excellent contact with laboratory experiments, see Chap. 4 of Chen (1974) or Chen (2016). For more sophisticated and detailed textbook treatments, we especially like the relevant chapters of Clemmow and Dougherty (1969), Krall and Trivelpiece (1973), Lifshitz and Pitaevskii (1981), Sturrock (1994), Boyd and Sanderson (2003), Bittencourt (2004), Bellan (2006), and Piel (2017). For treatments that focus on space plasmas, see Melrose (1980), Parks (2004), and Gurnett and Bhattarcharjee (2017).

CHAPTER TWENTY-TWO

Kinetic Theory of Warm Plasmas

Any complete theory of the kinetics of a plasma
cannot ignore the plasma oscillations.
IRVING LANGMUIR (1928)

22.1 Overview

At the end of Chap. 21, we showed how to generalize cold-plasma two-fluid theory to accommodate several distinct plasma beams, and thereby we discovered the two-stream instability. If the beams are not individually monoenergetic (i.e., cold), as we assumed there, but instead have broad velocity dispersions that overlap in velocity space (i.e., if the beams are *warm*), then the two-fluid approach of Chap. 21 cannot be used, and a more powerful, kinetic-theory description of the plasma is required.

Chapter 21's approach entailed specifying the positions and velocities of specific groups of particles (the "fluids"); this is an example of a *Lagrangian* description. It turns out that the most robust and powerful method for developing the kinetic theory of warm plasmas is an *Eulerian* one in which we specify how many particles are to be found in a fixed volume of one-particle phase space.

In this chapter, using this Eulerian approach, we develop the kinetic theory of plasmas. We begin in Sec. 22.2 by introducing kinetic theory's one-particle distribution function $f(\mathbf{v}, \mathbf{x}, t)$ and recovering its evolution equation (the collisionless Boltzmann equation, also called the Vlasov equation), which we have met previously in Chap. 3. We then use this *Vlasov equation* to derive the two-fluid formalism used in Chap. 21 and to deduce some physical approximations that underlie the two-fluid description of plasmas.

In Sec. 22.3, we explore the application of the Vlasov equation to Langmuir waves—the 1-dimensional electrostatic modes in an unmagnetized plasma that we studied in Chap. 21 using the two-fluid formalism. Using kinetic theory, we rederive Sec. 21.4.3's Bohm-Gross dispersion relation for Langmuir waves, and as a bonus we uncover a physical damping mechanism, called *Landau damping*, that did not and cannot emerge from the two-fluid analysis. This subtle process leads to the transfer of energy from a wave to those particles that can "surf" or "phase-ride" the wave (i.e., those whose velocity projected parallel to the wave vector is slightly less than the wave's phase speed). We show that Landau damping works because there are usually fewer particles traveling faster than the wave and losing energy to it than those traveling slower and extracting energy from it. However, in a collisionless plasma, the particle distributions need not be Maxwellian. In particular, it is possible for a plasma to possess an "inverted" particle distribution with more fast than slow ones; then there is a net injection of particle energy into the waves, which creates an instability. In Sec. 22.4, we use kinetic theory to derive a necessary and sufficient criterion for this instability.

In Sec. 22.5, we examine in greater detail the physics of Landau damping and show that it is an intrinsically nonlinear phenomenon; and we give a semi-quantitative discussion of *nonlinear Landau damping*, preparatory to a more detailed treatment of some other nonlinear plasma effects in the following chapter.

Although the kinetic-theory, Vlasov description of a plasma that is developed and used in this chapter is a great improvement on the two-fluid description of Chap. 21, it is still an approximation, and some situations require more accurate descriptions. We conclude this chapter in Sec. 22.6 by introducing greater accuracy via *N-particle distribution functions*, and as applications we use them (i) to explore the approximations underlying the Vlasov description, and (ii) to explore two-particle correlations that are induced in a plasma by Coulomb interactions and the influence of those correlations on a plasma's equation of state.

22.2 Basic Concepts of Kinetic Theory and Its Relationship to Two-Fluid Theory

22.2.1 Distribution Function and Vlasov Equation

In Chap. 3, we introduced the number density of particles in phase space, the distribution function $\mathcal{N}(\mathbf{p}, \mathbf{x}, t)$. We showed that this quantity is Lorentz invariant, and that it satisfies the collisionless Boltzmann equation (3.64) and (3.65). We interpreted this equation as $\mathcal{N}$ being constant along the phase-space trajectory of any freely moving particle.

To comply with the conventions of the plasma-physics community, we use the name *Vlasov equation* in place of *collisionless Boltzmann equation*.[1] We change notation in a manner described in Sec. 3.2.3: we use a particle's velocity $\mathbf{v}$ rather than its momentum $\mathbf{p}$ as an independent variable, and we define the distribution function f to be the number density of particles in physical and velocity space:

Vlasov equation

$$f(\mathbf{v}, \mathbf{x}, t) = \frac{dN}{d\mathcal{V}_x d\mathcal{V}_v} = \frac{dN}{dx\,dy\,dz\,dv_x\,dv_y\,dv_z}. \tag{22.1}$$

distribution function in plasma physics conventions

Note that the integral of f over velocity space is the number density $n(\mathbf{x}, t)$ of particles in physical space:

$$\boxed{\int f(\mathbf{v}, \mathbf{x}, t)\, d\mathcal{V}_v = n(\mathbf{x}, t),} \tag{22.2}$$

where $d\mathcal{V}_v \equiv dv_x\,dv_y\,dv_z$ is the 3-dimensional volume element of velocity space. (For simplicity, we also restrict ourselves to nonrelativistic speeds; the generalization to relativistic plasma theory is straightforward.)

This one-particle distribution function $f(\mathbf{v}, \mathbf{x}, t)$ and its resulting kinetic theory give a good description of a plasma in the regime of large Debye number, $N_D \gg 1$—which includes almost all plasmas that occur in the universe (cf. Sec. 20.3.2 and Fig. 20.1). The reason is that, when $N_D \gg 1$, we can define $f(\mathbf{v}, \mathbf{x}, t)$ by averaging over a physical-space volume that is large compared to the average interparticle spacing and that thus contains many particles, but is still small compared to the Debye length. By such an average—the starting point of kinetic theory—the electric fields of individual particles are made unimportant, as are the Coulomb-interaction-induced correlations between pairs of particles. We explore this issue in detail in Sec. 22.6.2, using a two-particle distribution function.

In Chap. 3, we showed that, in the absence of collisions (a good assumption for plasmas!), the distribution function evolves in accord with the Vlasov equation (3.64) and (3.65). We now rederive that Vlasov equation beginning with the law of conservation of particles for each species $s = e$ (electrons) and p (protons):

$$\frac{\partial f_s}{\partial t} + \boldsymbol{\nabla} \cdot (f_s \mathbf{v}) + \boldsymbol{\nabla}_v \cdot (f_s \mathbf{a}) \equiv \frac{\partial f_s}{\partial t} + \frac{\partial(f_s v_j)}{\partial x_j} + \frac{\partial(f_s a_j)}{\partial v_j} = 0. \tag{22.3}$$

particle conservation for species s (electron or proton)

---

1. This equation was introduced and explored in 1913 by James Jeans in the context of stellar dynamics, and then rediscovered and explored by Anatoly Alexandrovich Vlasov in 1938 in the context of plasma physics. Plasma physicists have honored Vlasov by naming the equation after him. For details of this history, see Hénon (1982).

Here

$$\mathbf{a} = \frac{d\mathbf{v}}{dt} = \frac{q_s}{m_s}(\mathbf{E} + \mathbf{v} \times \mathbf{B}) \tag{22.4}$$

is the electromagnetic acceleration of a particle of species s, which has mass m_s and charge q_s, and $\mathbf{E}$ and $\mathbf{B}$ are the electric and magnetic fields averaged over the same volume as is used in constructing f. Equation (22.3) has the standard form for a conservation law: the time derivative of a density (in this case density of particles in phase space, not just physical space), plus the divergence of a flux (in this case the spatial divergence of the particle flux, $f\mathbf{v} = f d\mathbf{x}/dt$, in the physical part of phase space, plus the velocity divergence of the particle flux, $f\mathbf{a} = f d\mathbf{v}/dt$, in velocity space) is equal to zero.

Now $\mathbf{x}$ and $\mathbf{v}$ are independent variables, so that $\partial x_i/\partial v_j = 0$ and $\partial v_i/\partial x_j = 0$. In addition, $\mathbf{E}$ and $\mathbf{B}$ are functions of $\mathbf{x}$ and t but not of $\mathbf{v}$, and the term $\mathbf{v} \times \mathbf{B}$ is perpendicular to $\mathbf{v}$. Therefore, we have

$$\mathbf{\nabla}_v \cdot (\mathbf{E} + \mathbf{v} \times \mathbf{B}) = 0. \tag{22.5}$$

These facts permit us to pull $\mathbf{v}$ and $\mathbf{a}$ out of the derivatives in Eq. (22.3), thereby obtaining

Vlasov equation for species s

$$\frac{\partial f_s}{\partial t} + (\mathbf{v} \cdot \mathbf{\nabla})f_s + (\mathbf{a} \cdot \mathbf{\nabla}_v)f_s \equiv \frac{\partial f_s}{\partial t} + \frac{dx_j}{dt}\frac{\partial f_s}{\partial x_j} + \frac{dv_j}{dt}\frac{\partial f_s}{\partial v_j} = 0. \tag{22.6}$$

We recognize this as the statement that f_s is a constant along the trajectory of a particle in phase space:

$$\frac{df_s}{dt} = 0, \tag{22.7}$$

which is the *Vlasov equation* for species s.

Equation (22.7) tells us that, when the space density near a given particle increases, the velocity-space density must decrease, and vice versa. Of course, if we find that other forces or collisions are important in some situation, we can represent them by extra terms added to the right-hand side of the Vlasov equation (22.7) in the manner of the Boltzmann transport equation (3.66) (cf. Sec. 3.6).

So far, we have treated the electromagnetic field as being somehow externally imposed. However, it is actually produced by the net charge and current densities associated with the two particle species. These are expressed in terms of the distribution functions by

charge density and current density as integrals over distribution functions

$$\rho_e = \sum_s q_s \int f_s \, d\mathcal{V}_v, \qquad \mathbf{j} = \sum_s q_s \int f_s \mathbf{v} \, d\mathcal{V}_v. \tag{22.8}$$

Equations (22.8), together with Maxwell's equations and the Vlasov equation (22.6), with $\mathbf{a} = d\mathbf{v}/dt$ given by the Lorentz force law (22.4), form a complete set of equations for the structure and dynamics of a plasma. They constitute the kinetic theory of plasmas.

22.2.2 Relation of Kinetic Theory to Two-Fluid Theory

Before developing techniques to solve the Vlasov equation, we first relate it to the two-fluid approach used in the previous chapter. We begin by constructing the moments of the distribution function f_s, defined by

$$n_s = \int f_s \, d\mathcal{V}_v,$$

$$\mathbf{u}_s = \frac{1}{n_s} \int f_s \mathbf{v} \, d\mathcal{V}_v,$$

macroscopic, two-fluid quantities as integrals over distribution functions

$$\mathbf{P}_s = m_s \int f_s (\mathbf{v} - \mathbf{u}_s) \otimes (\mathbf{v} - \mathbf{u}_s) \, \mathcal{V}_v. \qquad (22.9)$$

These are respectively the density, the mean fluid velocity, and the pressure tensor for species s. (Of course, $\mathbf{P}_s$ is just the 3-dimensional stress tensor $\mathbf{T}_s$ [Eq. (3.32d)] evaluated in the rest frame of the fluid.)

By integrating the Vlasov equation (22.6) over velocity space and using

$$\int (\mathbf{v} \cdot \boldsymbol{\nabla}) f_s \, d\mathcal{V}_v = \int \boldsymbol{\nabla} \cdot (f_s \mathbf{v}) \, d\mathcal{V}_v = \boldsymbol{\nabla} \cdot \int f_s \mathbf{v} \, d\mathcal{V}_v,$$

$$\int (\mathbf{a} \cdot \boldsymbol{\nabla}_v) f_s \, d\mathcal{V}_v = - \int (\boldsymbol{\nabla}_v \cdot \mathbf{a}) f_s \, d\mathcal{V}_v = 0, \qquad (22.10)$$

together with Eq. (22.9), we obtain the continuity equation,

$$\frac{\partial n_s}{\partial t} + \boldsymbol{\nabla} \cdot (n_s \mathbf{u}_s) = 0, \qquad (22.11)$$

continuity equation

for each particle species s. [It should not be surprising that the Vlasov equation implies the continuity equation, since the Vlasov equation is equivalent to the conservation of particles in phase space [Eq. (22.3)], while the continuity equation is just the conservation of particles in physical space.]

The continuity equation is the first of the two fundamental equations of two-fluid theory. The second is the equation of motion (i.e., the evolution equation for the fluid velocity $\mathbf{u}_s$). To derive this, we multiply the Vlasov equation (22.6) by the particle velocity $\mathbf{v}$ and then integrate over velocity space (i.e., we compute the Vlasov equation's first moment). The details are a useful exercise for the reader (Ex. 22.1); the result is

$$n_s m_s \left(\frac{\partial \mathbf{u}_s}{\partial t} + (\mathbf{u}_s \cdot \boldsymbol{\nabla}) \mathbf{u}_s \right) = -\boldsymbol{\nabla} \cdot \mathbf{P}_s + n_s q_s (\mathbf{E} + \mathbf{u}_s \times \mathbf{B}), \qquad (22.12)$$

equation of motion

which is identical with Eq. (21.13b).

constructing two-fluid
equations from kinetic
theory by method of
moments

A difficulty now presents itself in the two-fluid approximation to kinetic theory. We can use Eqs. (22.8) to compute the charge and current densities from n_s and $\mathbf{u}_s$, which evolve via the fluid equations (22.11) and (22.12). However, we do not yet know how to compute the pressure tensor $\mathbf{P}_s$ in the two-fluid approximation. We could derive a fluid equation for its evolution by taking the second moment of the Vlasov equation (i.e., multiplying it by $\mathbf{v} \otimes \mathbf{v}$ and integrating over velocity space), but that evolution equation would involve an unknown third moment of f_s on the right-hand side, $\mathbf{M}_3 = \int f_s \mathbf{v} \otimes \mathbf{v} \otimes \mathbf{v} \, d\mathcal{V}_v$, which is related to the heat-flux tensor. To determine the evolution of this $\mathbf{M}_3$, we would have to construct the third moment of the Vlasov equation, which would involve the fourth moment of f_s as a driving term, and so on. Clearly, this procedure will never terminate unless we introduce some additional relationship between the moments. Such a relationship, called a *closure relation*, permits us to build a self-contained theory involving only a finite number of moments.

For the two-fluid theory of Chap. 21, the closure relation that we implicitly used was the same idealization that one makes when regarding a fluid as perfect, namely, that the heat-flux tensor vanishes. This idealization is less well justified in a collisionless plasma, with its long mean free paths, than in a normal gas or liquid (which has short mean free paths).

An example of an alternative closure relation is one that is appropriate if radiative processes thermostat the plasma to a particular temperature, so $T_s = \text{const}$; then we can set $\mathbf{P}_s = n_s k_B T_s \mathbf{g} \propto n_s$, where $\mathbf{g}$ is the metric tensor. Clearly, a fluid theory of plasmas can be no more accurate than its closure relation.

22.2.3 Jeans' Theorem

Let us now turn to the difficult task of finding solutions to the Vlasov equation. There is an elementary (and, after the fact, obvious) method to write down a class of solutions that is often useful. This is based on Jeans' theorem (named after the astronomer who first drew attention to it in the context of stellar dynamics; Jeans, 1929).

Suppose that we know the particle acceleration $\mathbf{a}$ as a function of $\mathbf{v}$, $\mathbf{x}$, and t. (We assume this for pedagogical purposes; it is not necessary for our final conclusion.) Then for any particle with phase-space coordinates $(\mathbf{x}_0, \mathbf{v}_0)$ specified at time t_0, we can (at least in principle) compute the particle's future motion: $\mathbf{x} = \mathbf{x}(\mathbf{x}_0, \mathbf{v}_0, t)$, $\mathbf{v} = \mathbf{v}(\mathbf{x}_0, \mathbf{v}_0, t)$. These particle trajectories are the *characteristics* of the Vlasov equation, analogous to the characteristics of the equations for 1-dimensional supersonic fluid flow that we studied in Sec. 17.4 (see Fig. 17.7). For many choices of the acceleration $\mathbf{a}(\mathbf{v}, \mathbf{x}, t)$, there are *constants of the motion*, also known as *integrals of the motion*, that are preserved along the particle trajectories. Simple examples, familiar from elementary mechanics, include the energy (for a time-independent plasma) and the angular momentum (for a spherically symmetric plasma). These integrals can be expressed in terms of the initial coordinates $(\mathbf{x}_0, \mathbf{v}_0)$. If we know n constants of the motion, then only $6 - n$ additional variables need be chosen from $(\mathbf{x}_0, \mathbf{v}_0)$ to completely specify the motion of the particle.

The Vlasov equation tells us that f_s is constant along a trajectory in $\mathbf{x}$-$\mathbf{v}$ space. Therefore, in general f_s must be expressible as a function of $(\mathbf{x}_0, \mathbf{v}_0)$. Equivalently, it can be rewritten as a function of the n constants of the motion and the remaining $6 - n$ initial phase-space coordinates. However, there is no requirement that it actually depend on all of these variables. In particular, any function of the integrals of motion alone that is independent of the remaining initial coordinates will satisfy the Vlasov equation (22.6). This is *Jeans' theorem*. In words, *functions of constants of the motion take constant values along actual dynamical trajectories in phase space and therefore satisfy the Vlasov equation.*

Jeans' theorem: solution of Vlasov equation based on constants of particle motion

Of course, a situation may be so complex that no integrals of the particles' equation of motion can be found, in which case Jeans' theorem won't help us find solutions of the Vlasov equation. Alternatively, there may be integrals but the initial conditions may be sufficiently complex that extra variables are required to determine f_s. However, it turns out that in a wide variety of applications, particularly those with symmetries (e.g., time independence $\partial f_s/\partial t = 0$), simple functions of simple constants of the motion suffice to describe a plasma's distribution functions.

We have already met and used a variant of Jeans' theorem in our analysis of statistical equilibrium in Sec. 4.4. There the statistical mechanics distribution function ρ was found to depend only on the integrals of the motion.

We have also, unknowingly, used Jeans' theorem in our discussion of Debye shielding in a plasma (Sec. 20.3.1). To understand this, let us suppose that we have a single isolated positive charge at rest in a stationary plasma ($\partial f_s/\partial t = 0$), and we want to know the electron distribution function in its vicinity. Let us further suppose that the electron distribution at large distances from the charge is known to be Maxwellian with temperature T: $f_e(\mathbf{v}, \mathbf{x}, t) \propto \exp[-\frac{1}{2}m_e v^2/(k_B T)]$. Now, the electrons have an energy integral, $E = \frac{1}{2}m_e v^2 - e\Phi$, where Φ is the electrostatic potential. As Φ becomes constant at large distances from the charge, we can therefore write $f_e \propto \exp[-E/(k_B T)]$ at large distances. However, the particles near the charge must have traveled there from a large distance and so must have this same distribution function. Therefore, close to the charge, we have

$$f_e \propto e^{-E/k_B T} = e^{-[(m_e v^2/2 - e\Phi)/(k_B T)]},\tag{22.13}$$

and the electron density is obtained by integration over velocity:

$$n_e = \int f_e \, d\mathcal{V}_v \propto e^{[e\Phi/(k_B T)]}.\tag{22.14}$$

This is just the Boltzmann distribution that we asserted to be appropriate in Sec. 20.3.1.

Exercise 22.1 *Derivation: Two-Fluid Equation of Motion*
Derive the two-fluid equation of motion (22.12) by multiplying the Vlasov equation (22.6) by $\mathbf{v}$ and integrating over velocity space.

Exercise 22.2 *Example: Positivity of Distribution Function*
The particle distribution function $f(\mathbf{v}, \mathbf{x}, t)$ ought not to become negative if it is to remain physical. Show that this is guaranteed if it initially is everywhere nonnegative and it evolves by the collisionless Vlasov equation.

Exercise 22.3 *Problem and Challenge: Jeans' Theorem in Stellar Dynamics*
Jeans' theorem is of great use in studying the motion of stars in a galaxy. The stars are also almost collisionless and can be described by a distribution function $f(\mathbf{v}, \mathbf{x}, t)$. However, there is only one sign of "charge" and no possibility of screening. In this problem, we make a model of a spherical galaxy composed of identical-mass stars.[2]

(a) The simplest type of distribution function is a function of one integral of a star's motion, the energy per unit mass: $E = \frac{1}{2}v^2 + \Phi$, where Φ is the gravitational potential. A simple example is $f(E) \propto |-E|^{7/2}$ for negative energy and zero for positive energy. Verify that the associated mass density satisfies: $\rho \propto [1 + (r/s)^2]^{-5/2}$, where r is the radius, and s is a scale that you should identify.

(b) This density profile does not agree with observations, so we need to find an algorithm for computing the distribution function that gives a specified density profile. Show that for a general $f(E)$ satisfying appropriate boundary conditions,

$$\frac{d\rho}{d\Phi} = -8^{1/2}\pi\, m \int_{\Phi}^{0} dE\, \frac{f(E)}{(E - \Phi)^{1/2}}, \qquad (22.15a)$$

where m is the mass of a star.

(c) Equation (22.15a) is an Abel integral equation. Confirm that it can be inverted to give the desired algorithm:

$$f(E) = \frac{1}{8^{1/2}\pi^2\, m} \frac{d}{dE} \int_{E}^{0} d\Phi\, \frac{d\rho/d\Phi}{(\Phi - E)^{1/2}}. \qquad (22.15b)$$

(d) Compute the distribution function that is paired with the *Jaffe* profile, which looks more like a real galaxy:

$$\rho \propto \frac{1}{r^2(r + s)^2}. \qquad (22.16)$$

---

2. We now know that the outer parts of galaxies are dominated by *dark matter* (Sec. 28.3.2), which may comprise weakly interacting elementary particles that should satisfy the Vlasov equation just as stars do.

(e) We can construct *two-integral* spherical models using $f(E, L)$, where L is a star's total angular momentum per unit mass. What extra feature can we hope to capture using this broader class of distribution functions?

22.3 Electrostatic Waves in an Unmagnetized Plasma: Landau Damping

22.3

As our principal application of the kinetic theory of plasmas, we explore its predictions for the dispersion relations; stability; and damping of longitudinal, electrostatic waves in an unmagnetized plasma—Langmuir waves and ion-acoustic waves. When studying these waves in Sec. 21.4.3 using two-fluid theory, we alluded time and again to properties of the waves that could not be derived by fluid techniques. Our goal now is to elucidate those properties using kinetic theory. As we shall see, their origin lies in the plasma's velocity-space dynamics.

22.3.1 Formal Dispersion Relation

22.3.1

Consider an electrostatic wave propagating in the z direction. Such a wave is 1-dimensional in that the electric field points in the z direction, $\mathbf{E} = E\mathbf{e}_z$, and varies as $e^{i(kz-\omega t)}$, so it depends only on z and not on x or y; the distribution function similarly varies as $e^{i(kz-\omega t)}$ and is independent of x, y; and the Vlasov, Maxwell, and Lorentz force equations produce no coupling of particle velocities v_x and v_y to the z direction. These properties suggest the introduction of 1-dimensional distribution functions, obtained by integration over v_x and v_y:

$$\boxed{F_s(v, z, t) \equiv \int f_s(v_x, v_y, v = v_z, z, t) dv_x dv_y.}$$

(22.17)

1-dimensional distribution function for plasma interacting with a longitudinal electrostatic wave (Langmuir or ion-acoustic)

Here and throughout we suppress the subscript z on v_z.

Restricting ourselves to weak waves, so nonlinearities can be neglected, we linearize the 1-dimensional distribution functions:

$$\boxed{F_s(v, z, t) \simeq F_{s0}(v) + F_{s1}(v, z, t).}$$

(22.18)

Here $F_{s0}(v)$ is the distribution function of the unperturbed particles ($s = e$ for electrons and $s = p$ for protons) in the absence of the wave, and F_{s1} is the perturbation induced by and linearly proportional to the electric field E. The evolution of F_{s1} is governed by the linear approximation to the Vlasov equation (22.6):

$$\frac{\partial F_{s1}}{\partial t} + v \frac{\partial F_{s1}}{\partial z} + \frac{q_s E}{m_s} \frac{dF_{s0}}{dv} = 0.$$

(22.19)

Here E is a first-order quantity, so in its term we keep only the zero-order dF_{s0}/dv.

We seek a monochromatic plane-wave solution to this Vlasov equation, so $\partial/\partial t \rightarrow -i\omega$ and $\partial/\partial z \rightarrow ik$ in Eq. (22.19). Solving the resulting equation for F_{s1}, we obtain

solution of linearized
Vlasov equation

$$F_{s1} = \frac{-iq_s}{m_s(\omega - kv)} \frac{dF_{s0}}{dv} E. \tag{22.20}$$

This equation implies that the charge density associated with the wave is related to the electric field by

$$\rho_e = \sum_s q_s \int_{-\infty}^{+\infty} F_{s1}dv = \left(\sum_s \frac{-iq_s^2}{m_s} \int_{-\infty}^{+\infty} \frac{F_{s0}' \, dv}{\omega - kv} \right) E, \tag{22.21}$$

where the prime denotes a derivative with respect to v: $F_{s0}' = dF_{s0}/dv$.

A quick route from here to the waves' dispersion relation is to insert this charge density into Poisson's equation, $\nabla \cdot \mathbf{E} = ikE = \rho_e/\epsilon_0$, and note that both sides are proportional to E, so a solution is possible only if

$$1 + \sum_s \frac{q_s^2}{m_s \epsilon_0 k} \int_{-\infty}^{+\infty} \frac{F_{s0}' \, dv}{\omega - kv} = 0. \tag{22.22}$$

An alternative route, which makes contact with the general analysis of waves in a dielectric medium (Sec. 21.2), is developed in Ex. 22.4. This route reveals that the dispersion relation is given by the vanishing of the zz component of the dielectric tensor, which we denoted ϵ_3 in Chap. 21 [Eq. (21.45)], and it shows that ϵ_3 is given by expression (22.22):

$$\epsilon_3(\omega, k) = 1 + \sum_s \frac{q_s^2}{m_s \epsilon_0 k} \int_{-\infty}^{+\infty} \frac{F_{s0}' \, dv}{\omega - kv} = 0. \tag{22.23}$$

Since $\epsilon_3 = \epsilon_{zz}$ is the only component of the dielectric tensor that we meet in this chapter, we simplify notation henceforth by omitting the subscript 3 (i.e., by denoting $\epsilon_{zz} = \epsilon$).

The form of the dispersion relation (22.23) suggests that we combine the unperturbed electron and proton distribution functions $F_{e0}(v)$ and $F_{p0}(v)$ to produce a single, unified distribution function:

unified distribution
function

$$F(v) \equiv F_{e0}(v) + \frac{m_e}{m_p} F_{p0}(v), \tag{22.24}$$

in terms of which the dispersion relation takes the form

general dispersion
relation for electrostatic
waves—derived by Fourier
techniques

$$\epsilon(\omega, k) = 1 + \frac{e^2}{m_e \epsilon_0 k} \int_{-\infty}^{+\infty} \frac{F' \, dv}{\omega - kv} = 0. \tag{22.25}$$

Note that each proton is weighted less heavily than each electron by a factor $m_e/m_p = 1/1{,}836$ in the unified distribution function (22.24) and the dispersion relation (22.25). This is due to the protons' greater inertia and corresponding weaker response to an applied electric field; it causes the protons to be of no importance in Langmuir waves (Sec. 22.3.5). However, in ion-acoustic waves (Sec. 22.3.6), the protons can play an important role, because large numbers of them may move with thermal speeds that are close to the waves' phase velocity, and thereby they can interact resonantly with the waves.

EXERCISES

Exercise 22.4 *Example: Dielectric Tensor and Dispersion Relation for Longitudinal, Electrostatic Waves*

Derive expression (22.23) for the zz component of the dielectric tensor in a plasma excited by a weak electrostatic wave, and show that the wave's dispersion relation is $\epsilon_3 = 0$. [Hint: Notice that the z component of the plasma's electric polarization P_z is related to the charge density by $\nabla \cdot \mathbf{P} = ikP_z = -\rho_e$ [Eq. (21.1)]. Combine this with Eq. (22.21) to get a linear relationship between P_z and $E_z = E$. Argue that the only nonzero component of the plasma's electric susceptibility is χ_{zz}, and deduce its value by comparing the above result with Eq. (21.3). Then construct the dielectric tensor ϵ_{ij} from Eq. (21.5) and the algebratized wave operator L_{ij} from Eq. (21.9), and deduce that the dispersion relation $\det||L_{ij}|| = 0$ takes the form $\epsilon_{zz} \equiv \epsilon_3 = 0$, where ϵ_3 is given by Eq. (22.23).]

22.3.2 Two-Stream Instability

22.3.2

As a first application of the general dispersion relation (22.25), we use it to rederive the dispersion relation (21.74) associated with the cold-plasma two-stream instability of Sec. 21.6.

We begin by performing an integration by parts on the general dispersion relation (22.25), obtaining:

multi-stream approach to two-stream instability

$$\frac{e^2}{m_e \epsilon_0} \int_{-\infty}^{+\infty} \frac{F \, dv}{(\omega - kv)^2} = 1. \tag{22.26}$$

We then presume, as in Sec. 21.6, that the fluid consists of two or more streams of cold particles (protons or electrons) moving in the z direction with different fluid speeds $u_1, u_2, \ldots$, so $F(v) = n_1 \delta(v - u_1) + n_2 \delta(v - u_2) + \ldots$. Here n_j is the number density of particles in stream j if the particles are electrons, and m_e/m_p times the number density if they are protons. Inserting this $F(v)$ into Eq. (22.26) and noting that $n_j e^2/(m_e \epsilon_0)$ is the squared plasma frequency ω_{pj}^2 of stream j, we obtain the dispersion relation

$$\frac{\omega_{p1}^2}{(\omega - ku_1)^2} + \frac{\omega_{p2}^2}{(\omega - ku_2)^2} + \cdots = 1, \qquad (22.27)$$

which is identical to the dispersion relation (21.74) used in our analysis of the two-stream instability.

It should be evident that the general dispersion relation (22.26) [or equally well, Eq. (22.25)] provides us with a tool for exploring how the two-stream instability is influenced by a warming of the plasma (i.e., by a spread of particle velocities around the mean, fluid velocity of each stream). We explore this in Sec. 22.4.

22.3.3

22.3.3 The Landau Contour

The general dispersion relation (22.25) has a troubling feature: for real ω and k its integrand becomes singular at $v = \omega/k =$ (the waves' phase velocity) unless dF/dv vanishes there, which is generically unlikely. This tells us that if k is real (as we shall assume), then ω cannot be real, except perhaps for a nongeneric mode whose phase velocity happens to coincide with a velocity for which $dF/dv = 0$.

ambiguity in general dispersion relation

With ω/k complex, we must face the possibility of some subtlety in how the integral over v in the dispersion relation (22.25) is performed—the possibility that we may have to make v complex in the integral and follow some special route in the complex velocity plane from $v = -\infty$ to $v = +\infty$. Indeed, there is such a subtlety, as Landau (1946) has shown. Our simple derivation of the dispersion relation in Sec. 22.3.1 cannot reveal this subtlety—and, indeed, is suspicious, since in going from Eq. (22.19) to Eq. (22.20), our derivation entailed dividing by $\omega - kv$, which vanishes when $v = \omega/k$, and dividing by zero is always a suspicious practice. Faced with this conundrum, Landau developed a more sophisticated derivation of the dispersion

derivation of general dispersion relation by Landau's method

relation. It is based on posing generic initial data for electrostatic waves, then evolving those data forward in time and identifying the plasma's electrostatic modes by their late-time sinusoidal behaviors, and finally reading off the dispersion relation for the modes from the equations for the late-time evolution. In the remainder of this section, we present a variant of Landau's analysis.

This analysis is very important, including the portion (Ex. 22.5) assigned for the reader to work out. The reader is encouraged to read through this section slowly, with care, so as to understand clearly what is going on.

For simplicity, from the outset we restrict ourselves to plane waves propagating in the z direction with some fixed, real wave number k, so the linearized 1-dimensional distribution function and the electric field have the forms

$$F_s(v, z, t) = F_{s0}(v) + F_{s1}(v, t)e^{ikz}, \quad E(z, t) = E(t)e^{ikz}. \qquad (22.28)$$

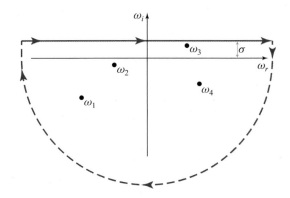

FIGURE 22.1 Contour of integration for evaluating $E(t)$ [Eq. (22.29)] as a sum over residues of the integrand's poles ω_n—the complex frequencies of the plasma's modes.

At $t = 0$ we pose initial data $F_{s1}(v, 0)$ for the electron and proton velocity distributions; these data determine the initial electric field $E(0)$ via Poisson's equation. We presume that these initial distributions [and also the unperturbed plasma's velocity distribution $F_{s0}(v)$] are analytic functions of velocity v, but aside from this constraint, the $F_{s1}(v, 0)$ are generic. (A Maxwellian distribution is analytic, and most any physically reasonable initial distribution can be well approximated by an analytic function.)

We then evolve these initial data forward in time. The ideal tool for such evolution is the Laplace transform, and *not* the Fourier transform. The power of the Laplace transform is much appreciated by engineers and is underappreciated by many physicists. Those readers who are not intimately familiar with evolution via Laplace transforms should work carefully through Ex. 22.5. That exercise uses Laplace transforms, followed by conversion of the final answer into Fourier language, to derive the following formula for the time-evolving electric field in terms of the initial velocity distributions $F_{s1}(v, 0)$:

key to the derivation— evolve arbitrary initial data using Laplace transform

$$E(t) = \int_{i\sigma-\infty}^{i\sigma+\infty} \frac{e^{-i\omega t}}{\epsilon(\omega, k)} \left[\sum_s \frac{q_s}{2\pi\epsilon_0 k} \int_{-\infty}^{+\infty} \frac{F_{s1}(v, 0)}{\omega - kv} dv \right] d\omega. \qquad (22.29)$$

Here the integral in frequency space is along the solid horizontal line at the top of Fig. 22.1, with the imaginary part of ω held fixed at $\omega_i = \sigma$ and the real part ω_r varying from $-\infty$ to $+\infty$. The Laplace techniques used to derive this formula are carefully designed to avoid any divergences and any division by zero. This careful design leads to the requirement that the height σ of the integration line above the real frequency axis be greater than the e-folding rate $\Im(\omega)$ of the plasma's most rapidly growing mode (or, if none grow, still larger than zero and thus larger than $\Im(\omega)$ for the most slowly decaying mode):

$$\sigma > p_o \equiv \max_n \Im(\omega_n), \quad \text{and} \quad \sigma > 0. \qquad (22.30)$$

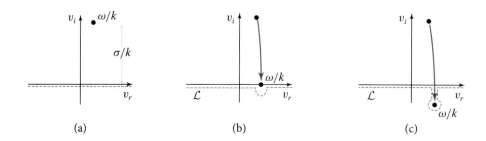

FIGURE 22.2 Derivation of the Landau contour $\mathcal{L}$. (a) The dielectric function $\epsilon(\omega, k)$ is originally defined, in Eq. (22.31), solely for the pole (the point labeled ω/k) lying in the upper half plane; that is, for $\omega_i/k = \sigma/k > 0$. (b,c) Since $\epsilon(\omega, k)$ must be an analytic function of ω at fixed k and thus must vary continuously as ω is continuously changed, the dashed contour of integration in Eq. (22.31) must be kept always below the pole $v = \omega/k$.

Here $n = 1, 2, \ldots$ labels the modes, ω_n is the complex frequency of mode n, and $\Im$ means take the imaginary part of. We shall see that the ω_n are the zeros of the dielectric function $\epsilon(\omega, k)$ that appears in Eq. (22.29), that is, they are poles of the integrand of the frequency integral.

Equation (22.29) also entails a velocity integral. In the Laplace-based analysis (Ex. 22.5) that leads to this formula, there is never any question about the nature of the velocity v: it is always real, so the integral is over real v running from $-\infty$ to $+\infty$. However, because all the frequencies ω appearing in Eq. (22.29) have imaginary parts $\omega_i = \sigma > 0$, there is no possibility in the velocity integral of any divergence of the integrand.

In Eq. (22.29) for the evolving field, $\epsilon(\omega, k)$ is the same dielectric function (22.25) as we deduced in our previous analysis (Sec. 22.3.1):

$$\epsilon(\omega, k) = 1 + \frac{e^2}{m_e \epsilon_0 k} \int_{-\infty}^{+\infty} \frac{F' \, dv}{\omega - kv}, \quad \text{where } F(v) = F_{e0}(v) + \frac{m_e}{m_p} F_{p0}. \quad (22.31)$$

However, by contrast with Sec. 22.3.1, our derivation here dictates unequivocally how to handle the v integration—the same way as in Eq. (22.29): v is strictly real, and the only frequencies appearing in the evolution equations have $\omega_i = \sigma > 0$, so the v integral, running along the real velocity axis, passes under the integrand's pole at $v = \omega/k$, as shown in Fig. 22.2a.

To read off the modal frequencies from the evolving field $E(t)$ at times $t > 0$, we use techniques from complex-variable theory. It can be shown that, because (by hypothesis) $F_{s1}(v, 0)$ and $F_{s0}(v)$ are analytic functions of v, the integrand of the ω integral in Eq. (22.29) is meromorphic: when the integrand is analytically continued throughout the complex frequency plane, its only singularities are poles. This permits us to evaluate the frequency integral, at times $t > 0$, by closing the integration contour in the lower-half frequency plane, as shown by the dashed curve in Fig. 22.1. Because of the exponential factor $e^{-i\omega t}$, the contribution from the dashed part of the contour vanishes, which means that the integral around the contour is equal to $E(t)$ (the contribution from the solid horizontal part). Complex-variable theory tells us that this integral is given by a sum over the residues R_n of the integrand at the poles (labeled $n = 1, 2, \ldots$):

$$E(t) = 2\pi i \sum_n R_n = \sum_n A_n e^{-i\omega_n t}. \tag{22.32}$$

Here ω_n is the frequency at pole n, and A_n is $2\pi i R_n$ with its time dependence $e^{-i\omega_n t}$ factored out. It is evident, then, that each pole of the analytically continued integrand of Eq. (22.29) corresponds to a mode of the plasma, and the pole's complex frequency is the mode's frequency.

Now, for special choices of the initial data $F_{s1}(v, 0)$, there may be poles in the square-bracketed term in Eq. (22.29), but for generic initial data there will be none, and the only poles will be the zeros of $\epsilon(\omega, k)$. Therefore, generically, the modes' frequencies are the zeros of $\epsilon(\omega, k)$—when that function (which was originally defined only for ω along the line $\omega_i = \sigma$) has been analytically extended throughout the complex frequency plane.

So how do we compute the analytically extended dielectric function $\epsilon(\omega, k)$? Imagine holding k fixed and real, and exploring the (complex) value of ϵ, thought of as a function of ω/k, by moving around the complex ω/k plane (same as the complex velocity plane). In particular, imagine computing ϵ from Eq. (22.31) at one point after another along the arrowed path shown in Fig. 22.2b,c. This path begins at an initial location ω/k, where $\omega_i/k = \sigma/k > 0$, and travels down to some other location below the real axis. At the starting point, the discussion in the paragraph following Eq. (22.30) tells us how to handle the velocity integral: just integrate v along the real axis. As ω/k is moved continuously (with k held fixed), $\epsilon(\omega, k)$ must vary continuously, because it is analytic. When ω/k crosses the real velocity axis, if the integration contour in Eq. (22.31) were to remain on the velocity axis, then the contour would jump over the integral's moving pole $v = \omega/k$, and the function $\epsilon(\omega, k)$ would jump discontinuously at the moment of crossing, which is not possible. To avoid such a discontinuous jump, it is necessary that the contour of integration be kept below the pole, $v = \omega/k$, as that pole moves into the lower half of the velocity plane (Fig. 22.2b,c).

The rule that the integration contour must always pass beneath the pole $v = \omega/k$ as shown in Fig. 22.2 is called the *Landau prescription*; the contour is called the *Landau contour* and is denoted $\mathcal{L}$. Our final formula for the dielectric function (and for its vanishing at the modal frequencies—the dispersion relation) is

Landau contour and the unambiguous general dispersion relation for electrostatic waves in an unmagnetized plasma

$$\boxed{\epsilon(\omega, k) = 1 + \frac{e^2}{m_e \epsilon_0 k} \int_{\mathcal{L}} \frac{F' dv}{\omega - kv} = 0,} \quad \text{where} \quad \boxed{F(v) = F_e(v) + \frac{m_e}{m_p} F_p(v).}$$

$$\tag{22.33}$$

For future use we have omitted the subscript 0 from the unperturbed distribution functions F_s, as there should be no confusion in future contexts. We refer to Eq. (22.33) as the general dispersion relation for electrostatic waves in an un-magnetized plasma.

Exercise 22.5 **Example: Electric Field for Electrostatic Wave Deduced Using Laplace Transforms*

Use Laplace-transform techniques to derive Eqs. (22.29)–(22.31) for the time-evolving electric field of electrostatic waves with fixed wave number k and initial velocity perturbations $F_{s1}(v, 0)$. A sketch of the solution follows.

(a) When the initial data are evolved forward in time, they produce $F_{s1}(v, t)$ and $E(t)$. Construct the Laplace transforms (see, e.g., Arfken, Weber, and Harris, 2013, Chap. 20, or Mathews and Walker, 1970, Sec. 4-3) of these evolving quantities:

$$\tilde{F}_{s1}(v, p) = \int_0^\infty dt\, e^{-pt} F_{s1}(v, t), \qquad \tilde{E}(p) = \int_0^\infty dt\, e^{-pt} E(t). \quad (22.34)$$

To ensure that the time integral is convergent, insist that $\Re(p)$ be greater than $p_0 \equiv \max_n \Im(\omega_n) \equiv$ (the e-folding rate of the most strongly growing mode—or, if none grow, then the most weakly damped mode). Also, to simplify the subsequent analysis, insist that $\Re(p) > 0$. Below, in particular, we need the Laplace transforms for $\Re(p) =$ (some fixed value σ that satisfies $\sigma > p_o$ and $\sigma > 0$).

(b) By constructing the Laplace transform of the 1-dimensional Vlasov equation (22.19) and integrating by parts the term involving $\partial F_{s1}/\partial t$, obtain an equation for a linear combination of $\tilde{F}_{s1}(v, p)$ and $\tilde{E}(p)$ in terms of the initial data $F_{s1}(v, t = 0)$. By then combining with the Laplace transform of Poisson's equation, show that

$$\tilde{E}(p) = \frac{1}{\epsilon(ip, k)} \sum_s \frac{q_s}{k\epsilon_0} \int_{-\infty}^\infty \frac{F_{s1}(v, 0)}{ip - kv} dv. \quad (22.35)$$

Here $\epsilon(ip, k)$ is the dielectric function (22.25) evaluated for frequency $\omega = ip$, with the integral running along the real v-axis, and [as we noted in part (a)] with $\Re(p)$ greater than p_0, the largest ω_i of any mode, and greater than 0. This situation for the dielectric function is the one depicted in Fig. 22.2a.

(c) Laplace-transform theory tells us that the time-evolving electric field (with wave number k) can be expressed in terms of its Laplace transform (22.35) by

$$E(t) = \int_{\sigma-i\infty}^{\sigma+i\infty} \tilde{E}(p)\, e^{pt} \frac{dp}{2\pi i}, \quad (22.36)$$

where σ [as introduced in part (a)] is any real number larger than p_0 and larger than 0. Combine this equation with expression (22.35) for $\tilde{E}(p)$, and set $p = -i\omega$. Thereby arrive at the desired result, Eq. (22.29).

In most practical situations, electrostatic waves are weakly damped or weakly unstable: $|\omega_i| \ll \omega_r$ (where ω_r and ω_i are the real and imaginary parts of the wave frequency ω), so the amplitude changes little in one wave period. In this case, the dielectric function (22.33) can be evaluated at $\omega = \omega_r + i\omega_i$ using the first term in a Taylor series expansion away from the real axis:

$$
\epsilon(\omega_r + i\omega_i, k) \simeq \epsilon(\omega_r, k) + \omega_i \left(\frac{\partial \epsilon_r}{\partial \omega_i} + i \frac{\partial \epsilon_i}{\partial \omega_i} \right)_{\omega_i = 0}
$$

$$
= \epsilon(\omega_r, k) + \omega_i \left(-\frac{\partial \epsilon_i}{\partial \omega_r} + i \frac{\partial \epsilon_r}{\partial \omega_r} \right)_{\omega_i = 0}
$$

$$
\simeq \epsilon(\omega_r, k) + i\omega_i \left(\frac{\partial \epsilon_r}{\partial \omega_r} \right)_{\omega_i = 0}. \tag{22.37}
$$

Here ϵ_r and ϵ_i are the real and imaginary parts of ϵ. In going from the first line to the second we have assumed that $\epsilon(\omega, k)$ is an analytic function of ω near the real axis and hence have used the Cauchy-Riemann equations for the derivatives (see, e.g., Arfken, Weber, and Harris, 2013). In going from the second line to the third we have used the fact that $\epsilon_i \to 0$ when the velocity distribution is one that produces $\omega_i \to 0$ [cf. Eq. (22.39)], so the middle term on the second line is second order in ω_i and can be neglected.

Equation (22.37) expresses the dielectric function slightly away from the real axis in terms of its value and derivative on and along the real axis. The on-axis value can be computed from Eq. (22.33) by breaking the Landau contour depicted in Fig. 22.2b into three pieces—two lines from $\pm\infty$ to a small distance δ from the pole, plus a semicircle of radius δ under and around the pole—and by then taking the limit $\delta \to 0$. The first two terms (the two straight lines) together produce the Cauchy principal value of the integral (denoted $\int_P$ below), and the third produces $2\pi i$ times half the residue of the pole at $v = \omega_r/k$, so Eq. (22.33) becomes:

$$
\epsilon(\omega_r, k) = 1 - \frac{e^2}{m_e \epsilon_0 k^2} \left[\int_P \frac{F' \, dv}{v - \omega_r/k} dv + i\pi F'(v = \omega_r/k) \right]. \tag{22.38}
$$

Inserting this equation and its derivative with respect to ω_r into Eq. (22.37), and setting the result to zero, we obtain

$$
\epsilon(\omega_r + i\omega_i, k) \simeq 1 - \frac{e^2}{m_e \epsilon_0 k^2} \left[\int_P \frac{F' \, dv}{v - \omega_r/k} + i\pi F'(\omega_r/k) \right.
$$

$$
\left. + i\omega_i \frac{\partial}{\partial \omega_r} \int_P \frac{F' \, dv}{v - \omega_r/k} \right] = 0. \tag{22.39}
$$

Notice that the vanishing of ϵ_r determines the real part ω_r of the frequency:

$$
\boxed{1 - \frac{e^2}{m_e \epsilon_0 k^2} \int_P \frac{F'}{v - \omega_r/k} dv = 0,} \tag{22.40a}
$$

and the vanishing of ϵ_i determines the imaginary part:

$$\boxed{\omega_i = \frac{\pi \, F'(\omega_r/k)}{-\frac{\partial}{\partial \omega_r} \int_P \frac{F'}{v - \omega_r/k} dv}.}$$ (22.40b)

general dispersion relation for weakly damped elastostatic waves in an unmagnetized plasma

Equations (22.40) are the dispersion relation in the limit $|\omega_i| \ll \omega_r$. We refer to this as the small-$|\omega_i|$ dispersion relation for electrostatic waves in an unmagnetized plasma.

Notice that the sign of ω_i is influenced by the sign of $F' = dF/dv$ at $v = \omega_r/k = V_{ph} =$ (the waves' phase velocity). As we shall see, this has a simple physical origin and important physical consequences. Usually, but not always, the denominator of Eq. (22.40b) is positive, so the sign of ω_i is the same as the sign of $F'(\omega_r/k)$.

22.3.5

22.3.5 Langmuir Waves and Their Landau Damping

We now apply the small-$|\omega_i|$ dispersion relation (22.40) to Langmuir waves in a thermalized plasma. Langmuir waves typically move so fast that the slow ions cannot interact with them, so their dispersion relation is influenced significantly only by the electrons. Therefore, we ignore the ions and include only the electrons in $F(v)$. We obtain $F(v)$ by integrating out v_y and v_z in the 3-dimensional Boltzmann distribution [Eq. (3.22d) with $E = \frac{1}{2} m_e (v_x^2 + v_y^2 + v_z^2)$]; the result, correctly normalized so that $\int F(v) dv = n$, is

$$F \simeq F_e = n \left(\frac{m_e}{2\pi k_B T} \right)^{1/2} e^{-[m_e v^2/(2 k_B T)]},$$ (22.41)

where T is the electron temperature.

Now, as we saw in Eq. (22.40b), ω_i is proportional to $F'(v = \omega_r/k)$ with a proportionality constant that is usually positive. Physically, this proportionality arises from the manner in which electrons surf on the waves. Those electrons moving slightly faster than the waves' phase velocity $V_{ph} = \omega_r/k$ (usually) lose energy to the waves on average, while those moving slightly slower (usually) extract energy from the waves on average. Therefore,

energy conservation in Landau damping

1. if there are more slightly slower particles than slightly faster [$F'(v = \omega_r/k) < 0$], then the particles on average gain energy from the waves and the waves are damped [$\omega_i < 0$];

2. if there are more slightly faster particles than slightly slower [$F'(v = \omega_r/k) > 0$], then the particles on average lose energy to the waves and the waves are amplified [$\omega_i > 0$]; and

3. the bigger the disparity between the number of slightly faster electrons and the number of slightly slower electrons [i.e., the bigger $|F'(\omega_r/k)|$], the larger will be the damping or growth of wave energy (i.e., the larger will be $|\omega_i|$).

Quantitatively, it turns out that if the waves' phase velocity ω_r/k is anywhere near the steepest point on the side of the electron velocity distribution (i.e., if ω_r/k is of order the electron thermal velocity $\sqrt{k_B T/m_e}$), then the waves will be strongly damped: $\omega_i \sim -\omega_r$. Since our dispersion relation (22.41) is valid only when the waves are weakly damped, we must restrict ourselves to waves with $\omega_r/k \gg \sqrt{k_B T/m_e}$ (a physically allowed regime) or to $\omega_r/k \ll \sqrt{k_B T/m_e}$ (a regime that does not occur in Langmuir waves; cf. Fig. 21.1).

Requiring, then, that $\omega_r/k \gg \sqrt{k_B T/m_e}$ and noting that the integral in Eq. (22.39) gets its dominant contribution from velocities $v \lesssim \sqrt{k_B T/m_e}$, we can expand $1/(v - \omega_r/k)$ in the integrand as a power series in vk/ω_r, obtaining

$$\int_P \frac{F' \, dv}{v - \omega_r/k} = -\int_{-\infty}^{\infty} dv \, F' \left[\frac{k}{\omega_r} + \frac{k^2 v}{\omega_r^2} + \frac{k^3 v^2}{\omega_r^3} + \frac{k^4 v^3}{\omega_r^4} + \cdots \right]$$

$$= \frac{nk^2}{\omega_r^2} + \frac{3n \langle v^2 \rangle k^4}{\omega_r^4} + \cdots$$

$$= \frac{nk^2}{\omega_r^2} \left(1 + \frac{3k_B T k^2}{m_e \omega_r^2} + \cdots \right)$$

$$\simeq \frac{nk^2}{\omega_r^2} \left(1 + 3k^2 \lambda_D^2 \frac{\omega_p^2}{\omega_r^2} \right). \tag{22.42}$$

Substituting Eqs. (22.41) and (22.42) into Eqs. (22.40a) and (22.40b), and noting that $\omega_r/k \gg \sqrt{k_B T/m_e} \equiv \omega_p \lambda_D$ implies $k\lambda_D \ll 1$ and $\omega_r \simeq \omega_p$, we obtain

$$\boxed{\omega_r = \omega_p(1 + 3k^2 \lambda_D^2)^{1/2},} \tag{22.43a}$$

$$\boxed{\omega_i = -\left(\frac{\pi}{8} \right)^{1/2} \frac{\omega_p}{k^3 \lambda_D^3} \exp \left(-\frac{1}{2k^2 \lambda_D^2} - \frac{3}{2} \right).} \tag{22.43b}$$

dispersion relation for Langmuir waves in an unmagnetized, thermalized plasma

The real part of this dispersion relation, $\omega_r = \omega_p \sqrt{1 + 3k^2 \lambda_D^2}$, reproduces the Bohm-Gross result (21.31) that we derived using the two-fluid theory in Sec. 21.4.3 and plotted in Fig. 21.1. The imaginary part reveals the damping of these Langmuir waves by surfing electrons—so-called *Landau damping*. The two-fluid theory could not predict this Landau damping, because the damping is a result of internal dynamics in the electrons' velocity space, of which that theory is oblivious.

Landau damping of Langmuir waves

Notice that, as the waves' wavelength is decreased (i.e., as k increases), the waves' phase velocity decreases toward the electron thermal velocity and the damping becomes stronger, as is expected from our discussion of the number of electrons that can surf on the waves. In the limit $k \to 1/\lambda_D$ (where our dispersion relation has broken down and so is only an order-of-magnitude guide), the dispersion relation predicts that $\omega_r/k \sim \sqrt{k_B T/m_e}$ and $\omega_i/\omega_r \sim 1/10$.

In the opposite regime of large wavelength $k\lambda_D \ll 1$ (where our dispersion relation should be quite accurate), the Landau damping is very weak—so weak that ω_i decreases to zero with increasing k faster than any power of k.

22.3.6 Ion-Acoustic Waves and Conditions for Their Landau Damping to Be Weak

As we saw in Sec. 21.4.3, ion-acoustic waves are the analog of ordinary sound waves in a fluid: they occur at low frequencies where the mean (fluid) electron velocity is very nearly locked to the mean (fluid) proton velocity, so the electric polarization is small; the restoring force is due to thermal pressure and not to the electrostatic field; and the inertia is provided by the heavy protons. It was asserted in Sec. 21.4.3 that to avoid these waves being strongly damped, the electron temperature must be much higher than the proton temperature: $T_e \gg T_p$. We can now understand this condition in terms of particle surfing and Landau damping.

requirement that $T_e \gg T_p$ for ion-acoustic waves

Suppose that the electrons and protons have Maxwellian velocity distributions but possibly with different temperatures. Because of their far greater inertia, the protons will have a far smaller mean thermal speed than the electrons, $\sqrt{k_B T_p/m_p} \ll \sqrt{k_B T_e/m_e}$, so the net 1-dimensional distribution function $F(v) = F_e(v) + (m_e/m_p)F_p(v)$ [Eq. (22.24)] that appears in the kinetic-theory dispersion relation has the form shown in Fig. 22.3. Now, if $T_e \sim T_p$, then the contributions of the electron pressure and proton pressure to the waves' restoring force will be comparable, and the waves' phase velocity will therefore be $\omega_r/k \sim \sqrt{k_B(T_e + T_p)/m_p} \sim \sqrt{k_B T_p/m_p} = v_{\mathrm{th},p}$, which is the thermal proton velocity and also is the speed at which the proton contribution to $F(v)$ has its steepest slope (see the left tick mark on the horizontal axis in Fig. 22.3); so $|F'(v = \omega_r/k)|$ is large. This means large numbers of protons can surf on the waves and a disparity develops between the number moving slightly slower than the waves (which extract energy from the waves) and the number moving slightly faster (which give energy to the waves). The result will be strong Landau damping by the protons.

ion-acoustic waves' strong Landau damping if electrons and ions are thermalized at same temperature

This strong Landau damping is avoided if $T_e \gg T_p$. Then the waves' phase velocity will be $\omega_r/k \sim \sqrt{k_B T_e/m_p}$, which is large compared to the proton thermal velocity $v_{\mathrm{th},p} = \sqrt{k_B T_p/m_p}$ and so is way out on the tail of the proton velocity distribution, where there are few protons that can surf and damp the waves; see the right tick mark on the horizontal axis in Fig. 22.3. Thus Landau damping by protons has been shut down by raising the electron temperature.

What about Landau damping by electrons? The phase velocity $\omega_r/k \sim \sqrt{k_B T_e/m_p}$ is small compared to the electron thermal velocity $v_{\mathrm{th},e} = \sqrt{k_B T_e/m_e}$, so the waves reside near the peak of the electron velocity distribution, where $F_e(v)$ is large, so many electrons can surf with the waves. But $F_e'(v)$ is small, so there are nearly equal numbers of faster and slower electrons, and the surfing produces little net Landau damping. Thus $T_e/T_p \gg 1$ leads to successful propagation of ion acoustic waves.

A detailed computation, based on our small-ω_i kinetic-theory dispersion relation (22.40) makes this physical argument quantitative. The details are carried out in

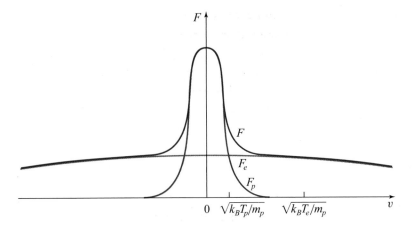

FIGURE 22.3 Electron and ion contributions to the net distribution function $F(v)$ in a thermalized plasma. When $T_e \sim T_p$, the phase speed of ion acoustic waves is near the left tick mark on the horizontal axis—a speed at which surfing protons have maximum ability to Landau-damp the waves, and the waves are strongly damped. When $T_e \gg T_p$, the phase speed is near the right tick mark—far out on the tail of the proton velocity distribution—so few protons can surf and damp the waves. The phase speed is near the peak of the electron distribution, so the number of electrons moving slightly slower than the waves is nearly the same as the number moving slightly faster, and there is little net damping by the electrons. In this case the waves can propagate.

Ex. 22.6 under the assumptions that $T_e \gg T_p$ and $\sqrt{k_B T_p/m_p} \ll \omega_r/k \ll \sqrt{k_B T_e/m_e}$ (corresponding to the above discussion). The result is

$$
\boxed{\frac{\omega_r}{k} = \sqrt{\frac{k_B T_e/m_p}{1 + k^2\lambda_D^2}},}
$$
(22.44a)

dispersion relation for ion-acoustic waves in an unmagnetized plasma with electrons and ions thermalized at different temperatures, $T_e \gg T_p$

$$
\boxed{\frac{\omega_i}{\omega_r} = -\frac{\sqrt{\pi/8}}{(1 + k^2\lambda_D^2)^{3/2}} \left[\sqrt{\frac{m_e}{m_p}} + \left(\frac{T_e}{T_p}\right)^{3/2} \exp\left(\frac{-T_e/T_p}{2(1 + k^2\lambda_D^2)} - \frac{3}{2}\right) \right].}
$$
(22.44b)

The real part of this dispersion relation was plotted in Fig. 21.1. As is shown there and in the above formulas, for $k\lambda_D \ll 1$ the waves' phase speed is $\sqrt{k_B T_e/m_p}$, and the waves are only weakly damped: they can propagate for roughly $\sqrt{m_p/m_e} \sim 43$ periods before damping has a strong effect. This damping is independent of the proton temperature, so it must be due to surfing electrons. When the wavelength is decreased (k is increased) into the regime $k\lambda_D \gtrsim 1$, the waves' frequency asymptotes toward $\omega_r = \omega_{pp}$, the proton plasma frequency. Then the phase velocity decreases, so more protons can surf the waves, and the Landau damping increases. Equations (22.44) show us that the damping becomes very strong when $k\lambda_D \sim \sqrt{T_e/T_p}$, and that this is also the point at which ω_r/k has decreased to the proton thermal velocity $\sqrt{k_B T_p/m_p}$—which is in accord with our physical arguments about proton surfing.

22.3 Electrostatic Waves in an Unmagnetized Plasma: Landau Damping   **1089**

When T_e/T_p is decreased from $\gg 1$ toward unity, the ion damping becomes strong regardless of how small may be k [cf. the second term of Eq. (22.44b)]. This is also in accord with our physical reasoning.

Ion-acoustic waves are readily excited at Earth's bow shock, where Earth's magnetosphere impacts the solar wind. It is observed that these waves are not able to propagate very far away from the shock, by contrast with Alfvén waves, which are much less rapidly damped.

Exercise 22.6 *Derivation: Ion-Acoustic Dispersion Relation*
Consider a plasma in which the electrons have a Maxwellian velocity distribution with temperature T_e, the protons are Maxwellian with temperature T_p, and $T_p \ll T_e$. Consider an ion acoustic mode in this plasma for which $\sqrt{k_B T_p/m_p} \ll \omega_r/k \ll \sqrt{k_B T_e/m_e}$ (i.e., the wave's phase velocity is between the left and right tick marks in Fig. 22.3). As was argued in the text, for such a mode it is reasonable to expect weak damping: $|\omega_i| \ll \omega_r$. Making approximations based on these "$\ll$" inequalities, show that the small-$|\omega_i|$ dispersion relation (22.40) reduces to Eqs. (22.44).

Exercise 22.7 *Problem: Dispersion Relations for a Non-Maxwellian Distribution Function*
Consider a plasma with cold protons and hot electrons with a 1-dimensional distribution function proportional to $1/(v_0^2 + v^2)$, so the full 1-dimensional distribution function is

$$F(v) = \frac{n v_0}{\pi (v_0^2 + v^2)} + n \frac{m_e}{m_p} \delta(v). \tag{22.45}$$

(a) Show that the dispersion relation for Langmuir waves, with phase speeds large compared to v_0, is

$$\omega \simeq \omega_{pe} - i k v_0. \tag{22.46}$$

(b) Compute the dispersion relation for ion-acoustic waves, assuming that their phase speeds are much less than v_0 but are large compared to the cold protons' thermal velocities (so the contribution from proton surfing can be ignored). Your result should be

$$\omega \simeq \frac{k v_0 (m_e/m_p)^{1/2}}{[1 + (k v_0/\omega_{pe})^2]^{1/2}} - \frac{i k v_0 (m_e/m_p)}{[1 + (k v_0/\omega_{pe})^2]^2}. \tag{22.47}$$

22.4

22.4 Stability of Electrostatic Waves in Unmagnetized Plasmas

Our small-ω_i dispersion relation (22.40) implies that the sign of F' at resonance dictates the sign of the imaginary part of ω. This raises the interesting possibility that distribution functions that increase with velocity over some range of positive v might be unstable to the exponential growth of electrostatic waves. In fact, the criterion for instability turns out to be a little more complex than this [as one might suspect from the fact that the sign of the denominator of Eq. (22.40b) is not obvious], and deriving it is an elegant exercise in complex-variable theory.

We carry out our derivation in two steps. We first introduce a general method, due to Harry Nyquist, for diagnosing instabilities of dynamical systems. Then we apply Nyquist's method explicitly to electrostatic waves in an unmagnetized plasma and thereby deduce an instability criterion due to Oliver Penrose.

22.4.1 Nyquist's Method

22.4.1

Consider any dynamical system whose modes of oscillation have complex eigen-freqencies ω that are zeros of some function $\mathcal{D}(\omega)$. [In our case the dynamical system is electrostatic waves in an unmagnetized plasma with some chosen wave number k, and because the waves' dispersion relation is $\epsilon(\omega, k) = 0$, Eq. (22.33), the function $\mathcal{D}$ can be chosen as $\mathcal{D}(\omega) = \epsilon(\omega, k)$.] Unstable modes are zeros of $\mathcal{D}(\omega)$ that lie in the upper half of the complex-ω plane.

Assume that $\mathcal{D}(\omega)$ is analytic in the upper half of the ω plane. Then a well-known theorem in complex-variable theory[3] says that the number N_z of zeros of $\mathcal{D}(\omega)$ in the upper half-plane, minus the number N_p of poles, is equal to the number of times that $\mathcal{D}(\omega)$ encircles the origin counterclockwise, in the complex-$\mathcal{D}$ plane, as ω travels counterclockwise along the closed path $\mathcal{C}$ that encloses the upper half of the frequency plane; see Fig. 22.4.

If one knows the number of poles of $\mathcal{D}(\omega)$ in the upper half of the frequency plane, then one can infer, from the Nyquist diagram, the number of unstable modes of the dynamical system.

In Sec. 22.4.2, we use this Nyquist method to derive the Penrose criterion for instability of electrostatic modes of an unmagnetized plasma. As a second example, in Box 22.2, we show how it can be used to diagnose the stability of a feedback control system.

Nyquist diagram and method for deducing the number of unstable modes of a dynamical system

22.4.2 Penrose's Instability Criterion

22.4.2

The straightforward way to apply Nyquist's method to electrostatic waves would be to set $\mathcal{D}(\omega) = \epsilon(\omega, k)$. However, to reach our desired instability criterion more quickly, we set $\mathcal{D} = k^2\epsilon$; then the zeros of $\mathcal{D}$ are still the electrostatic waves' modes. From Eq. (22.33) for ϵ, we see that

$$\mathcal{D}(\omega) = k^2 - Z(\omega/k), \qquad (22.48a)$$

---

3. This is variously called "the principle of the argument" or "Cauchy's theorem," and it follows from the theorem of residues (e.g., Copson, 1935, Sec. 6.2; Arfken, Weber, and Harris, 2013, Chap. 11).

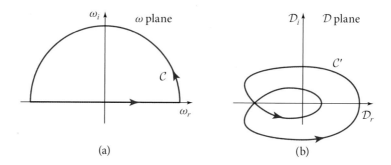

(a)　　　　　　　　　　　　　(b)

FIGURE 22.4 Nyquist diagram for stability of a dynamical system. (a) The curve C in the complex-ω plane extends along the real frequency axis from $\omega_r = -\infty$ to $\omega_r = +\infty$, then closes up along the semicircle at $|\omega| = \infty$, so it encloses the upper half of the frequency plane. (b) As ω travels along C, $\mathcal{D}(\omega)$ travels along the closed curve C' in the complex-$\mathcal{D}$ plane, which counterclockwise encircles the origin twice. Thus the number of zeros of the analytic function $\mathcal{D}(\omega)$ in the upper half of the frequency plane minus the number of poles is $N_z - N_p = 2$.

where

$$Z(\zeta) \equiv \frac{e^2}{m_e \epsilon_0} \int_{\mathcal{L}} \frac{F'}{v - \zeta} dv, \tag{22.48b}$$

with $\zeta = \omega/k$ the waves' phase velocity.

These equations have several important consequences.

1. For ζ in the upper half-plane—the region that concerns us—we can choose the Landau contour $\mathcal{L}$ to travel along the real v-axis from $-\infty$ to $+\infty$, and the resulting $\mathcal{D}(\omega)$ is analytic in the upper half of the frequency plane, as required for Nyquist's method.

2. For all ζ in the upper half-plane, and for all distribution functions $F(v)$ that are nonnegative and normalizable and thus physically acceptable, the velocity integral is finite, so there are no poles of $\mathcal{D}(\omega)$ in the upper half-plane. Thus there is an unstable mode, for fixed k, if and only if $\mathcal{D}(\omega)$ encircles the origin at least once, as ω travels around the curve C of Fig. 22.4. Note that the encircling is guaranteed to be counterclockwise, since there are no poles.

3. The wave frequency ω traveling along the curve C in the complex-frequency plane is equivalent to ζ traveling along the same curve in the complex-phase-velocity plane; and $\mathcal{D}$ encircling the origin of the complex-$\mathcal{D}$ plane is equivalent to $Z(\zeta)$ encircling the point $Z = k^2$ on the positive real axis of the complex-Z plane.

4. For every point on the semicircular segment of the curve C at $|\zeta| \to \infty$ (Fig. 22.4), $Z(\zeta)$ vanishes, so the curve C can be regarded as going just along the real axis from $-\infty$ to $+\infty$, during which $Z(\zeta)$ emerges from the origin, travels around some curve, and returns to the origin.

BOX 22.2.   STABILITY OF A FEEDBACK-CONTROL SYSTEM: ANALYSIS BY NYQUIST'S METHOD T2

A control system can be described quite generally by the following block diagram.

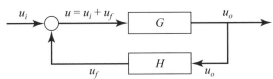

An input signal $u_i(t)$ and the feedback signal $u_f(t)$ are added, then fed through a filter G to produce an output signal $u_o(t) = \int_{-\infty}^{+\infty} G(t - t')u(t')dt'$; or, in the Fourier domain, $\tilde{u}_o(\omega) = \tilde{G}(\omega)\tilde{u}(\omega)$. [See Sec. 6.7 for filtering of signals. Here, for consistency with this plasma-physics chapter, we adopt the opposite sign convention for Fourier transforms from that in Sec. 6.7.] Then the output signal is fed through a filter H to produce the feedback signal $u_f(t)$.

As an example, consider the following simple model for an automobile's cruise control. The automobile's speed v is to be locked to some chosen value V by measuring v and applying a suitable feedback acceleration. To simplify the analysis, we focus on the difference $u \equiv v - V$, which is to be locked to zero. The input to the control system is the speed $u_i(t)$ that the automobile would have in the absence of feedback, plus the speed change $u_f(t)$ due to the feedback acceleration. Their sum, $u = u_i + u_f$, is measured by averaging over a short time interval, with the average exponentially weighted toward the present (in our simple model), so the output of the measurement is $u_o(t) = (1/\tau) \int_{-\infty}^{t} e^{(t'-t)/\tau} u(t')dt'$. By comparing with $u_o = \int_{-\infty}^{+\infty} G(t - t')u(t')dt'$ to infer the measurement filter's kernel $G(t)$, then Fourier transforming, we find that $\tilde{G}(\omega) \equiv \int_{-\infty}^{+\infty} G(t)e^{i\omega t}dt = 1/(1 - i\omega\tau)$. If $u_o(t)$ is positive, we apply to the automobile a negative feedback acceleration $a_f(t)$ proportional to it; if u_o is negative, we apply a positive feedback acceleration; so in either case, $a_f = -Ku_o$ for some positive constant K. The feedback speed u_f is the time integral of this acceleration: $u_f(t) = -K \int_{-\infty}^{t} u_o(t')dt'$. Setting this to $\int_{-\infty}^{\infty} H(t - t')u_o(t')dt'$, reading off the kernel H, and computing its Fourier transform, we find $\tilde{H}(\omega) = -K/(i\omega)$.

From the block diagram, we see, fully generally, that in the Fourier domain $\tilde{u}_o = \tilde{G}(\tilde{u}_i + \tilde{u}_f) = \tilde{G}(\tilde{u}_i + \tilde{H}\tilde{u}_o)$; so the output in terms of the input is

$$\tilde{u}_o = \frac{\tilde{G}}{1 + \tilde{G}\tilde{H}}\tilde{u}_i. \tag{1}$$

(continued)

BOX 22.2. (continued)

Evidently, the feedback system will undergo self-excited oscillations, with no input, at any complex frequency ω that is a zero of $\mathcal{D}(\omega) \equiv 1 + \tilde{G}(\omega)\tilde{H}(\omega)$. If that ω is in the lower half of the complex frequency plane, the oscillations will die out and so are not a problem; but if it is in the upper half-plane, they will grow exponentially with time. Thus the zeros of $\mathcal{D}(\omega)$ in the upper half of the ω plane represent unstable modes of self-excitation and must be avoided in the design of any feedback-control system.

For our cruise-control example, $\mathcal{D}$ is $1 + \tilde{G}\tilde{H} = 1 - K[i\omega(1 - i\omega\tau)]^{-1}$, which can be brought into a more convenient form by introducing the dimensionless frequency $z = \omega\tau$ and dimensionless feedback strength $\kappa = K\tau$: $\mathcal{D} = 1 - \kappa[iz(1 - iz)]^{-1}$. The Nyquist diagram for this $\mathcal{D}$ has the following form.

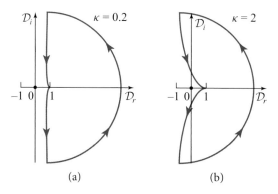

(a)   (b)

As $z = \omega\tau$ travels around the upper half of the frequency plane (curve $\mathcal{C}$ in Fig. 22.4a), $\mathcal{D}$ travels along the left curve (for feedback strength $\kappa = 0.2$), or the right curve (for $\kappa = 2$) in the above diagram. These curves do not encircle the origin at all—nor does the curve for any other $\kappa > 0$, so the number of zeros minus the number of poles in the upper half-plane is $N_z - N_p = 0$. Moreover, $\mathcal{D} = 1 - \kappa[iz(1 - iz)]^{-1}$ has no poles in the upper half-plane, so $N_p = 0$ and $N_z = 0$: our cruise-control feedback system is stable. For further details, see Ex. 22.10.

When designing control systems, it is important to have a significant margin of protection against instability. As an example, consider a control system for which $\tilde{G}\tilde{H} = -\kappa(1 + iz)[iz(1 - iz)]^{-1}$ (Ex. 22.11). The Nyquist diagrams take the common form shown here.

(continued)

BOX 22.2. (continued)

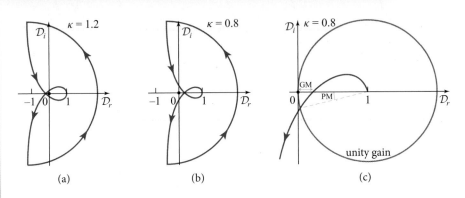

There are no poles in the upper half-plane; and for $\kappa > 1$ (drawing a) the origin is encircled twice, while for $\kappa < 1$ it is not encircled at all (drawing b). Therefore, the control system is unstable for $\kappa > 1$ and stable for $\kappa < 1$. One often wants to push κ as high as possible to achieve one's stabilization goals but must maintain a margin of safety against instability. That margin is quantified by either or both of two quantities: (i) The *phase margin* (labeled PM in diagram c): the amount by which the phase of $\tilde{G}\tilde{H} = \mathcal{D} - 1$ exceeds $180°$ at the unity gain point, $|\tilde{G}\tilde{H}| = 1$ (red curve). (ii) The *gain margin* GM: the amount by which the gain $|\tilde{G}\tilde{H}|$ is less than 1 when the phase of $\tilde{G}\tilde{H}$ reaches $180°$. As κ is increased toward the onset of instability, $\kappa = 1$, both PM and GM approach zero.

For the theory of control systems, see, for example, Franklin, Powell, and Emami-Naeini (2005); Dorf and Bishop (2012).

In view of these facts, Nyquist's method tells us the following: *there will be an unstable mode, for one or more values of k, if and only if, as ζ travels from $-\infty$ to $+\infty$, $Z(\zeta)$ encloses one or more points on the positive real Z-axis. In addition, the wave numbers for any resulting unstable modes are $k = \pm\sqrt{Z}$, for all Z on the positive real axis that are enclosed.*

In Fig. 22.5, we show three examples of $Z(\zeta)$ curves. For Fig. 22.5a, no points on the positive real axis are enclosed, so all electrostatic modes are stable for all wave numbers k. For Fig. 22.5b,c, a segment of the positive real axis is enclosed, so there are unstable modes; those unstable modes have $k = \pm\sqrt{Z}$ for all Z on the enclosed line segment.

Notice that in Fig. 22.5a,b, the rightmost crossing of the real axis is at positive Z, and the curve $\mathcal{C}'$ moves upward as it crosses. A little thought reveals that this must

Nyquist's method for diagnosing unstable electrostatic modes in an unmagnetized plasma

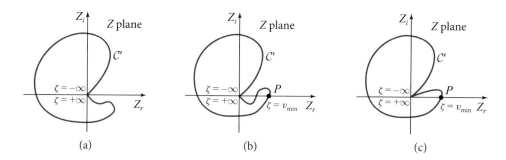

FIGURE 22.5 Nyquist diagrams for electrostatic waves. As a mode's real phase velocity ζ increases from $-\infty$ to $+\infty$, $Z(\zeta)$ travels, in the complex-Z plane, around the closed curve $\mathcal{C}'$, which always begins and ends at the origin. (a) The curve $\mathcal{C}'$ encloses no points on the positive real axis, so there are no unstable electrostatic modes. (b,c) The curve does enclose a set of points on the positive real axis, so there are unstable modes.

always be the case: $Z(\zeta)$ will encircle, counterclockwise, points on the positive-Z axis if and only if it somewhere crosses the positive-Z axis traveling upward.

alternative Nyquist criterion for instability

Therefore, *an unstable electrostatic mode exists in an unmagnetized plasma if and only if, as ζ travels along its real axis from $-\infty$ to $+\infty$, $Z(\zeta)$ crosses some point $\mathcal{P}$ on its positive real axis, traveling upward.* This version of the Nyquist criterion enables us to focus on the small-ω_i domain (while still treating the general case)—for which $\epsilon(\zeta, k)$ is given by Eq. (22.39). From that expression and real ζ (our case), we infer that $Z(\zeta) = k^2[1 - \epsilon]$ is given by

$$Z(\zeta) = \frac{e^2}{m_e \epsilon_0} \left[\int_{\mathrm{P}} \frac{F'\, dv}{v - \zeta} + i\pi F'(\zeta) \right], \tag{22.49a}$$

where P means the Cauchy principal value of the integral. This means that $Z(\zeta)$ crosses its real axis at any ζ where $F'(\zeta) = 0$, and it crosses moving upward if and only if $F''(\zeta) > 0$ at that crossing point. These two conditions together say that, at the crossing point $\mathcal{P}$, $\zeta = v_{\min}$, a particle speed at which $F(v)$ has a minimum. Moreover, Eq. (22.49a) says that $Z(\zeta)$ crosses its positive real axes (rather than negative) if and only if $\int_{\mathrm{P}}[F'/(v - v_{\min})]\, dv > 0$. We can evaluate this integral using an integration by parts:

$$\int_{\mathrm{P}} \frac{F'}{v - v_{\min}} dv = \int_{\mathrm{P}} \frac{d[F(v) - F(v_{\min})]/dv}{v - v_{\min}} dv$$

$$= \int_{\mathrm{P}} \frac{[F(v) - F(v_{\min})]}{(v - v_{\min})^2} dv + \lim_{\delta \to 0} \left[\frac{F(v_{\min} - \delta) - F(v_{\min})}{-\delta} \right.$$

$$\left. - \frac{F(v_{\min} + \delta) - F(v_{\min})}{\delta} \right]. \tag{22.49b}$$

The $\lim_{\delta \to 0}$ terms inside the square bracket vanish since $F'(v_{\min}) = 0$, and in the first $\int_{\mathrm{P}}$ term we do not need the Cauchy principal value, because F is a minimum at $v_{\min}$. Therefore, our requirement is that

$$\boxed{\int_{-\infty}^{+\infty} \frac{[F(v) - F(v_{\min})]}{(v - v_{\min})^2} dv > 0.} \qquad (22.50)$$

Penrose's criterion for instability of an electrostatic wave in an unmagnetized plasma

Thus, *a necessary and sufficient condition for an unstable mode is that there exist some velocity $v_{\min}$ at which the distribution function $F(v)$ has a minimum, and that in addition the minimum be deep enough that the integral (22.50) is positive.* This is called the *Penrose criterion* for instability (Penrose, 1960).

For a more in-depth, pedagogical derivation and discussion of the Penrose criterion, see, for example, Krall and Trivelpiece (1973, Sec. 9.6).

EXERCISES

Exercise 22.8 *Example: Penrose Criterion*

Consider an unmagnetized electron plasma with a 1-dimensional distribution function:

$$F(v) \propto [(v - v_0)^2 + u^2]^{-1} + [(v + v_0)^2 + u^2]^{-1}, \qquad (22.51)$$

where v_0 and u are constants. Show that this distribution function possesses a minimum if $v_0 > 3^{-1/2}u$, but the minimum is not deep enough to cause instability unless $v_0 > u$.

Exercise 22.9 *Problem: Range of Unstable Wave Numbers*

Consider a plasma with a distribution function $F(v)$ that has precisely two peaks, at $v = v_1$ and $v = v_2$ [with $F(v_2) \geq F(v_1)$], and a minimum between them at $v = v_{\min}$, and assume that the minimum is deep enough to satisfy the Penrose criterion for instability [Eq. (22.50)]. Show that there will be at least one unstable mode for every wave number k in the range $k_{\min} < k < k_{\max}$, where

$$k_{\min}^2 = \frac{e^2}{\epsilon_0 m_e} \int_{-\infty}^{+\infty} \frac{F(v) - F(v_1)}{(v - v_1)^2} dv, \quad k_{\max}^2 = \frac{e^2}{\epsilon_0 m_e} \int_{-\infty}^{+\infty} \frac{F(v) - F(v_{\min})}{(v - v_{\min})^2} dv.$$

$$(22.52)$$

Show, further, that the marginally unstable mode at $k = k_{\max}$ has phase velocity $\omega/k = v_{\min}$, and the marginally unstable mode at $k = k_{\min}$ has $\omega/k = v_1$. [Hint: Use a Nyquist diagram like those in Fig. 22.5.]

Exercise 22.10 *Example and Derivation: Cruise-Control System* **T2**

(a) Show that the cruise-control feedback system described at the beginning of Box 22.2 has $\tilde{G}(z) = 1/(1 - iz)$ and $\tilde{H} = -\kappa/(iz)$, with $z = \omega\tau$ and $\kappa = K\tau$, as claimed.

(b) Show that the Nyquist diagram has the forms shown in the second figure in Box 22.2, and that this control system is stable for all feedback strengths $\kappa > 0$.

(c) Solve explicitly for the zeros of $\mathcal{D} = 1 + \tilde{G}(z)\tilde{H}(z)$, and verify that none are in the upper half of the frequency plane.

(d) To understand the stability from a different viewpoint, imagine that the automobile's speed v is oscillating with an amplitude δv and a real frequency ω around the desired speed V, $v = V + \delta v \, \sin(\omega t)$, and that the feedback is turned off. Show that the output of the control system is $u_o = [\delta v / \sqrt{1 + \omega^2 \tau^2}] \, \sin(\omega t - \Delta\varphi)$, with a phase delay $\Delta\varphi = \arctan(\omega\tau)$ relative to the oscillations of v. Now turn on the feedback, but at a low strength, so it only weakly changes the speed's oscillations in one period. Show that, because $\Delta\varphi < \pi/2$, $d(\delta v^2)/dt$ is negative, so the feedback damps the oscillations. Show that an instability would arise if the phase delay were in the range $\pi/2 < |\Delta\varphi| < 3\pi/2$. For high-frequency oscillations, $\omega\tau \gg 1$, $\Delta\varphi$ approaches $\pi/2$, so the cruise-control system is only marginally stable.

Exercise 22.11 *Derivation: Phase Margin and Gain Margin for a Feedback-Control System* T2

Consider the control system discussed in the last two long paragraphs of Box 22.2. It has $\tilde{G}\tilde{H} = -\kappa(1 + iz)[iz(1 - iz)]^{-1}$, with $z = \omega\tau$ a dimensionless frequency and τ some time constant.

(a) Show that there are no poles of $\mathcal{D} = 1 + \tilde{G}\tilde{H}$ in the upper half of the complex frequency plane (z plane).

(b) Construct the Nyquist diagram for various feedback strengths κ. Show that for $\kappa > 1$ the curve encircles the origin twice (diagram a in the third figure in Box 22.2), so the control system is unstable, while for $\kappa < 1$, it does not encircle the origin (diagram b in the same figure), so the control system is stable.

(c) Show that the phase margin and gain margin, defined in diagram c in the third figure in Box 22.2, approach zero as κ increases toward the instability point, $\kappa = 1$.

(d) Compute explicitly the zeros of $\mathcal{D} = 1 + \tilde{G}\tilde{H}$, and plot their trajectories in the complex frequency plane as κ increases from zero through one to ∞. Verify that two zeros enter the upper half of the frequency plane as κ increases through one, and they remain in the upper half-plane for all $\kappa > 1$, as is guaranteed by the Nyquist diagrams.

22.5 Particle Trapping

We now return to the description of Landau damping. Our treatment so far has been essentially linear in the wave amplitude (or equivalently, in the perturbation to the distribution function). What happens when the wave amplitude is not infinitesimally small?

Consider a single Langmuir wave mode as observed in a frame moving with the mode's phase velocity and, for the moment, ignore its growth or damping. Then in this frame the electrostatic field oscillates spatially, $E = E_0 \sin kz$, but has no time

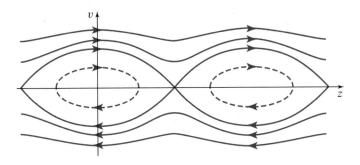

FIGURE 22.6 Phase-space orbits for trapped (dashed lines) and untrapped (solid lines) electrons.

dependence. Figure 22.6 shows some phase-space orbits of electrons in this oscillatory potential. The solid curves are orbits of particles that move fast enough to avoid being trapped in the potential wells at $kz = 0, 2\pi, 4\pi, \ldots$. The dashed curves are orbits of trapped particles. As seen in another frame, these trapped particles are surfing on the wave, with their velocities performing low-amplitude oscillations around the wave's phase velocity ω/k.

<aside>trapped and untrapped electrons in a Langmuir wave</aside>

The equation of motion for an electron trapped in the minimum $z = 0$ has the form

$$\ddot{z} = \frac{-eE_0 \sin kz}{m_e}$$

$$\simeq -\omega_b^2 z, \tag{22.53}$$

where we have assumed small-amplitude oscillations and approximated $\sin kz \simeq kz$, and where

$$\omega_b = \left(\frac{eE_0 k}{m_e}\right)^{1/2} \tag{22.54}$$

is known as the *bounce frequency*. Since the potential well is actually anharmonic, the trapped particles will mix in phase quite rapidly.

The growth or damping of the wave is characterized by a growth or damping of E_0, and correspondingly by a net acceleration or deceleration of untrapped particles, when averaged over a wave cycle. It is this net feeding of energy into and out of the untrapped particles that causes the wave's Landau damping or growth.

<aside>Landau damping associated with accelerating or decelerating untrapped particles</aside>

Now suppose that the amplitude E_0 of this particular wave mode is changing on a time scale τ due to interactions with the electrons, or possibly (as we shall see in Chap. 23) due to interactions with other waves propagating through the plasma. The potential well then changes on this same timescale, and we expect that τ is also a measure of the maximum length of time a particle can be trapped in the potential well. Evidently, nonlinear wave-trapping effects should only be important when the bounce period $\sim \omega_b^{-1}$ is short compared with τ, that is, when $E_0 \gg m_e/(ek\tau^2)$.

Electron trapping can cause particles to be bunched together at certain preferred phases of a growing wave. This can have important consequences for the radiative properties of the plasma. Suppose, for example, that the plasma is magnetized. Then the electrons gyrate around the magnetic field and emit cyclotron radiation. If their gyrational phases are random, then the total power that they radiate will be the sum of their individual particle powers. However, if N electrons are localized at the same gyrational phase due to being trapped in a potential well of a wave, then they will radiate like one giant electron with a charge $-Ne$. As the radiated power is proportional to the square of the charge carried by the radiating particle, the total power radiated by the bunched electrons will be N times the power radiated by the same number of unbunched electrons. Very large amplification factors are thereby possible both in the laboratory and in Nature, for example, in the Jovian magnetosphere.

radiation from bunched, trapped electrons: example of a nonlinear plasma effect

This brief discussion suggests that there may be much more richness in plasma waves than is embodied in our dispersion relations with their simple linear growth and decay, even when the wave amplitude is small enough that the particle motion is only slightly perturbed by its interaction with the wave. This motivates us to discuss more systematically nonlinear plasma physics, which is the topic of our next chapter.

EXERCISES

Exercise 22.12 *Challenge: BGK Waves*

Consider a steady, 1-dimensional, large-amplitude electrostatic wave in an unmagnetized, proton-electron plasma. Write down the Vlasov equation for each particle species in a frame moving with the wave [i.e., a frame in which the electrostatic potential is a time-independent function of z, $\Phi = \Phi(z)$, not necessarily precisely sinusoidal].

(a) Use Jeans' theorem to argue that proton and electron distribution functions that are just functions of the energy,

$$F_s = F_s(W_s), \qquad W_s = \frac{m_s v_s^2}{2} + q_s \Phi(z), \qquad (22.55a)$$

satisfy the Vlasov equation; here $s = p, e$, and as usual $q_p = e$, $q_e = -e$. Then show that Poisson's equation for the potential Φ can be rewritten in the form

$$\frac{1}{2}\left(\frac{d\Phi}{dz}\right)^2 + V(\Phi) = \text{const}, \qquad (22.55b)$$

where the potential V is $-2/\epsilon_0$ times the kinetic-energy density of the particles:

$$V = \frac{-2}{\epsilon_0} \sum_s \int \frac{1}{2} m_s v_s^2 F_s \, dv_s, \qquad (22.55c)$$

which depends on Φ. (Yes, it is weird to construct a potential V from kinetic energy, and a "kinetic energy" term from the electrical potential Φ, but it also is very clever.)

(b) It is possible to find many electrostatic potential profiles $\Phi(z)$ and distribution functions $F_s[W_s(v, \Phi)]$ that satisfy Eqs. (22.55). These are called BGK waves after

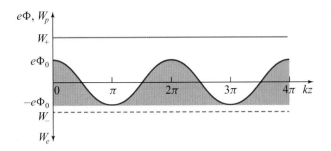

FIGURE 22.7 BGK waves as analyzed in Ex. 22.12. Two quantitites are plotted upward: $e\Phi$, where $\Phi(z)$ is the 1-dimensional electrostatic potential, and W_p, the proton total energy. Values $e\Phi_0$ and $-e\Phi_0$ of $e\Phi$, and W_+ of W_p are marked on the vertical axis. The electron total energy W_e is plotted downward, and its value W_- is marked on the axis. The shaded region shows the range of bound electron energies as a function of location z; the solid line W_+ and dashed line W_- are the energies of the unbound protons and electrons.

Bernstein, Greene, and Kruskal (1957), who first analyzed them. Explain how, in principle, one can solve for (nonunique) BGK distribution functions F_s in a large-amplitude wave of given electrostatic potential profile $\Phi(z)$.

(c) Carry out this procedure, assuming that the potential profile is of the form $\Phi(z) = \Phi_0 \cos kz$, with $\Phi_0 > 0$. Assume also that the protons are monoenergetic, with $W_p = W_+ > e\Phi_0$, and move along the positive z direction. In addition, assume that there are both monoenergetic (with $W_e = W_- > -e\Phi_0$), untrapped electrons (also moving along the positive z direction), and trapped electrons with distribution $F_e(W_e)$, $-e\Phi_0 \leq W_e < e\Phi_0$; see Fig. 22.7. Show that the density of trapped electrons must vanish at the wave troughs [at $z = (2n+1)\pi/k$; $n = 0, 1, 2, 3\ldots$]. Let the proton density at the troughs be n_{p0}, and assume that there is no net current. Show that the total electron density can then be written as

$$n_e(z) = \left[\frac{m_e(W_+ + e\Phi_0)}{m_p(W_- + e\Phi)}\right]^{1/2} n_{p0} + \int_{-e\Phi}^{e\Phi_0} \frac{dW_e F_e(W_e)}{[2m_e(W_e + e\Phi)]^{1/2}}. \tag{22.56}$$

(d) Use Poisson's equation to show that

$$\int_{-e\Phi}^{e\Phi_0} \frac{dW_e F_e(W_e)}{[2m_e(W_e + e\Phi)]^{1/2}}$$
$$= -\frac{\epsilon_0 k^2 \Phi}{e} + n_{p0}\left[\left(\frac{W_+ + e\Phi_0}{W_+ - e\Phi}\right)^{1/2} - \left(\frac{m_e(W_+ + e\Phi_0)}{m_p(W_- + e\Phi)}\right)^{1/2}\right]. \tag{22.57}$$

(e) Solve the integral equation in part (d) for $F_e(W_e)$. [Hint: It is of Abel type.]

(f) Exhibit some solutions of Eq. (22.57) graphically.

particle correlations are
missed by the Vlasov
formalism

Before turning to nonlinear phenomena in plasmas (the next chapter), let us digress briefly and explore ways to study correlations among particles, of which our Vlasov formalism is oblivious.

The Vlasov formalism treats each particle as independent, migrating through phase space in response to the local mean electromagnetic field and somehow un-affected by individual electrostatic interactions with neighboring particles. As we discussed in Chap. 20, this approximation is likely to be good in a collisionless plasma because of the influence of Debye screening—except in the tiny "independent-particle" region of Fig. 20.1. However, we would like to have some means of quantify-ing this and of deriving an improved formalism that takes into account the correlations among individual particles.

One environment where this formalism may be relevant is the interior of the Sun. Here, although the gas is fully ionized, the Debye number is not particularly large (i.e., one is near the independent-particle region of Fig. 20.1). As a result, Coulomb corrections to the perfect-gas equation of state may be responsible for measurable changes in the Sun's internal structure, as deduced, for example, using helioseismological analysis (cf. Sec. 16.2.4). In this application our task is simpler than in the general case, because the gas will locally be in thermodynamic equilibrium at some temperature T. It turns out that the general case, where the plasma departs significantly from thermal equilibrium, is extremely hard to treat.

The one-particle distribution function that we use in the Vlasov formalism is the first member of a hierarchy of k-particle distribution functions: $f^{(k)}(\mathbf{x}_1, \mathbf{x}_2, \ldots, \mathbf{x}_k, \mathbf{v}_1, \mathbf{v}_2, \ldots, \mathbf{v}_k, t)$. For example, $f^{(2)}(\mathbf{x}_1, \mathbf{x}_2, \mathbf{v}_1, \mathbf{v}_2, t)d\mathbf{x}_1 d\mathbf{v}_1 d\mathbf{x}_2 d\mathbf{v}_2 \equiv$ [the probability of finding a particle (any particle) in a volume $d\mathbf{x}_1 d\mathbf{v}_1 \equiv dx_1 dy_1 dz_1 dv_{x_1} dv_{y_1} dv_{z_1}$ of phase space, and another particle (any particle) in volume $d\mathbf{x}_2 d\mathbf{v}_2$ of phase space].[4] This definition and its obvious generalization dictate the normalization

$$\int f^{(k)} d\mathbf{x}_1 d\mathbf{v}_1 \cdots d\mathbf{x}_k d\mathbf{v}_k = N^k, \tag{22.58}$$

where $N \gg k \geq 1$ is the total number of particles.

N-particle distribution
function f_{all} in *N*-particle
plasma

It is useful to relate the distribution functions $f^{(k)}$ to the concepts of statistical mechanics, which we developed in Chap. 4. Suppose we have an ensemble of N-electron plasmas, and let the probability that a member of this ensemble is in a volume $d^N\mathbf{x}\, d^N\mathbf{v}$ of the $6N$-dimensional phase space of all its particles be $f_{\text{all}} d^N\mathbf{x}\, d^N\mathbf{v}$ (N is

---

4. In this section—and only in this section—we adopt the volume notation commonly used in this multi-particle subject: we use $d\mathbf{x}_j \equiv dx_j dy_j dz_j$ and $d\mathbf{v}_j \equiv dv_{x_j} dv_{y_j} dv_{z_j}$ to denote the volume elements for particle j. Warning: Despite the boldface notation, $d\mathbf{x}_j$ and $d\mathbf{v}_j$ are not vectors! Also, in this section we do *not* study waves, so k represents the number of particles in a distribution function, rather than a wave number.

a very large number!). Of course, f_{all} satisfies the Liouville equation

$$\frac{\partial f_{\text{all}}}{\partial t} + \sum_{i=1}^{N} \left[(\mathbf{v}_i \cdot \boldsymbol{\nabla}_i) f_{\text{all}} + (\mathbf{a}_i \cdot \boldsymbol{\nabla}_{\mathbf{v}i}) f_{\text{all}} \right] = 0, \tag{22.59}$$

Liouville equation for f_{all}

where $\mathbf{a}_i$ is the electromagnetic acceleration of the ith particle, and $\boldsymbol{\nabla}_i$ and $\boldsymbol{\nabla}_{\mathbf{v}i}$ are gradients with respect to the position and velocity of particle i, respectively. We can construct the k-particle "reduced" distribution function from the statistical-mechanics distribution function f_{all} by integrating over all but k of the particles:

$$f^{(k)}(\mathbf{x}_1, \mathbf{x}_2, \ldots, \mathbf{x}_k, \mathbf{v}_1, \mathbf{v}_2, \ldots, \mathbf{v}_k, t)$$
$$= N^k \int d\mathbf{x}_{k+1} \ldots d\mathbf{x}_N d\mathbf{v}_{k+1} \ldots d\mathbf{v}_N f_{\text{all}}(\mathbf{x}_1, \ldots, \mathbf{x}_N, \mathbf{v}_1, \ldots, \mathbf{v}_N). \tag{22.60}$$

k-particle distribution function constructed from f_{all}

(Note that k is typically a very small number, by contrast with N; below we shall only be concerned with $k = 1, 2, 3$.) The reason for the prefactor N^k in Eq. (22.60) is that, whereas f_{all} refers to the probability of finding particle 1 in $d\mathbf{x}_1 d\mathbf{v}_1$, particle 2 in $d\mathbf{x}_2 d\mathbf{v}_2$, and so forth, the reduced distribution function $f^{(k)}$ describes the probability of finding *any* of the N identical (though distinguishable) particles in $d\mathbf{x}_1 d\mathbf{v}_1$ and so on. (As long as we are dealing with nondegenerate plasmas we can count the electrons classically.) As $N \gg k$, the number of possible ways we can choose k particles for k locations in phase space is approximately N^k.

For simplicity, suppose that the protons are immobile and form a uniform, neutralizing background of charge, so we need only consider the electron distribution and its correlations. Let us further suppose that the forces associated with the mean electromagnetic fields produced by external charges and currents can be ignored. We can then restrict our attention to direct electron-electron electrostatic interactions. The acceleration of an electron is then

$$\mathbf{a}_i = \frac{e}{m_e} \sum_j \boldsymbol{\nabla}_i \Phi_{ij}, \tag{22.61}$$

where $\Phi_{ij}(x_{ij}) = -e/(4\pi \epsilon_0 x_{ij})$ is the electrostatic potential of electron i at the location of electron j, and $x_{ij} \equiv |\mathbf{x}_i - \mathbf{x}_j|$.

22.6.1 BBGKY Hierarchy T2

We can now derive the so-called BBGKY hierarchy of kinetic equations (Bogolyubov, 1962; Born and Green, 1949; Kirkwood, 1946; Yvon, 1935) which relate the k-particle distribution function to integrals over the $(k + 1)$-particle distribution function. The first equation in this hierarchy is given by integrating the Liouville equation (22.59) over $d\mathbf{x}_2 \ldots d\mathbf{x}_N d\mathbf{v}_2 \ldots d\mathbf{v}_N$. If we assume that the distribution function decreases to zero at large distances, then integrals of the type $\int d\mathbf{x}_i \mathbf{\nabla}_i f_{\text{all}}$ vanish, and the one-particle distribution function evolves according to

$$\frac{\partial f^{(1)}}{\partial t} + (\mathbf{v}_1 \cdot \mathbf{\nabla}_1) f^{(1)} = \frac{-eN}{m_e} \int d\mathbf{x}_2 \dots d\mathbf{x}_N d\mathbf{v}_2 \dots d\mathbf{v}_N \Sigma_j \mathbf{\nabla}_1 \Phi_{1j} \cdot \mathbf{\nabla}_{\mathbf{v}_1} f_{\text{all}}$$

$$= \frac{-eN^2}{m_e} \int d\mathbf{x}_2 \dots d\mathbf{x}_N d\mathbf{v}_2 \dots d\mathbf{v}_N \mathbf{\nabla}_1 \Phi_{12} \cdot \mathbf{\nabla}_{\mathbf{v}_1} f_{\text{all}}$$

$$= \frac{-e}{m_e} \int d\mathbf{x}_2 d\mathbf{v}_2 \left(\mathbf{\nabla}_{\mathbf{v}1} f^{(2)} \cdot \mathbf{\nabla}_1 \right) \Phi_{12}, \tag{22.62}$$

where, in the second line, we have replaced the probability of having any particle at a location in phase space by N times the probability of having one specific particle there. The left-hand side of Eq. (22.62) describes the evolution of independent particles, and the right-hand side takes account of their pairwise mutual correlation.

The evolution equation for $f^{(2)}$ can similarly be derived by integrating the Liouville equation (22.59) over $d\mathbf{x}_3 \dots d\mathbf{x}_N d\mathbf{v}_3 \dots d\mathbf{v}_N$:

$$\frac{\partial f^{(2)}}{\partial t} + (\mathbf{v}_1 \cdot \mathbf{\nabla}_1) f^{(2)} + (\mathbf{v}_2 \cdot \mathbf{\nabla}_2) f^{(2)} + \frac{e}{m_e} \left[\left(\mathbf{\nabla}_{\mathbf{v}1} f^{(2)} \cdot \mathbf{\nabla}_1 \right) \Phi_{12} + \left(\mathbf{\nabla}_{\mathbf{v}2} f^{(2)} \cdot \mathbf{\nabla}_2 \right) \Phi_{12} \right]$$

$$= \frac{-e}{m_e} \int d\mathbf{x}_3 d\mathbf{v}_3 \left[\left(\mathbf{\nabla}_{\mathbf{v}1} f^{(3)} \cdot \mathbf{\nabla}_1 \right) \Phi_{13} + \left(\mathbf{\nabla}_{\mathbf{v}2} f^{(3)} \cdot \mathbf{\nabla}_2 \right) \Phi_{23} \right]. \tag{22.63}$$

BBGKY hierarchy of evolution equations for k-particle distribution functions, derived by integrating the Liouville equation

Generalizing and allowing for the presence of a mean electromagnetic field (in addition to the inter-electron electrostatic field) causing an acceleration $\mathbf{a}^{\text{ext}} = -(e/m_e)(\mathbf{E} + \mathbf{v} \times \mathbf{B})$, we obtain the BBGKY hierarchy of kinetic equations:

$$\frac{\partial f^{(k)}}{\partial t} + \sum_{i=1}^{k} \left[(\mathbf{v}_i \cdot \mathbf{\nabla}_i) f^{(k)} + (\mathbf{a}_i^{\text{ext}} \cdot \mathbf{\nabla}_{\mathbf{v}i}) f^{(k)} + \frac{e}{m_e} (\mathbf{\nabla}_{\mathbf{v}_i} f^{(k)} \cdot \mathbf{\nabla}_i) \sum_{j \neq i}^{k} \Phi_{ij} \right]$$

$$= \frac{-e}{m_e} \int d\mathbf{x}_{k+1} d\mathbf{v}_{k+1} \sum_{i=1}^{k} \left(\mathbf{\nabla}_{\mathbf{v}_i} f^{(k+1)} \cdot \mathbf{\nabla}_i \right) \Phi_{ik+1}. \tag{22.64}$$

This kth equation in the hierarchy shows explicitly how we require knowledge of the $(k+1)$-particle distribution function to determine the evolution of the k-particle distribution function.

22.6.2 22.6.2 Two-Point Correlation Function T2

It is convenient to define the *two-point correlation function*, $\xi_{12}(\mathbf{x}_1, \mathbf{v}_1, \mathbf{x}_2, \mathbf{v}_2, t)$ for particles 1 and 2, by

two-point correlation function

$$\boxed{f^{(2)}(\mathbf{x}_1, \mathbf{v}_1, \mathbf{x}_2, \mathbf{v}_2, t) = f_1 f_2 (1 + \xi_{12}),} \tag{22.65}$$

where we introduce the notation $f_1 = f^{(1)}(\mathbf{x}_1, \mathbf{v}_1, t)$, and $f_2 = f^{(1)}(\mathbf{x}_2, \mathbf{v}_2, t)$. For the analysis that follows to be accurate, we require $|\xi_{12}| \ll 1$. We restrict attention to a plasma in thermal equilibrium at temperature T. In this case, f_1 and f_2 are

Maxwellian distribution functions, independent of $\mathbf{x}$ and t. Now, let us make an ansatz, namely, that ξ_{12} is just a function of the electrostatic interaction energy between the two electrons, and therefore it does not involve the electron velocities. (It is, actually, possible to justify this directly for an equilibrium distribution of particles interacting electrostatically, but we make do with showing that our final answer for ξ_{12} is just a function of $x_{12} = |\mathbf{x}_1 - \mathbf{x}_2|$, in accord with our ansatz.) As Debye screening should be effective at large distances, we anticipate that $\xi_{12} \to 0$ as $x_{12} \to \infty$.

Now turn to Eq. (22.62), and introduce the simplest imaginable closure relation: $\xi_{12} = 0$. In other words, completely ignore all correlations. We can then replace $f^{(2)}$ by $f_1 f_2$ and perform the integral over $\mathbf{x}_2$ and $\mathbf{v}_2$ to recover the collisionless Vlasov equation (22.6). Therefore, we see explicitly that particle-particle correlations are indeed ignored in the simple Vlasov approach.

For the three-particle distribution function, we expect that, when electron 1 is distant from both electrons 2 and 3, then $f^{(3)} \sim f_1 f_2 f_3 (1 + \xi_{23})$, and so forth. Summing over all three pairs, we write:

$$f^{(3)} = f_1 f_2 f_3 (1 + \xi_{23} + \xi_{31} + \xi_{12} + \chi_{123}), \qquad (22.66)$$

three-particle distribution function and three-point correlation function

where χ_{123} is the *three-point correlation function* that ought to be significant when all three particles are close together. The function χ_{123} is, of course, determined by the next equation in the BBGKY hierarchy.

We next make the closure relation $\chi_{123} = 0$, that is to say, we ignore the influence of third bodies on pair interactions. This is plausible because close, three-body encounters are even less frequent than close two-body encounters. We can now derive an equation for ξ_{12} by seeking a steady-state solution to Eq. (22.63), that is, a solution with $\partial f^{(2)}/\partial t = 0$. We substitute Eqs. (22.65) and (22.66) into Eq. (22.63) (with $\chi_{123} = 0$) to obtain

closure relation for computing two-point correlation function

$$f_1 f_2 \left[(\mathbf{v}_1 \cdot \mathbf{\nabla}_1)\xi_{12} + (\mathbf{v}_2 \cdot \mathbf{\nabla}_2)\xi_{12} - \frac{e(1 + \xi_{12})}{k_B T} \left\{ (\mathbf{v}_1 \cdot \mathbf{\nabla}_1)\Phi_{12} + (\mathbf{v}_2 \cdot \mathbf{\nabla}_2)\Phi_{12} \right\} \right]$$

$$= \frac{e f_1 f_2}{k_B T} \int d\mathbf{x}_3 d\mathbf{v}_3 f_3 \left(1 + \xi_{23} + \xi_{31} + \xi_{12} \right) \left[(\mathbf{v}_1 \cdot \mathbf{\nabla}_1)\Phi_{13} + (\mathbf{v}_2 \cdot \mathbf{\nabla}_2)\Phi_{23} \right],$$

$$(22.67)$$

where we have used the relation

$$\mathbf{\nabla}_{\mathbf{v}1} f_1 = -\frac{m_e \mathbf{v}_1 f_1}{k_B T}, \qquad (22.68)$$

valid for an unperturbed Maxwellian distribution function. We can rewrite this equation using the relations

$$\mathbf{\nabla}_1 \Phi_{12} = -\mathbf{\nabla}_2 \Phi_{12}, \quad \mathbf{\nabla}_1 \xi_{12} = -\mathbf{\nabla}_2 \xi_{12}, \quad \xi_{12} \ll 1, \quad \int d\mathbf{v}_3 f_3 = n, \qquad (22.69)$$

to obtain

$$(\mathbf{v}_1 - \mathbf{v}_2) \cdot \nabla_1 \left(\xi_{12} - \frac{e\Phi_{12}}{k_B T} \right)$$

$$= \frac{ne}{k_B T} \int d\mathbf{x}_3 (1 + \xi_{23} + \xi_{31} + \xi_{12}) [(\mathbf{v}_1 \cdot \nabla_1)\Phi_{13} + (\mathbf{v}_2 \cdot \nabla_2)\Phi_{23}]. \qquad (22.70)$$

Now, symmetry considerations tell us that

$$\int d\mathbf{x}_3 (1 + \xi_{31})\nabla_1 \Phi_{13} = 0, \qquad \int d\mathbf{x}_3 (1 + \xi_{23})\nabla_2 \Phi_{23} = 0, \qquad (22.71)$$

and, in addition,

$$\int d\mathbf{x}_3 \xi_{12}\nabla_1 \Phi_{13} = -\xi_{12} \int d\mathbf{x}_3 \nabla_3 \Phi_{13} = 0,$$

$$\int d\mathbf{x}_3 \xi_{12}\nabla_2 \Phi_{23} = -\xi_{12} \int d\mathbf{x}_3 \nabla_3 \Phi_{23} = 0. \qquad (22.72)$$

Therefore, we end up with

$$(\mathbf{v}_1 - \mathbf{v}_2) \cdot \nabla_1 \left(\xi_{12} - \frac{e\Phi_{12}}{k_B T} \right) = \frac{ne}{k_B T} \int d\mathbf{x}_3 [\xi_{23}(\mathbf{v}_1 \cdot \nabla_1)\Phi_{13} + \xi_{31}(\mathbf{v}_2 \cdot \nabla_2)\Phi_{23}].$$

$$(22.73)$$

As this equation must be true for arbitrary velocities, we can set $\mathbf{v}_2 = 0$ and obtain

$$\nabla_1 (k_B T \xi_{12} - e\Phi_{12}) = ne \int d\mathbf{x}_3 \xi_{23}\nabla_1 \Phi_{13}. \qquad (22.74)$$

We take the divergence of Eq. (22.74) and use Poisson's equation, $\nabla_1^2 \Phi_{12} = e\delta(\mathbf{x}_{12})/\epsilon_0$, to obtain

$$\nabla_1^2 \xi_{12} - \frac{\xi_{12}}{\lambda_D^2} = \frac{e^2}{\epsilon_0 k_B T} \delta(\mathbf{x}_{12}), \qquad (22.75)$$

where $\lambda_D = [k_B T \epsilon_0/(ne^2)]^{1/2}$ is the Debye length [Eq. (20.10)]. The solution of Eq. (22.75) is

two-point correlation
function for electrons
in an unmagnetized,
thermalized plasma

$$\boxed{\xi_{12} = \frac{-e^2}{4\pi\epsilon_0 k_B T} \frac{e^{-x_{12}/\lambda_D}}{x_{12}}.} \qquad (22.76)$$

Note that the sign is negative, because the electrons repel one another. Note also that, to order of magnitude, $\xi_{12}(x_{12} = \lambda_D) \sim -N_D^{-1}$, which is $\ll 1$ in magnitude if the Debye number N_D is much greater than unity. At the mean interparticle spacing, we have $\xi_{12}(x_{12} = n^{-1/3}) \sim -N_D^{-2/3}$. Only for distances $x_{12} \lesssim e^2/(\epsilon_0 k_B T)$ will the correlation effects become large and our expansion procedure and truncation ($\chi_{123} = 0$) become invalid. This analysis justifies the use of the Vlasov equation when $N_D \gg 1$; see the discussion at the end of Sec. 22.6.3.

Exercise 22.13 *Problem: Correlations in a Tokamak Plasma* T2

For a tokamak plasma, compute, to order of magnitude, the two-point correlation function for two electrons separated by

(a) a Debye length, and

(b) the mean interparticle spacing.

22.6.3 Coulomb Correction to Plasma Pressure T2

22.6.3

Let us now turn to the problem of computing the Coulomb correction to the pressure of a thermalized ionized gas. It is easiest to begin by computing the Coulomb correction to the internal energy density. Once again ignoring the protons, this is simply given by

$$U_c = \frac{-e}{2} \int d\mathbf{x}_1 n_1 n_2 \xi_{12} \Phi_{12}, \tag{22.77}$$

where the factor 1/2 compensates for double counting the interactions. Substituting Eq. (22.76) and performing the integral, we obtain

$$U_c = \frac{-ne^2}{8\pi\epsilon_0 \lambda_D} = -\frac{(e^2 n/\epsilon_0)^{3/2}}{8\pi(k_B T)^{1/2}}, \tag{22.78}$$

where n is the number density of electrons. The pressure can be obtained from this energy density using elementary thermodynamics. From the definition (5.32) of the physical free energy converted to a per-unit-volume basis and the first law of thermodynamics [Eq. (5.33)], the volume density of Coulomb free energy, $\mathcal{F}_c$, is given by integrating

$$U_c = -T^2 \left(\frac{\partial (\mathcal{F}_c/T)}{\partial T} \right)_n. \tag{22.79}$$

From this expression, we obtain $\mathcal{F}_c = \frac{2}{3} U_c$. The Coulomb contribution to the pressure is then given by $P = -\partial F/\partial V$ [Eq. (5.33)], rewritten as

$$P_c = n^2 \left(\frac{\partial (\mathcal{F}_c/n)}{\partial n} \right)_T = \frac{1}{3} U_c. \tag{22.80}$$

Therefore, including the Coulomb interaction decreases the pressure at a given density and temperature.

So far we have only allowed the electrons to move and have kept the protons fixed, and we found a Coulomb pressure and energy density independent of the mass of the electron. Therefore, if we had only allowed the protons to move, we would have gotten the same answer. If (as is the case in reality) we allow both protons and electrons to move, we must include the proton-electron attractions as well; but because U_c and P_c are proportional to e^2, these contribute with identical magnitude and sign to the electron-electron repulsions. Therefore the electrons and protons play completely equivalent roles, and the full electron-proton U_c and P_c are simply obtained by replacing n by the total density of particles, $2n$. Inserting $\lambda_D = \sqrt{\epsilon_0 k_B T/(ne^2)}$, the end result for the Coulomb correction to the pressure is

Coulomb correction to pressure in a thermalized plasma

$$P_c = \frac{-n^{3/2}e^3}{2^{3/2}3\pi \epsilon_0^{3/2}(k_B T)^{1/2}}, \tag{22.81}$$

where n is still the number density of electrons. Numerically, the gas pressure for a perfect electron-proton gas is

$$P = 1.6 \times 10^{13}(\rho/1{,}000\,\text{kg m}^{-3})(T/10^6\,\text{K})\,\text{N m}^{-2}, \tag{22.82}$$

and the Coulomb correction to this pressure is

$$P_c = -7.3 \times 10^{11}(\rho/1{,}000\,\text{kg m}^{-3})^{3/2}(T/10^6\,\text{K})^{-1/2}\,\text{N m}^{-2}. \tag{22.83}$$

In the interior of the Sun this is about 1% of the total pressure. In denser, cooler stars, it is significantly larger.

By contrast, for most of the plasmas that one encounters, our analysis implies that the order of magnitude of the two-point correlation function ξ_{12} is $\sim N_D^{-1}$ across a Debye sphere and only $\sim N_D^{-2/3}$ at the distance of the mean interparticle spacing (see end of Sec. 22.6.2). Only those particles that are undergoing large deflections, through angles ~ 1 radian, are close enough together for $\xi_{12} = O(1)$. This is the ultimate justification for treating plasmas as collisionless and for using mean electromagnetic fields in the Vlasov description.

EXERCISES

Exercise 22.14 *Derivation: Thermodynamic Identities* T2
Verify Eqs. (22.79) and (22.80).

Exercise 22.15 *Problem: Thermodynamics of a Coulomb Plasma* T2
Compute the entropy of a proton-electron plasma in thermal equilibrium at temperature T including the Coulomb correction.

Bibliographic Note

The kinetic theory of warm plasmas and its application to electrostatic waves and their stability are treated in nearly all texts on plasma physics. For maximum detail and good pedagogy, we particularly like Krall and Trivelpiece (1973, Chaps. 7, 8) (but beware of typographical errors). We also recommend, in Bellan (2006): the early parts of Chap. 2, and all of Chap. 5, and for the extension to magnetized plasmas, Chap. 8.

Also useful are Schmidt (1979, Chaps. 3, 7), Lifshitz and Pitaevskii (1981, Chap. 3), Stix (1992, Chaps. 8, 10), Boyd and Sanderson (2003, Chap. 8), and Swanson (2003, Chap. 4).

For brief discussions of the BBGKY hierarchy of N-particle distribution functions and their predicted correlations in a plasma, see Boyd and Sanderson (2003, Chap. 12) and Swanson (2003, Sec. 4.1.3). For detailed discussions, see the original literature cited in Sec. 22.6.1.

The theory of galaxy dynamics is very clearly developed in Binney and Tremaine (2003). The reader will find many parallels between warm plasmas and self-gravitating stellar distributions.

23

Nonlinear Dynamics of Plasmas

Plasma seems to have the kinds of properties one would like for life. It's somewhat like liquid water—unpredictable and thus able to behave in an enormously complex fashion. It could probably carry as much information as DNA does. It has at least the potential for organizing itself in interesting ways.

FREEMAN DYSON (1986)

23.1 Overview

In Sec. 21.6 we met our first example of a velocity-space instability: the two-stream instability, which illustrated the general principle that in a collisionless plasma, departures from Maxwellian equilibrium in velocity space can be unstable and lead to exponential growth of small-amplitude waves. This is similar to the Kelvin-Helmholtz instability in a fluid (Sec. 14.6.1), where a spatially varying fluid velocity produces an instability. In Chap. 22, we derived the dispersion relation for electrostatic waves in a warm, unmagnetized plasma. We discovered Landau damping of the waves when the phase-space density of the resonant particles diminishes with increasing speed, and we showed that in the opposite case of an increasing phase-space density, the waves can grow at the expense of the energies of near-resonant particles (a generalization of the two-stream instability). In this chapter, we explore the back-reaction of the waves on the near-resonant particles. This back-reaction is a (weakly) nonlinear process, so we have to extend our analysis of wave-particle interactions to include the leading nonlinearity.

This extension is called *quasilinear theory* or *the theory of weak plasma turbulence*, and it allows us to follow the time development of the waves and the near-resonant particles simultaneously. We develop this formalism in Sec. 23.2 and verify that it enforces the laws of particle conservation, energy conservation, and momentum conservation. Our initial development of the formalism is entirely in classical language and meshes nicely with the kinetic theory of electrostatic waves as presented in Chap. 22. In Sec. 23.3, we reformulate the theory in terms of the emission, absorption, and scattering of wave quanta, which are called *plasmons*. Although waves in plasmas almost always entail large quantum occupation numbers and thus are highly classical, this quantum formulation of the classical theory has great computational and heuristic power. And, as one might expect, despite the presence of Planck's reduced constant $\hbar$ in the formalism, $\hbar$ nowhere appears in the final answers to problems.

Our initial derivation and development of the formalism is restricted to the interaction of electrons with electrostatic waves, but in Sec. 23.3 we also describe how the formalism can be generalized to describe a multitude of wave modes and particle species interacting with one another.

We also describe circumstances in which this formalism can fail because the resonant particles couple strongly, *not* to a broad-band distribution of incoherent waves (as the formalism presumes) but instead to one or a few individual, coherent modes. (In Sec. 23.6, we explore an example.)

In Sec. 23.4, we give two illustrative applications of quasilinear theory. For a warm electron beam that propagates through a stable plasma, generating Langmuir plasmons, we show how the plasmons act back on the beam's particle distribution function so as to shut down the plasmon production. For our second application, we describe how galactic cosmic rays (relativistic charged particles) generate Alfvén waves, which in turn scatter the cosmic rays, isotropizing their distribution functions and confining them to our galaxy's interior much longer than one might expect.

In the remainder of this chapter, we describe a few of the many other nonlinear phenomena that can be important in plasmas. In Sec. 23.5, we consider parametric instabilities in plasmas (analogs of the optical parametric amplification that we met in nonlinear crystals in Sec. 10.7). These instabilities are important for the absorption of laser light in experimental studies of the compression of small deuterium-tritium pellets—a possible forerunner of a commercial nuclear fusion reactor. In Sec. 23.6, we return to ion-acoustic solitons (which we studied, for small amplitudes, in Ex. 21.6) and explore how they behave when their amplitudes are so large that nonlinearities are strong. We discover that dissipation can convert such a soliton into a collisionless shock, similar to the bow shock that forms where Earth's magnetic field meets the solar wind.

23.2 Quasilinear Theory in Classical Language

23.2.1 Classical Derivation of the Theory

In Chap. 22, we discovered that hot, thermalized electrons or ions can Landau damp a wave mode. We also showed that some nonthermal particle velocity distributions lead to exponential growth of the waves. Either way there is energy transfer between the waves and the particles. We now turn to the back-reaction of the waves on the near-resonant particles that damp or amplify them. For simplicity, we derive and explore the back-reaction equations ("quasilinear theory") in the special case of electrons interacting with electrostatic Langmuir waves. Then we assert the (rather obvious) generalization to protons or other ions and (less obviously) to interactions with other types of wave modes.

TWO-LENGTHSCALE EXPANSION

We begin with the electrons' 1-dimensional distribution function $F_e(v, z, t)$ [Eq. (22.17)]. As in Sec. 22.3.1, we split F_e into two parts, but we must do so more carefully here than we did there. The foundation for our split is a two-lengthscale expansion of the same sort as we used in developing geometric optics (Sec. 7.3). We introduce two disparate lengthscales, the short one being the typical reduced wavelength of a Langmuir wave $\lambda \sim 1/k$, and the long one being a scale $L \gg \lambda$ over which we perform spatial averages. Later, when applying our formalism to an inhomogeneous plasma, L must be somewhat shorter than the spatial inhomogeneity scale but still $\gg \lambda$. In our present situation of a homogeneous background plasma, there can still be large-scale spatial inhomogeneities caused by the growth or damping of the wave modes via their interaction with the electrons. We must choose L somewhat smaller than the growth or damping length but still large compared to λ.

Our split of F_e is into the spatial average of F_e over the length L (denoted F_0) plus a rapidly varying part that averages to zero (denoted F_1):

$$\boxed{F_0 \equiv \langle F_e \rangle, \quad F_1 \equiv F_e - F_0, \quad F_e = F_0 + F_1.} \tag{23.1}$$

derivation of quasilinear theory

two-lengthscale expansion of electron distribution function

[For simplicity, we omit the subscript e from F_0 and F_1. Therefore—by contrast with Chap. 22, where $F_0 = F_{e0} + (m_e/m_p)F_{p0}$—we here include only F_{e0} in F_0.]

The time evolution of F_e, and hence of F_0 and F_1, is governed by the 1-dimensional Vlasov equation, in which we assume a uniform neutralizing ion (proton) background, no background magnetic field, and interaction with 1-dimensional (planar) electrostatic waves. We cannot use the linearized Vlasov equation (22.19), which formed the foundation for all of Chap. 22, because the processes we wish to study are nonlinear. Rather, we must use the fully nonlinear Vlasov equation [Eq. (22.6) integrated over the irrelevant components v_x and v_y of velocity, as in Eq. (22.17)]:

$$\frac{\partial F_e}{\partial t} + v\frac{\partial F_e}{\partial z} - \frac{eE}{m_e}\frac{\partial F_e}{\partial v} = 0. \tag{23.2}$$

Here E is the rapidly varying electric field (pointing in the z direction) associated with the waves.

Inserting $F_e = F_0 + F_1$ into this Vlasov equation, and arranging the terms in an order that will make sense below, we obtain

$$\frac{\partial F_0}{\partial t} + v\frac{\partial F_0}{\partial z} - \frac{e}{m_e}\frac{\partial F_1}{\partial v}E + \frac{\partial F_1}{\partial t} + v\frac{\partial F_1}{\partial z} - \frac{e}{m_e}\frac{\partial F_0}{\partial v}E = 0. \tag{23.3}$$

We then split this equation into two parts: its average over the large lengthscale L and its remaining time-varying part.

The averaged part gets contributions only from the first three terms (since the last three are linear in F_1 and E, which have vanishing averages):

$$\frac{\partial F_0}{\partial t} + v\frac{\partial F_0}{\partial z} - \frac{e}{m_e}\left\langle\frac{\partial F_1}{\partial v}E\right\rangle = 0. \tag{23.4}$$

This is an evolution equation for the averaged distribution F_0; the third, nonlinear term drives the evolution. This driving term is the only nonlinearity that we keep in the quasilinear Vlasov equation.

RAPID EVOLUTION OF F_1

The rapidly varying part of the Vlasov equation (23.3) is just the last three terms of Eq. (23.3):

$$\frac{\partial F_1}{\partial t} + v\frac{\partial F_1}{\partial z} - \frac{e}{m_e}\frac{\partial F_0}{\partial v}E = 0, \tag{23.5}$$

plus a nonlinear term

$$-\frac{e}{m_e}\left(\frac{\partial F_1}{\partial v}E - \left\langle\frac{\partial F_1}{\partial v}E\right\rangle\right), \tag{23.6}$$

which we discard as being far smaller than the linear ones. If we were to keep this term, we would find that it can produce a "three-wave mixing" (analogous to that for light in a nonlinear crystal, Sec. 10.6), in which two electrostatic waves with different wave numbers k_1 and k_2 interact weakly to try to generate a third electrostatic wave with wave number $k_3 = k_1 \pm k_2$. We discuss such three-wave mixing in Sec. 23.3.6; for the moment we ignore it and correspondingly discard the nonlinearity (23.6).

Equation (23.5) is the same linear evolution equation for F_1 as we developed and studied in Chap. 22. Here, as there, we bring its physics to the fore by decomposing into monochromatic modes; but here we are dealing with many modes and sums over the effects of many modes, so we must do the decomposition a little more carefully

evolution of slowly varying part F_0 of electron distribution function

evolution of rapidly varying part F_1 of electron distribution function

three-wave mixing

than in Chap. 22. The foundation for the decomposition is a spatial Fourier transform inside our averaging "box" of length L:

$$\tilde{F}_1(v, k, t) = \int_0^L e^{-ikz} F_1(v, z, t) dz, \quad \tilde{E}(k, t) = \int_0^L e^{-ikz} E(z, t) dz. \quad (23.7)$$

We take F_1 and E to represent the physical quantities and thus to be real; this implies that $\tilde{F}_1(-k) = \tilde{F}_1^*(k)$ and similarly for $\tilde{E}$, so the inverse Fourier transforms are

$$F_1(v, z, t) = \int_{-\infty}^{\infty} e^{ikz} \tilde{F}_1(v, k, t) \frac{dk}{2\pi},$$
$$E(z, t) = \int_{-\infty}^{\infty} e^{ikz} \tilde{E}(v, k, t) \frac{dk}{2\pi}, \quad \text{for } 0 < z < L. \quad (23.8)$$

(This choice of how to do the mathematics corresponds to idealizing F_1 and E as vanishing outside the box; alternatively, we could treat them as though they were periodic with period L and replace Eq. (23.8) by a sum over discrete values of k—multiples of $2\pi/L$.)

From our study of linearized waves in Chap. 21, we know that a mode with real wave number k will oscillate in time with some complex frequency $\omega(k)$:

$$\tilde{F}_1 \propto e^{-i\omega(k)t}, \quad \tilde{E} \propto e^{-i\omega(k)t}. \quad (23.9)$$

For simplicity, we assume that the mode propagates in the $+z$ direction; when studying modes traveling in the opposite direction, we just turn our z-axis around. (In Sec. 23.2.4, we generalize to 3-dimensional situations and include all directions of propagation simultaneously.) For simplicity, we also assume that for each wave number k there is at most one mode type present (i.e., only a Langmuir wave or only an ion-acoustic wave). With these simplifications, $\omega(k)$ is a unique function with its real part positive ($\omega_r > 0$) when $k > 0$. Notice that the reality of $E(z, t)$ implies [from the second of Eqs. (23.7)] that $\tilde{E}(-k, t) = \tilde{E}^*(k, t)$ for all t; in other words [cf. Eqs. (23.9)], $\tilde{E}(-k, 0)e^{-i\omega(-k)t} = \tilde{E}^*(k, 0)e^{+i\omega^*(k)t}$ for all t, which in turn implies

$$\omega(-k) = -\omega^*(k); \quad \text{or} \quad \omega_r(-k) = -\omega_r(k), \quad \omega_i(-k) = \omega_i(k). \quad (23.10)$$

These equations should be obvious physically: they say that, for our chosen conventions, both the negative k and positive k contributions to Eq. (23.8) propagate in the $+z$ direction, and both grow or are damped in time at the same rate. In general, $\omega(k)$ is determined by the Landau-contour dispersion relation (22.33). However, throughout Secs. 23.2–23.4, we specialize to weakly damped or growing Langmuir waves with phase velocities ω_r/k large compared to the rms electron speed:

$$v_{\text{ph}} = \frac{\omega_r}{k} \gg v_{\text{rms}} = \sqrt{\frac{1}{n} \int v^2 F_0(v) dv}. \quad (23.11)$$

For such waves, from Eqs. (22.40a), (22.40b), and the first two lines of Eq. (22.42), we deduce the following explicit forms for the real and imaginary parts of ω:

dispersion relation for slowly evolving Langmuir waves

$$\omega_r^2 = \omega_p^2 \left(1 + 3\frac{v_{\text{rms}}^2}{(\omega_p/k)^2}\right), \quad \text{for } k > 0, \tag{23.12a}$$

$$\omega_i = \frac{\pi e^2}{2\epsilon_0 m_e}\frac{\omega_r}{k^2}F_0'(\omega_r/k), \quad \text{for } k > 0. \tag{23.12b}$$

The linearized Vlasov equation (23.5) implies that the modes' amplitudes $\tilde{F}_1(v, k, t)$ and $\tilde{E}(k, t)$ are related by

$$\tilde{F}_1 = \frac{ie}{m_e}\frac{\partial F_0/\partial v}{(\omega - kv)}\tilde{E}. \tag{23.13}$$

This is just Eq. (22.20) with d/dv replaced by $\partial/\partial v$, because F_0 now varies slowly in space and time as well as varying with v.

SPECTRAL ENERGY DENSITY $\mathcal{E}_K$

Now leave the rapidly varying quantities F_1 and E and their Vlasov equation (23.5), dispersion relation (23.12a), and damping rate (23.12b), and turn to the spatially averaged distribution function F_0 and its spatially averaged Vlasov equation (23.4). We shall bring this Vlasov equation's nonlinear term into a more useful form. The key quantity in this nonlinear term is the average of the product of the rapidly varying quantities F_1 and E. Parseval's theorem permits us to rewrite this as

$$\langle EF_1\rangle = \frac{1}{L}\int_0^L EF_1 dz = \frac{1}{L}\int_{-\infty}^{\infty} EF_1 dz = \int_{-\infty}^{\infty}\frac{dk}{2\pi}\frac{\tilde{E}^*\tilde{F}_1}{L}, \tag{23.14}$$

where in the second step we have used our mathematical idealization that F_1 and E vanish outside our averaging box, and the third equality is Parseval's theorem. Inserting Eq. (23.13), we bring Eq. (23.14) into the form

$$\langle EF_1\rangle = \frac{e}{m_e}\int_{-\infty}^{\infty}\frac{dk}{2\pi}\frac{\tilde{E}^*\tilde{E}}{L}\frac{i}{\omega - kv}\frac{\partial F_0}{\partial v}. \tag{23.15}$$

The quantity $\tilde{E}^*\tilde{E}/L$ is a function of wave number k, time t, and also the location and size L of the averaging box. For Eq. (23.15) to be physically and computationally useful, it is essential that this quantity not fluctuate wildly as k, t, L, and the box location are varied. In most circumstances, if the box is chosen to be far larger than $\lambda = 1/k$, then $\tilde{E}^*\tilde{E}/L$ indeed will not fluctuate wildly. When one develops the quasilinear theory with greater care and rigor than we can do in so brief a treatment, one discovers

random-phase approximation (RPA)

that this nonfluctuation is a consequence of the *random-phase approximation* or RPA for short—an approximation that the phase of $\tilde{E}$ varies randomly with k, t, L, and the box location on suitably short lengthscales.[1] Like ergodicity (Secs. 4.6 and 6.2.3), although the RPA is often valid, sometimes it can fail. Sometimes there is an organized bunching of the particles in phase space that induces nonrandom phases on the plasma waves. Quasilinear theory requires that RPA be valid, and for the moment we assume it is, but in Sec. 23.6 we meet an example for which it fails: strong ion-acoustic solitons.

The RPA implies that, as we increase the length L of our averaging box, $\tilde{E}^*\tilde{E}/L$ will approach a well-defined limit. This limit is half the spectral density $S_E(k)$ of the random process $E(z, t)$ at fixed time t [cf. Eq. (6.25)]. Correspondingly, it is natural to express quasilinear theory in the language of spectral densities. We shall do so, but with a normalization of the spectral density that is tied to the physical energy density and differs slightly from that used in Chap. 6: in place of $S_E(k)$, we use the *Langmuir-wave spectral energy density* $\mathcal{E}_k$. We follow plasma physicists' conventions by defining this quantity to include the oscillatory kinetic energy in the electrons, as well as the electrical energy to which it is, on average, equal. As in the theory of random processes (Chap. 6), we add the energy at $-k$ to that at $+k$, so that all the energy is regarded as residing at positive wave number, and $\int_0^\infty dk\, \mathcal{E}_k$ is the total wave energy per unit volume in the plasma, averaged over length L.

Invoking the RPA, we can use Parseval's theorem to compute the electrical energy density:

$$\frac{\epsilon_0 \langle E^2 \rangle}{2} = \int_{-\infty}^{\infty} \frac{dk}{2\pi} \epsilon_0 \left\langle \frac{\tilde{E}\tilde{E}^*}{2L} \right\rangle = \int_0^{\infty} \frac{dk}{2\pi} \epsilon_0 \left\langle \frac{\tilde{E}\tilde{E}^*}{L} \right\rangle, \tag{23.16}$$

where we have used $\tilde{E}(k)\tilde{E}^*(k) = \tilde{E}(-k)\tilde{E}^*(-k)$. We double this quantity to account for the wave energy in the oscillating electrons and then read off the spectral energy density as the integrand:

$$\boxed{\mathcal{E}_k = \frac{\epsilon_0 \langle \tilde{E}\tilde{E}^* \rangle}{\pi L}.} \tag{23.17}$$

spectral energy density of Langmuir waves

This wave energy density can be regarded as a function either of wave number k or wave phase velocity $v_{\rm ph} = \omega_r/k$. It is useful to plot $\mathcal{E}_k(v_{\rm ph})$ on the same graph as the averaged electron velocity distribution $F_0(v)$. Figure 23.1 is such a plot. It shows the physical situation we are considering: approximately thermalized electrons with (possibly) a weak beam (bump) of additional electrons at velocities $v \gg v_{\rm rms}$, and a distribution of Langmuir waves with phase velocities $v_{\rm ph} \gg v_{\rm rms}$.

EVOLUTION OF WAVE SPECTRAL ENERGY DENSITY $\mathcal{E}_K$

There is an implicit time dependence associated with the growth or decay of the waves, so $\mathcal{E}_k \propto e^{2\omega_i t}$. Now, the waves' energy density $\mathcal{E}_k$ travels through phase space (physical

---

1. For detailed discussions, see Davidson (1972) and Pines and Schrieffer (1962).

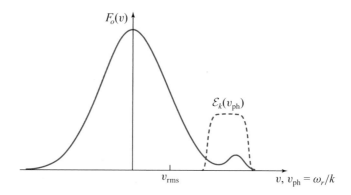

FIGURE 23.1 The spatially averaged electron velocity distribution $F_0(v)$ (solid curve) and wave spectral energy density $\mathcal{E}_k(v_{\text{ph}})$ (dashed curve) for the situation treated in Secs. 23.2–23.4.

space and wave-vector space) on the same trajectories as a wave packet (i.e., along the geometric-optics rays discussed in Sec. 7.3), which for our waves propagating in the z direction are given by

derivatives moving with Langmuir-wave rays (wave packets)

$$\left(\frac{dz}{dt}\right)_{\text{wp}} = \frac{\partial \omega_r}{\partial k_j}, \qquad \left(\frac{dk}{dt}\right)_{\text{wp}} = -\frac{\partial \omega_r}{\partial z} \tag{23.18}$$

[Eqs. (7.25)]. Correspondingly, the waves' growth or decay actually occurs along these rays (with the averaging boxes also having to be carried along the rays). Thus the equation of motion for the waves' energy density is

evolution of Langmuir waves' spectral energy density along a ray

$$\frac{d\mathcal{E}_k}{dt} \equiv \frac{\partial \mathcal{E}_k}{\partial t} + \left(\frac{dz}{dt}\right)_{\text{wp}} \frac{\partial \mathcal{E}_k}{\partial z} + \left(\frac{dk}{dt}\right)_{\text{wp}} \frac{\partial \mathcal{E}_k}{\partial k} = 2\omega_i \mathcal{E}_k. \tag{23.19}$$

Here we have used the fact that the electrostatic waves are presumed to propagate in the z direction, so the only nonzero component of $\mathbf{k}$ is $k_z \equiv k$, and the only nonzero component of the group velocity is $V_{gz} = (dz/dt)_{\text{wp}} = (\partial \omega_r/\partial k)_z$. For weakly damped, high-phase-speed Langmuir waves, $\omega_r(\mathbf{x}, k)$ is given by Eq. (23.12a), with the $\mathbf{x}$ dependence arising from the slowly spatially varying electron density $n(\mathbf{x})$, which induces a slow spatial variation in the plasma frequency: $\omega_p = \sqrt{ne^2/(\epsilon_0 m_e)}$.

SLOW EVOLUTION OF F_0

evolution of slowly varying part F_0 of electron distribution function

The context in which the quantity $\tilde{E}(k)^* \tilde{E}(k) = (\pi L/\epsilon_0)\mathcal{E}_k$ arose was our evaluation of the nonlinear term in the Vlasov equation (23.4) for the electrons' averaged distribution function F_0. By inserting Eqs. (23.15) and (23.17) into Eq. (23.4), we bring that nonlinear Vlasov equation into the form

$$\frac{\partial F_0}{\partial t} + v \frac{\partial F_0}{\partial z} = \frac{\partial}{\partial v}\left(D \frac{\partial F_0}{\partial v}\right), \tag{23.20}$$

where

$$D(v) = \frac{e^2}{2\epsilon_0 m_e^2} \int_{-\infty}^{\infty} dk\, \mathcal{E}_k \frac{i}{\omega - kv}$$

diffusion coefficient for electrons

$$= \frac{e^2}{\epsilon_0 m_e^2} \int_0^{\infty} dk\, \mathcal{E}_k \frac{\omega_i}{(\omega_r - kv)^2 + \omega_i^2}. \tag{23.21}$$

In the second equality we have used Eq. (23.10).

Equation (23.20) says that $F_0(v, z, t)$ is transported in physical space with the electron speed v and diffuses in velocity space with the *velocity diffusion coefficient* $D(v)$. Notice that $D(v)$ is manifestly real, and a major contribution to it comes from waves whose phase speeds ω_r/k nearly match the particle speed v (i.e., from resonant waves).

The two-lengthscale approximation that underlies quasilinear theory requires that the waves grow or damp on a lengthscale long compared to a wavelength, and equivalently, that $|\omega_i|$ be much smaller than ω_r. This allows us, for each v, to split the integral in Eq. (23.21) into a piece due to modes that can resonate with electrons of that speed because $\omega_r/k \simeq v$ plus a piece that cannot so resonate. We consider these two pieces in turn.

The resonant piece of the diffusion coefficient can be written, in the limit $|\omega_i| \ll \omega_r$, by approximating the resonance in Eq. (23.21) as a delta function:

$$\boxed{D^{\mathrm{res}} \simeq \frac{e^2 \pi}{\epsilon_0 m_e^2} \int_0^{\infty} dk\, \mathcal{E}_k \delta(\omega_r - kv).} \tag{23.22a}$$

diffusion coefficient for electrons that are resonant with the waves

In the diffusion equation (23.20) this influences $F_0(v)$ only at those velocities v of resonating waves with substantial wave energy (i.e., on the tail of the electron velocity distribution, under the dashed $\mathcal{E}_k$ curve of Fig. 23.1). We refer to electrons in this region as the *resonant electrons*. In Sec. 23.4, we explore the dynamical influence of this resonant diffusion coefficient on the velocity distribution $F_0(v)$ of the resonant electrons.

Nearly all the electrons reside at velocities $|v| \lesssim v_{\mathrm{rms}}$, where there are no waves (because waves there get damped very quickly). For these *nonresonant electrons*, the denominator in Eq. (23.21) for the diffusion coefficient is approximately equal to $\omega_r^2 \simeq \omega_p^2 = e^2 n/(\epsilon_0 m_e)$, and correspondingly, the diffusion coefficient has the form

$$\boxed{D^{\mathrm{non\text{-}res}} \simeq \frac{1}{nm_e} \int_0^{\infty} \omega_i \mathcal{E}_k dk,} \tag{23.22b}$$

diffusion coefficient for nonresonant electrons

which is independent of the electron velocity v.

The nonresonant electrons at $v \lesssim v_{\mathrm{rms}}$ are the ones that participate in the wave motions and account for the waves' oscillating charge density (via F_1). The time-averaged kinetic energy of these nonresonant electrons (embodied in F_0) thus must include a conserved part independent of the waves plus a part associated with the waves. The wave part must be equal to the waves' electrical energy and so to half the waves' total energy, $\frac{1}{2} \int_0^\infty dk \mathcal{E}_k$, and it thus must change at a rate $2\omega_i$ times that energy. Correspondingly, we expect the nonresonant piece of $D(v)$ to produce a rate change of the time-averaged electron energy (embodied in F_0) given by

nonresonant energy
conservation

$$\frac{\partial U_e}{\partial t} + \frac{\partial \mathcal{F}_{ez}}{\partial z} = \frac{1}{2} \int_0^\infty dk \, 2\omega_i \mathcal{E}_k, \tag{23.23}$$

where $\mathcal{F}_{ez}$ is the electron energy flux. Indeed, this is the case; see Ex. 23.1. Because we have already accounted for this electron contribution to the wave energy in our definition of $\mathcal{E}_k$, we ignore it henceforth in the evolution of $F_0(v)$, and correspondingly, for weakly damped or growing waves we focus solely on the resonant part of the diffusion coefficient, Eq. (23.22a).

EXERCISES

Exercise 23.1 *Problem: Nonresonant Particle Energy in Wave*
Show that the nonresonant part of the diffusion coefficient in velocity space, Eq. (23.22b), produces a rate of change of electron kinetic energy given by Eq. (23.23).

23.2.2

23.2.2 Summary of Quasilinear Theory

All the fundamental equations of quasilinear theory are now in hand. They are:

summary: fundamental
equations of quasilinear
theory for electrons
interacting with Langmuir
waves

1. The general dispersion relation (22.40) for the (real part of the) waves' frequency $\omega_r(k)$ and their growth rate $\omega_i(k)$ [which, for the high-speed Langmuir waves on which we are focusing, reduces to Eqs. (23.12)]; this dispersion relation depends on the electrons' slowly evolving, time-averaged velocity distribution $F_0(v, z, t)$.

2. The equation of motion (23.19) for the waves' slowly evolving spectral energy density $\mathcal{E}_k(k, z, t)$, in which appear $\omega_r(k)$ and $\omega_i(k)$.

3. Equations (23.21) or (23.22) for the diffusion coefficient $D(v)$ in terms of $\mathcal{E}_k$.

4. The diffusive evolution equation (23.20) for the slow evolution of $F_0(v, z, t)$.

The fundamental functions in this theory are $\mathcal{E}_k(k, z, t)$ for the waves and $F_0(v, z, t)$ for the electrons.

Quasilinear theory sweeps under the rug and ignores the details of the oscillating electric field $E(z, t)$ and the oscillating part of the distribution function $F_1(v, z, t)$. Those quantities were needed in deriving the quasilinear equations, but they are needed no longer—except, sometimes, as an aid to physical understanding.

It is instructive to verify that the quasilinear equations enforce the conservation of particles, momentum, and energy.

We begin with particles (electrons). The number density of electrons is $n = \int F_0 dv$, and the z component of particle flux is $S_z = \int nv dv$ (where F_1 contributes nothing because it is oscillatory, and here and below all velocity integrals go from $-\infty$ to $+\infty$). Therefore, by integrating the diffusive evolution equation (23.20) for F_0 over velocity, we obtain

$$\frac{\partial n}{\partial t} + \frac{\partial S_z}{\partial z} = \int \left(\frac{\partial}{\partial t} F_0 + v \frac{\partial}{\partial z} F_0 \right) dv = \int \frac{\partial}{\partial v} \left(D \frac{\partial F_0}{\partial v} \right) dv = 0, \quad (23.24)$$

particle conservation

which is the law of particle conservation for our 1-dimensional situation where nothing depends on x or y.

The z component of electron momentum density is $G_z^e \equiv \int m_e v F_0 dv$, and the zz component of electron momentum flux (stress) is $T_{zz}^e = \int m_e v^2 F_0 dv$; so evaluating the first moment of the evolution equation (23.20) for F_0, we obtain

$$\frac{\partial G_z^e}{\partial t} + \frac{\partial T_{zz}^e}{\partial z} = m_e \int v \left(\frac{\partial F_0}{\partial t} + v \frac{\partial F_0}{\partial z} \right) dv$$

$$= m_e \int v \frac{\partial}{\partial v} \left(D \frac{\partial F_0}{\partial v} \right) dv = -m_e \int D \frac{\partial F_0}{\partial v} dv, \quad (23.25)$$

where we have used integration by parts in the last step. The waves influence the momentum of the resonant electrons through the delta-function part of the diffusion coefficient D [Eq. (23.22a)], and the momentum of the nonresonant electrons through the nonresonant part of the diffusion coefficient (23.22b). Because we have included the evolving part of the nonresonant electrons' momentum and energy as part of the waves' momentum and energy, we must restrict attention in Eq. (23.25) to the resonant electrons (see last sentence in Sec. 23.2.1). We therefore insert the delta-function part of D [Eq. (23.22a)] into Eq. (23.25), thereby obtaining

$$\frac{\partial G_z^e}{\partial t} + \frac{\partial T_{zz}^e}{\partial z} = -\frac{\pi e^2}{\epsilon_0 m_e} \int dv \int dk \mathcal{E}_k \delta(\omega_r - kv) \frac{\partial F_0}{\partial v} = -\frac{\pi e^2}{\epsilon_0 m_e} \int \mathcal{E}_k F_0'(\omega_r/k) \frac{dk}{k}.$$

$$(23.26)$$

Here we have changed the order of integration, integrated out v, and set $F_0'(v) \equiv \partial F_0/\partial v$. Assuming, for definiteness, high-speed Langmuir waves, we can rewrite the last expression in terms of ω_i with the aid of Eq. (23.12b):

$$\frac{\partial G_z^e}{\partial t} + \frac{\partial T_{zz}^e}{\partial z} = -2 \int dk \omega_i \frac{\mathcal{E}_k k}{\omega_r} = -\frac{\partial}{\partial t} \int dk \frac{\mathcal{E}_k k}{\omega_r} - \frac{\partial}{\partial z} \int dk \frac{\mathcal{E}_k k}{\omega_r} \frac{\partial \omega_r}{\partial k}$$

conservation of total momentum in quasilinear theory

$$= -\frac{\partial G_z^w}{\partial t} - \frac{\partial T_{zz}^w}{\partial z}. \quad (23.27)$$

The second equality follows from Eq. (23.19) together with the fact that for high-speed Langmuir waves, $\omega_r \simeq \omega_p$ so $\partial\omega_r/\partial z \simeq \frac{1}{2}(\omega_p/n_e)\partial n_e/\partial z$ is independent of k. After the second equality in Eq. (23.19), $\int dk\, \mathcal{E}_k k/\omega_r = G_z^w$ is the waves' density of the z component of momentum [as one can see from the fact that each plasmon (wave quantum) carries a momentum $p_z = \hbar k$ and an energy $\hbar\omega_r$; cf. Secs. 7.3.2 and 23.3]. Similarly, since the waves' momentum and energy travel with the group velocity $d\omega_r/dk$, $\int dk\, \mathcal{E}_k(k/\omega_r)(\partial\omega_r/\partial k) = T_{zz}^w$ is the waves' flux of momentum. Obviously, Eq. (23.27) represents the conservation of total momentum, that of the resonant electrons plus that of the waves (which includes the evolving part of the nonresonant electron momentum).

conservation of total energy in quasilinear theory

Energy conservation can be handled in a similar fashion; see Ex. 23.2.

EXERCISES

Exercise 23.2 *Problem: Energy Conservation*
Show that the quasilinear evolution equations guarantee conservation of total energy —that of the resonant electrons plus that of the waves. Pattern your analysis after that for momentum, Eqs. (23.25)–(23.27).

23.2.4

23.2.4 Generalization to 3 Dimensions

quasilinear equations in 3 dimensions

So far, we have restricted our attention to Langmuir waves propagating in one direction, $+\mathbf{e}_z$. The generalization to 3 dimensions is straightforward. The waves' wave number k is replaced by the wave vector $\mathbf{k} = k\hat{\mathbf{k}}$, where $\hat{\mathbf{k}}$ is a unit vector in the direction of the phase velocity. The waves' spectral energy density becomes $\mathcal{E}_\mathbf{k}$, which depends on $\mathbf{k}$, varies slowly in space $\mathbf{x}$ and time t, and is related to the waves' total energy density by

waves' spectral energy density

$$U_w = \int \mathcal{E}_\mathbf{k}\, d\mathcal{V}_k, \tag{23.28}$$

where $d\mathcal{V}_k \equiv dk_x dk_y dk_z$ is the volume integral in wave-vector space.

Because the plasma is isotropic, the dispersion relation $\omega(k) = \omega(|\mathbf{k}|)$ has the same form as in the 1-dimensional case, and the group and phase velocities point in the same direction $\hat{\mathbf{k}}$: $\mathbf{V}_{\mathrm{ph}} = (\omega/k)\hat{\mathbf{k}}$, $\mathbf{V}_\mathrm{g} = (d\omega_r/dk)\hat{\mathbf{k}}$. The evolution equation (23.19) for the waves' spectral energy density moving along a ray (a wave-packet trajectory) becomes

evolution of waves' spectral energy density

$$\frac{d\mathcal{E}_k}{dt} \equiv \frac{\partial\mathcal{E}_k}{\partial t} + \left(\frac{dx_j}{dt}\right)_{\mathrm{wp}}\frac{\partial\mathcal{E}_j}{\partial x_j} + \left(\frac{dk_j}{dt}\right)_{\mathrm{wp}}\frac{\partial\mathcal{E}_\mathbf{k}}{\partial k_j} = 2\omega_i\mathcal{E}_\mathbf{k}, \tag{23.29a}$$

and the rays themselves are given by

rays

$$\left(\frac{dx_j}{dt}\right)_{\mathrm{wp}} = \frac{\partial\omega_r}{\partial k_j} = V_{\mathrm{g}\,j}, \quad \left(\frac{dk_j}{dt}\right)_{\mathrm{wp}} = -\frac{\partial\omega_r}{\partial x_j}. \tag{23.29b}$$

We shall confine ourselves to the resonant part of the diffusion coefficient, as the nonresonant part is only involved in coupling a wave's electrostatic part to its oscillating, nonresonant-electron part, and we already understand that issue [Eqs. (23.22b), (23.23), and associated discussion] and shall not return to it.

The 3-dimensional velocity diffusion coefficient acts only along the direction of the waves; that is, its $\hat{\mathbf{k}} \otimes \hat{\mathbf{k}}$ component has the same resonant form [Eq. (23.22a)] as in the 1-dimensional case, and components orthogonal to $\hat{\mathbf{k}}$ vanish, so

$$\mathbf{D} = \frac{\pi e^2}{\epsilon_0 m_e^2} \int \mathcal{E}_{\mathbf{k}}\, \hat{\mathbf{k}} \otimes \hat{\mathbf{k}}\, \delta(\omega_r - \mathbf{k} \cdot \mathbf{v})d\mathcal{V}_k.$$

(23.29c)

tensorial diffusion coefficient for resonant electrons

Because the waves will generally propagate in a variety of directions, the net $\mathbf{D}$ is not unidirectional. This diffusion coefficient enters into the obvious generalization of the Fokker-Planck-like evolution equation (23.20) for the averaged distribution function $f_0(\mathbf{v}, \mathbf{x}, t)$—in which we suppress the subscript 0 for ease of notation:

$$\frac{\partial f}{\partial t} + \mathbf{v} \cdot \nabla f = \nabla_v \cdot (\mathbf{D} \cdot \nabla_v f), \quad \text{or} \quad \frac{\partial f}{\partial t} + v_j \frac{\partial f}{\partial x_j} = \frac{\partial}{\partial v_i}\left(D_{ij}\frac{\partial f}{\partial v_j}\right).$$

evolution of electrons' averaged distribution function

(23.29d)

Here ∇_v (in index notation, $\partial/\partial v_j$) is the gradient in velocity space, not to be confused with the spatial derivative along a vector $\mathbf{v}$, $\nabla_\mathbf{v} \equiv \mathbf{v} \cdot \nabla = v_j \partial/\partial x_j$.

23.3 Quasilinear Theory in Quantum Mechanical Language

23.3

The attentive reader will have noticed a familiar structure to our quasilinear theory. It is reminiscent of the geometric-optics formalism that we introduced in Chap. 7. Here, as there, we can reinterpret the formalism in terms of quanta carried by the waves. At the most fundamental level, we could (second) quantize the field of plasma waves into quanta (usually called "plasmons"), and describe their creation and annihilation using quantum mechanical transition probabilities. However, there is no need to go through the rigors of the quantization procedure, since the basic concepts of creation, annihilation, and transition probabilities should already be familiar to most readers in the context of photons coupled to atomic systems. Those concepts can be carried over essentially unchanged to plasmons, and by doing so, we recover our quasilinear theory rewritten in quantum language.

plasmons: quantized plasma wave's particles

23.3.1 Plasmon Occupation Number η

23.3.1

A major role in the quantum theory is played by the *occupation number* for electrostatic wave modes, which are just the single-particle states of quantum theory. Our electrostatic waves have spin zero, since there is no polarization freedom (the direction of $\mathbf{E}$ is unique: it must point along $\mathbf{k}$). In other words, there is only one polarization state for each $\mathbf{k}$, so the number of modes (i.e., of quantum states) in a volume

$dV_x dV_k = dx\,dy\,dz\,dk_x\,dk_y\,dk_z$ of phase space is $dN_{\text{states}} = dV_x dV_k / (2\pi)^3$, and correspondingly, the number density of states in phase space is

$$dN_{\text{states}}/dV_x dV_k = 1/(2\pi)^3 \quad (23.30)$$

(cf. Sec. 3.2.5 with $\mathbf{p} = \hbar\mathbf{k}$). The density of energy in phase space is $\mathcal{E}_{\mathbf{k}}$, and the energy of an individual plasmon is $\hbar\omega_r$, so the number density of plasmons in phase space is

$$dN_{\text{plasmons}}/dV_x dV_k = \mathcal{E}_{\mathbf{k}}/\hbar\omega_r. \quad (23.31)$$

Therefore, the states' occupation number is given by

$$\boxed{\eta(\mathbf{k}, \mathbf{x}, t) = \frac{dN_{\text{plasmons}}/dV_x dV_k}{dN_{\text{states}}/dV_x dV_k} = \frac{dN_{\text{plasmons}}}{dV_x\,dV_k/(2\pi)^3} = \frac{(2\pi)^3 \mathcal{E}_{\mathbf{k}}}{\hbar\omega_r}.} \quad (23.32)$$

This is actually the mean occupation number; the occupation numbers of individual states will fluctuate statistically around this mean. In this chapter (as in most of our treatment of statistical physics in Part II of this book), we do not deal with the individual occupation numbers, since quasilinear theory is oblivious to them and deals only with the mean. Thus, without any danger of ambiguity, we simplify our terminology by suppressing the word "mean."

Equation (23.32) says that $\eta(\mathbf{k}, \mathbf{x}, t)$ and $\mathcal{E}_{\mathbf{k}}$ are the same quantity, aside from normalization. In the classical formulation of quasilinear theory we use $\mathcal{E}_{\mathbf{k}}$; in the equivalent quantum formulation we use η. We can think of η equally well as a function of the state's wave number $\mathbf{k}$ or of the momentum $\mathbf{p} = \hbar\mathbf{k}$ of the individual plasmons that reside in the state.

The third expression in Eq. (23.32) allows us to think of η as the number density in $\mathbf{x}$-$\mathbf{k}$ phase space, with the relevant phase-space volume renormalized from dV_k to $dV_k/(2\pi)^3 = (dk_x/2\pi)(dk_y/2\pi)(dk_z/2\pi)$. The factors of 2π appearing here are the same ones as appear in the relationship between a spatial Fourier transform and its inverse [e.g., Eq. (23.8)]. We shall see the quantity $dV_k/(2\pi)^3$ appearing over and over again in the quantum mechanical theory, and it can generally be traced to that Fourier transform relationship.

23.3.2 23.3.2 Evolution of η for Plasmons via Interaction with Electrons

CLASSICAL FORMULA FOR EVOLUTION OF η

The motion of individual plasmons is governed by Hamilton's equations with the hamiltonian determined by their dispersion relation:

$$H(\mathbf{p}, \mathbf{x}, t) = \hbar\omega_r(\mathbf{k} = \mathbf{p}/\hbar, \mathbf{x}, t); \quad (23.33)$$

see Sec. 7.3.2. The plasmon trajectories in phase space are, of course, identical to the wave-packet trajectories (the rays) of the classical wave theory, since

$$\frac{dx_j}{dt} = \frac{\partial H}{\partial p_j} = \frac{\partial \omega_r}{\partial k_j}, \qquad \frac{dp_j}{dt} = \hbar\frac{\partial k_j}{\partial t} = -\frac{\partial H}{\partial x_j} = -\hbar\frac{\partial \omega_r}{\partial x_j}. \quad (23.34)$$

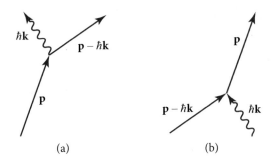

(a)   (b)

FIGURE 23.2 Feynman diagrams showing (a) creation and (b) annihilation of a plasmon (with momentum $\hbar\mathbf{k}$) by an electron (with momentum $\mathbf{p}$).

Interactions between the waves and resonant electrons cause $\mathcal{E}_{\mathbf{k}}$ to evolve as $d\mathcal{E}_{\mathbf{k}}/dt = 2\omega_i \mathcal{E}_{\mathbf{k}}$ [Eq. (23.29a)]. Therefore, the plasmon occupation number will also vary as

$$\frac{d\eta}{dt} \equiv \frac{\partial\eta}{\partial t} + \frac{dx_j}{dt}\frac{\partial\eta}{\partial x_j} + \frac{dp_j}{dt}\frac{\partial\eta}{\partial p_j} = 2\omega_i\eta, \qquad (23.35)$$

evolution of plasmon (mean) occupation number

where dx_j/dt and dp_j/dt are given by Hamilton's equations (23.34).

QUANTUM FORMULA FOR EVOLUTION OF η

The fundamental process that we are dealing with in Eq. (23.35) is the creation (or annihilation) of a plasmon by an electron (Fig. 23.2). The kinematics of this process is simple: energy and momentum must be conserved in the interaction. In plasmon creation (Fig. 23.2a), the plasma gains one quantum of energy $\hbar\omega_r$ so the electron must lose this same energy: $\hbar\omega_r = -\Delta(m_e v^2/2) \simeq -\Delta\mathbf{p}\cdot\mathbf{v}$, where $\Delta\mathbf{p}$ is the electron's change of momentum, and $\mathbf{v}$ is its velocity; by "$\simeq$" we assume that the electron's fractional change of energy is small (an assumption inherent in quasilinear theory). Since the plasmon momentum change, $\hbar\mathbf{k}$, is minus the electron momentum change, we conclude that

$$\hbar\omega_r = -\Delta(m_e v^2/2) \simeq -\Delta\mathbf{p}\cdot\mathbf{v} = \hbar\mathbf{k}\cdot\mathbf{v}. \qquad (23.36)$$

conservation of energy and momentum in the creation of plasmons: electron resonance and surfing

This is just the resonance condition contained in the delta function $\delta(\omega_r - \mathbf{k}\cdot\mathbf{v})$ of Eq. (23.29c). Thus, energy and momentum conservation in the fundamental plasmon creation process imply that the electron producing the plasmon must resonate with the plasmon's mode (i.e., the component of the electron's velocity along $\mathbf{k}$ must be the same as the mode's phase speed). In other words, the electron must "surf" with the wave mode in which it is creating the plasmon, always remaining in the same trough or crest of the mode.

A fundamental quantity in the quantum description is the probability per unit time for an electron with velocity $\mathbf{v}$ to *spontaneously* emit a Langmuir plasmon into a volume $\Delta\mathcal{V}_k$ in $\mathbf{k}$ space (i.e., the number of $\mathbf{k}$ plasmons emitted by a single velocity-$\mathbf{v}$

electron per unit time into the volume $\Delta \mathcal{V}_k$ centered on some wave vector $\mathbf{k}$). This probability is expressed in the following form:

spontaneous emission
of Langmuir plasmons:
fundamental probability W

$$\left(\frac{dN_{\text{plasmons}}}{dt} \right)_{\text{from one electron}} \equiv W(\mathbf{v}, \mathbf{k}) \frac{\Delta \mathcal{V}_k}{(2\pi)^3}. \tag{23.37}$$

In a volume $\Delta \mathcal{V}_x$ of physical space and $\Delta \mathcal{V}_v$ of electron-velocity space, there are $dN_e = f(\mathbf{v})\Delta \mathcal{V}_x \Delta \mathcal{V}_v$ electrons. Equation (23.37) tells us that these electrons increase the number of plasmons in $\Delta \mathcal{V}_x$, and with $\mathbf{k}$ in $\Delta \mathcal{V}_k$, by

$$\left(\frac{dN_{\text{plasmons}}}{dt} \right)_{\text{from all electrons in } \Delta \mathcal{V}_x \Delta \mathcal{V}_v} \equiv W(\mathbf{v}, \mathbf{k}) \frac{\Delta \mathcal{V}_k}{(2\pi)^3} f(\mathbf{v})\Delta \mathcal{V}_x \Delta \mathcal{V}_v. \tag{23.38}$$

Dividing by $\Delta \mathcal{V}_x$ and by $\Delta \mathcal{V}_k/(2\pi)^3$, and using the second expression for η in Eq. (23.32), we obtain for the rate of change of the plasmon occupation number produced by electrons with velocity $\mathbf{v}$ in $\Delta \mathcal{V}_v$:

$$\left(\frac{d\eta(\mathbf{k})}{dt} \right)_{\text{from all electrons in } \Delta \mathcal{V}_v} = W(\mathbf{v}, \mathbf{k}) f(\mathbf{v})\Delta \mathcal{V}_v. \tag{23.39}$$

Integrating over all of velocity space, we obtain a final expression for the influence of spontaneous plasmon emission, by electrons of all velocities, on the plasmon occupation number:

spontaneous plasmon
emission

$$\left(\frac{d\eta(\mathbf{k})}{dt} \right)_s = \int W(\mathbf{v}, \mathbf{k}) f(\mathbf{v}) d\mathcal{V}_v. \tag{23.40}$$

(Here and below the subscript s means "spontaneous.") Our introduction of the factor $(2\pi)^3$ in the definition (23.37) of $W(\mathbf{v}, \mathbf{k})$ was designed to avoid a factor $(2\pi)^3$ in this equation for the evolution of the occupation number.

Below [Eq. (23.45)] we deduce the fundamental emission rate W for high-speed Langmuir plasmons by comparing with our classical formulation of quasilinear theory.

Because the plasmons have spin zero, they obey Bose-Einstein statistics, which means that the rate for *induced* emission of plasmons is larger than that for spontaneous emission by the occupation number η of the state that receives the plasmons. Furthermore, the principle of detailed balance ("unitarity" in quantum mechanical language) tells us that W is also the relevant transition probability for the inverse process of absorption of a plasmon in a transition between the same two electron momentum states (Fig. 23.2b). This permits us to write down a *master equation* for the evolution of the plasmon occupation number in a homogeneous plasma:

master equation for
evolution of plasmon
occupation number

$$\frac{d\eta}{dt} = \int W(\mathbf{v}, \mathbf{k}) \left\{ f(\mathbf{v})[1 + \eta(\mathbf{k})] - f(\mathbf{v} - \hbar\mathbf{k}/m_e)\eta(\mathbf{k}) \right\} d\mathcal{V}_v. \tag{23.41}$$

The $f(\mathbf{v})$ term in the integrand is the contribution from spontaneous emission, the $f(\mathbf{v})\eta(\mathbf{k})$ term is that from induced emission, and the $f(\mathbf{v} - \hbar\mathbf{k}/m_e)\eta(\mathbf{k})$ term is from absorption.

COMPARISON OF CLASSICAL AND QUANTUM FORMULAS FOR $d\eta/dt$; THE FUNDAMENTAL QUANTUM EMISSION RATE W

The master equation (23.41) is actually the evolution law (23.35) for η in disguise, with the e-folding rate ω_i written in a fundamental quantum mechanical form. To make contact with Eq. (23.35), we first notice that in our classical development of quasilinear theory, for which $\eta \gg 1$, we neglected spontaneous emission, so we drop it from Eq. (23.41). In the absorption term, the momentum of the plasmon is so much smaller than the electron momentum that we can make a Taylor expansion:

$$f(\mathbf{v} - \hbar\mathbf{k}/m_e) \simeq f(\mathbf{v}) - (\hbar/m_e)(\mathbf{k} \cdot \nabla_v)f. \tag{23.42}$$

Inserting this into Eq. (23.41) and removing the spontaneous-emission term, we obtain

$$\boxed{\frac{d\eta}{dt} \simeq \eta \int W \frac{\hbar}{m_e}(\mathbf{k} \cdot \nabla_v)f\, d\mathcal{V}_v.} \tag{23.43}$$

For comparison, Eq. (23.35), with ω_i given by the classical high-speed Langmuir relation (23.12b) and converted to 3-dimensional notation, becomes

$$\frac{d\eta}{dt} = \eta \int \frac{\pi e^2 \omega_r}{\epsilon_0 k^2 m_e} \delta(\omega_r - \mathbf{k} \cdot \mathbf{v})\mathbf{k} \cdot \nabla_v f\, d\mathcal{V}_v. \tag{23.44}$$

Comparing Eqs. (23.43) and (23.44), we infer that the fundamental quantum emission rate for plasmons must be

$$\boxed{W = \frac{\pi e^2 \omega_r}{\epsilon_0 k^2 \hbar} \delta(\omega_r - \mathbf{k} \cdot \mathbf{v}).} \tag{23.45}$$

fundamental plasmon emission rate deduced from classical limit

Note that this emission rate is inversely proportional to $\hbar$ and is therefore a very large number under the classical conditions of our quasilinear theory.

This computation has shown that the classical absorption rate $-\omega_i$ is the difference between the quantum mechanical absorption rate and induced emission rate. Under normal conditions, when $\mathbf{k} \cdot \nabla_v f < 0$ ($\partial F_0/\partial v < 0$ in 1-dimensional language), the absorption dominates over emission, so the absorption rate $-\omega_i$ is positive, describing Landau damping. However (as we saw in Chap. 22), when this inequality is reversed, there can be wave growth [subject of course to there being a suitable mode into which the plasmons can be emitted, as guaranteed when the Penrose criterion (22.50) is fulfilled].

SPONTANEOUS EMISSION OF LANGMUIR PLASMONS AS CERENKOV RADIATION

Although spontaneous emission was absent from our classical development of quasi-linear theory, it nevertheless can be a classical process and therefore must be added

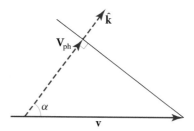

FIGURE 23.3 Geometry for Cerenkov emission. An electron moving with velocity $\mathbf{v}$ and speed $v = |\mathbf{v}|$ emits waves with phase speed $V_{ph} < v$ along a direction $\hat{\mathbf{k}}$ that makes an angle $\alpha = \cos^{-1}(V_{ph}/v)$ with the direction of the electron's motion. The slanted solid line is a phase front.

plasmon emission as Cerenkov radiation

to the quasilinear formalism. Classically or quantum mechanically, the spontaneous emission is a form of *Cerenkov radiation*, since (as for Cerenkov light emitted by electrons moving through a dielectric medium), the plasmons are produced when an electron moves through the plasma faster than the waves' phase velocity. More specifically, only when $v > V_{ph} = \omega_r/k$ can there be an angle α of $\mathbf{k}$ relative to $\mathbf{v}$ along which the resonance condition is satisfied: $\mathbf{v} \cdot \hat{\mathbf{k}} = v \cos\alpha = \omega_r/k$. The plasmons are emitted at this angle α to the electron's direction of motion (Fig. 23.3).

The spontaneous Cerenkov emission rate (23.40) takes the following form when we use the Langmuir expression (23.45) for W:

$$\left(\frac{d\eta}{dt}\right)_s = \frac{\pi e^2}{\epsilon_0 \hbar} \frac{\omega_r}{k^2} \int f(\mathbf{v})\delta(\omega_r - \mathbf{k} \cdot \mathbf{v})d\mathcal{V}_v. \tag{23.46}$$

Translated into classical language via Eq. (23.32), this Cerenkov emission rate is

$$\boxed{\left(\frac{d\mathcal{E}_\mathbf{k}}{dt}\right)_s = \frac{e^2}{8\pi^2\epsilon_0} \frac{\omega_r^2}{k^2} \int f(\mathbf{v})\delta(\omega_r - \mathbf{k} \cdot \mathbf{v})d\mathcal{V}_v.} \tag{23.47}$$

Note that Planck's constant is absent from the classical expression, but present in the quantum one.

In the above analysis, we computed the fundamental emission rate W by comparing the quantum induced emission rate minus absorption rate with the classical growth rate for plasma energy. An alternative route to Eq. (23.45) for W would have been to use classical plasma considerations to compute the classical Cerenkov emission rate (23.47), then convert to quantum language using $\eta = (2\pi)^3 \mathcal{E}_\mathbf{k}/\hbar\omega_r$, thereby obtaining Eq. (23.46), and then compare with the fundamental formula (23.40).

conditions for spontaneous Cerenkov emission to be ignorable

By comparing Eqs. (23.44) and (23.46) and assuming a thermal (Maxwellian) distribution for the electron velocities, we see that the spontaneous Cerenkov emission is ignorable in comparison with Landau damping when the electron temperature T_e is smaller than $\eta\hbar\omega_r/k_B$. Sometimes it is convenient to define a classical brightness temperature $T_B(\mathbf{k})$ for the plasma waves given implicitly by

$$\eta(\mathbf{k}) \equiv [e^{\hbar\omega_r/[k_B T_B(\mathbf{k})]} - 1]^{-1} \tag{23.48}$$

$$\simeq \frac{k_B T_B(\mathbf{k})}{\hbar\omega_r} \quad \text{when } \eta(\mathbf{k}) \gg 1, \text{ as in the classical regime.}$$

In this language, spontaneous emission of plasmons with wave vector $\mathbf{k}$ is generally ignorable when the wave brightness temperature exceeds the electron kinetic temperature—as one might expect on thermodynamic grounds. In a plasma in strict thermal equilibrium, Cerenkov emission is balanced by Landau damping, so as to maintain a thermal distribution of Langmuir waves with a temperature equal to that of the electrons: $T_B(\mathbf{k}) = T_e$ for all $\mathbf{k}$.

the balance of Cerenkov
emission by Landau
damping in thermal
equilibrium

EXERCISES

Exercise 23.3 *Problem: Cerenkov Power in Electrostatic Waves*
Show that the Langmuir wave power radiated by an electron moving with speed v in a plasma with plasma frequency ω_p is given by

$$P \simeq \frac{e^2\omega_p^2}{4\pi\epsilon_0 v} \ln\left(\frac{k_{\max}v}{\omega_p}\right), \tag{23.49}$$

where $k_{\max}$ is the largest wave number at which the waves can propagate. (For larger k the waves are strongly Landau damped.)

23.3.3 Evolution of f for Electrons via Interaction with Plasmons

23.3.3

Now turn from the evolution of the plasmon distribution $\eta(\mathbf{k}, \mathbf{x}, t)$ to that of the particle distribution $f(\mathbf{v}, \mathbf{x}, t)$. Classically, f evolves via the velocity-space diffusion equation (23.29d). We shall write down a fundamental quantum mechanical evolution equation (the "kinetic equation") that appears at first sight to differ remarkably from Eq. (23.29d), but then we will recover (23.29d) in the classical limit.

To derive the electron kinetic equation, we must consider three electron velocity states, $\mathbf{v}$ and $\mathbf{v} \pm \hbar\mathbf{k}/m_e$ (Fig. 23.4). Momentum conservation says that an electron can move between these states by emission or absorption of a plasmon with wave vector $\mathbf{k}$. The fundamental probability for these transitions is the same one, W, as for plasmon

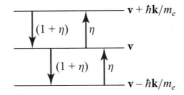

FIGURE 23.4 Three-level system for understanding the electron kinetic equation.

emission, since these transitions are merely plasmon emissions and absorptions as seen from the electron's viewpoint. Therefore, the electron kinetic equation must take the form

evolution of averaged electron distribution function due to plasmon emission and absorption— fundamental form of equation

$$\frac{df(\mathbf{v})}{dt} = \int \frac{d\mathcal{V}_k}{(2\pi)^3} \left\{ (1+\eta)[W(\mathbf{v}+\hbar\mathbf{k}/m_e, \mathbf{k})f(\mathbf{v}+\hbar\mathbf{k}/m_e) - W(\mathbf{v}, \mathbf{k})f(\mathbf{v})] \right.$$
$$\left. -\eta[W(\mathbf{v}+\hbar\mathbf{k}/m_e, \mathbf{k})f(\mathbf{v}) - W(\mathbf{v}, \mathbf{k})f(\mathbf{v}-\hbar\mathbf{k}/m_e)] \right\}. \qquad (23.50)$$

The four terms can be understood by inspection of Fig. 23.4. The two downward transitions in that diagram entail plasmon emission and thus are weighted by $(1+\eta)$, where the 1 is the spontaneous contribution and the η is the induced emission. In the first of these $(1+\eta)$ terms in Eq. (23.50), the $\mathbf{v}$ electron state gets augmented, so the sign is positive; in the second it gets depleted, so the sign is negative. The two upward transitions entail plasmon absorption and thus are weighted by η; the sign in Eq. (23.50) is plus when the final electron state has velocity $\mathbf{v}$, and minus when the initial state is $\mathbf{v}$.

In the domain of classical quasilinear theory, the momentum of each emitted or absorbed plasmon must be small compared to that of the electron, so we can expand the terms in Eq. (23.50) in powers of $\hbar\mathbf{k}/m_e$. Carrying out that expansion to second order and retaining those terms that are independent of $\hbar$ and therefore classical, we obtain (Ex. 23.4) the *quasilinear electron kinetic equation*:

evolution of averaged electron distribution in quasilinear domain— Fokker-Planck equation

$$\boxed{\frac{df}{dt} = \nabla_v \cdot [\mathbf{R}(\mathbf{v})f + \mathbf{D}(\mathbf{v}) \cdot \nabla_v f],} \qquad (23.51a)$$

where ∇_v is the gradient in velocity space (and not $\mathbf{v} \cdot \nabla$), and where

$$\mathbf{R}(\mathbf{v}) = \int \frac{d\mathcal{V}_k}{(2\pi)^3} \frac{W(\mathbf{v}, \mathbf{k})\hbar\mathbf{k}}{m_e}, \qquad (23.51b)$$

$$\mathbf{D}(\mathbf{v}) = \int \frac{d\mathcal{V}_k}{(2\pi)^3} \frac{\eta(\mathbf{k})W(\mathbf{v}, \mathbf{k})\hbar\mathbf{k} \otimes \hbar\mathbf{k}}{m_e^2}. \qquad (23.51c)$$

Note that $\mathbf{R}$ is a resistive coefficient associated with spontaneous emission, and $\mathbf{D}$ is a diffusive coefficient associated with plasmon absorption and induced emission in the limit that electron recoil due to spontaneous emission can be ignored. Following Eq. (6.106c), we can rewrite Eq. (23.51c) for $\mathbf{D}$ as

$$\mathbf{D} = \left\langle \frac{\Delta\mathbf{v} \otimes \Delta\mathbf{v}}{\Delta t} \right\rangle \qquad (23.51d)$$

with $\Delta\mathbf{v} = -\hbar\mathbf{k}/m_e$.

Our electron kinetic equation (23.51a) has the standard Fokker-Planck form (6.94) derived in Sec. 6.9.1 (generalized to three dimensions), except that the resistive co-efficient there is $A = -R + \partial D/\partial v$ and the diffusion coefficient there is $B = 2D$. Note that the diffusive term here has the form $(\partial/\partial v)(D\,\partial f/\partial v)$ discussed in Ex. 6.21 [Eq. (6.102)], as it should, because our formalism does not admit electron recoil (so the probability of a change in $\mathbf{v}$ due to plasmon absorption is the same as for the opposite change due to plasmon emission).

In our classical, quasilinear analysis, we ignored spontaneous emission and thus had no resistive term $\mathbf{R}$ in the evolution equation (23.29d) for f. We can recover that evolution equation and its associated $\mathbf{D}$ by dropping the resistive term from the quantum kinetic equation (23.51a) and inserting expression (23.45) for W into Eq. (23.51c) for the quantum $\mathbf{D}$. The results agree with the classical equations (23.29c) and (23.29d).

EXERCISES

Exercise 23.4 *Derivation: Electron Fokker-Planck Equation*
Fill in the missing details in the derivation of the electron Fokker-Planck equation (23.51a). It may be helpful to use index notation.

23.3.4 Emission of Plasmons by Particles in the Presence of a Magnetic Field

23.3.4

Let us return briefly to Cerenkov emission by an electron or ion. In the presence of a background magnetic field $\mathbf{B}$, the resonance condition for Cerenkov emission must be modified. Only the momentum parallel to the magnetic field need be conserved, not the total vectorial momentum. The unbalanced components of momentum perpendicular to $\mathbf{B}$ are compensated by a reaction force from $\mathbf{B}$ itself and thence by the steady currents that produce $\mathbf{B}$; correspondingly, the unbalanced perpendicular momentum ultimately does work on those currents. In this situation, one can show that the Cerenkov resonance condition $\omega_r - \mathbf{k} \cdot \mathbf{v} = 0$ is modified to

$$\omega_r - \mathbf{k}_\parallel \cdot \mathbf{v}_\parallel = 0, \tag{23.52}$$

where $\parallel$ means the component parallel to $\mathbf{B}$. If we allow for the electron gyrational motion as well, then some number N of gyrational (cyclotron) quanta can be fed into each emitted plasmon, so Eq. (23.52) gets modified to read

$$\omega_r - \mathbf{k}_\parallel \cdot \mathbf{v}_\parallel = N\omega_{ce}, \tag{23.53}$$

Cerenkov resonance condition in presence of a magnetic field

where N is an integer. For nonrelativistic electrons, the strongest resonance is for $N = 1$. We use this modified Cerenkov resonance condition in Sec. 23.4.1.

23.3.5 Relationship between Classical and Quantum Mechanical Formalisms

23.3.5

We have demonstrated how the structure of the classical quasilinear equations is mandated by quantum mechanics. In developing the quantum equations, we had to rely on one classical calculation, Eq. (23.44), which gave us the emission rate W. However, even this was not strictly necessary, since with significant additional effort we could have calculated the relevant quantum mechanical matrix elements and then computed W directly from Fermi's golden rule.[2] This has to be the case, because

---

2. See, for example, Cohen-Tannoudji, Diu, and Laloë (1977), and, in the plasma physics context, Melrose (2008, 2012).

quantum mechanics is the fundamental physical theory and thus must be applicable to plasma excitations just as it is applicable to atoms. Of course, if we are only interested in classical processes, as is usually the case in plasma physics, then we end up taking the limit $\hbar \to 0$ in all observable quantities, and the classical rate is all we need.

This raises an important point of principle. Should we perform our calculations classically or quantum mechanically? The best answer is to be pragmatic. Many calculations in nonlinear plasma theory are so long and arduous that we need all the help we can get to complete them. We therefore combine both classical and quantum considerations (confident that both must be correct throughout their overlapping domain of applicability), in whatever proportion minimizes our computational effort.

using quantum
calculations in the
classical domain

23.3.6

23.3.6 Evolution of η via Three-Wave Mixing

We have discussed plasmon emission and absorption both classically and quantum mechanically. Our classical and quantum formalisms can be generalized straightforwardly to encompass other nonlinear processes.

Among the most important other processes are three-wave interactions (in which two waves coalesce to form a third wave or one wave splits up into two) and scattering processes (in which waves are scattered off particles without creating or destroying plasmons). In this section we focus on three-wave mixing. We present the main ideas in the text but leave most of the details to Exs. 23.5 and 23.6.

three-wave mixing in
general

In three-wave mixing, where waves A and B combine to create wave C (Fig. 23.5a), the equation for the growth of the amplitude of wave C will contain nonlinear driving terms that combine the harmonic oscillations of waves A and B, that is, driving terms proportional to $\exp[i(\mathbf{k}_A \cdot \mathbf{x} - \omega_A t)]\exp[i(\mathbf{k}_B \cdot \mathbf{x} - \omega_B t)]$. For wave C to build up coherently over many oscillation periods, it is necessary that the spacetime dependence of these driving terms be the same as that of wave C, $\exp[i(\mathbf{k}_C \cdot \mathbf{x} - \omega_C t)]$; in other words, it is necessary that

$$\mathbf{k}_C = \mathbf{k}_A + \mathbf{k}_B, \qquad \omega_C = \omega_A + \omega_B. \tag{23.54}$$

Quantum mechanically, these expressions can be recognized as momentum and energy conservation for the waves' plasmons. We have met three-wave mixing previously, for electromagnetic waves in a nonlinear dielectric crystal (Sec. 10.6). There, as here, the conservation laws (23.54) were necessary for the mixing to proceed; see Eqs. (10.29) and associated discussion.

When the three waves are all electrostatic, the three-wave mixing arises from the nonlinear term (23.6) in the rapidly varying part [wave part, Eq. (23.5)] of the Vlasov equation, which we discarded in our quasilinear analysis. Generalized to 3 dimensions with this term treated as a driver, that Vlasov equation takes the form

three-wave mixing for
electrostatic waves:
driving term from
quasilinear analysis

$$\frac{\partial f_1}{\partial t} + \mathbf{v} \cdot \nabla f_1 - \frac{e}{m_e}\mathbf{E} \cdot \nabla_v f_0 = \frac{e}{m_e}\left(\mathbf{E} \cdot \nabla_v f_1 - \langle \mathbf{E} \cdot \nabla_v f_1 \rangle\right). \tag{23.55}$$

In the driving term (right-hand side of this equation), $\mathbf{E}$ could be the electric field of wave A and f_1 could be the perturbed velocity distribution of wave B or vice versa,

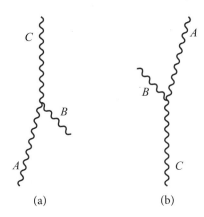

FIGURE 23.5 (a) A three-wave process in which two plasmons A and B interact nonlinearly to create a third plasmon C. Conserving energy and linear momentum, we obtain $\omega_C = \omega_A + \omega_B$ and $\mathbf{k}_C = \mathbf{k}_A + \mathbf{k}_B$. For example, A and C might be transverse electromagnetic plasmons satisfying the dispersion relation (21.24), and B might be a longitudinal plasmon (Langmuir or ion acoustic); or A and C might be Langmuir plasmons, and B might be an ion-acoustic plasmon—the case treated in the text and in Exs. 23.5 and 23.6. (b) The time-reversed three-wave process in which plasmon C generates plasmon B by an analog of Cerenkov emission, and while doing so recoils into plasmon state A.

and the $\mathbf{E}$ and f_1 terms on the left side could be those of wave C. If the wave vectors and frequencies are related by Eq. (23.54), then via this equation waves A and B will coherently generate wave C.

The dispersion relations for Langmuir and ion-acoustic waves permit the conservation law (23.54) to be satisfied if A is Langmuir (so $\omega_A \sim \omega_{pe}$), B is ion acoustic (so $\omega_B \lesssim \omega_{pp} \ll \omega_A$), and C is Langmuir. By working out the detailed consequences of the driving term (23.55) in the quasilinear formalism and comparing with the quantum equations for three-wave mixing (Ex. 23.5), one can deduce the fundamental rate for the process $A + B \to C$ [Fig. 23.5a; Eqs. (23.56)]. Detailed balance (unitarity) guarantees that the time-reversed process $C \to A + B$ (Fig. 23.5b) will have identically the same fundamental rate. This time-reversed process has a physical interpretation analogous to the emission of a Cerenkov plasmon by a high-speed, resonant electron: C is a "high-energy" Langmuir plasmon ($\omega_C \sim \omega_{pe}$) that can be thought of as Cerenkov-emitting a "low-energy" ion-acoustic plasmon ($\omega_B \lesssim \omega_{pp} \ll \omega_C$) and in the process recoiling slightly into Langmuir state A.

electrostatic three-wave mixing—two Langmuir waves (A and C) and one ion-acoustic wave (B)

The fundamental rate that one obtains for this wave-wave Cerenkov process $A + B \to C$ and its time reversal $C \to A + B$, when the plasma's electrons are thermalized at temperature T_e, is (Ex. 23.5)[3]

$$\boxed{W_{AB \leftrightarrow C} = R_{AB \leftrightarrow C}(\mathbf{k}_A, \mathbf{k}_B, \mathbf{k}_C)\delta(\mathbf{k}_A + \mathbf{k}_B - \mathbf{k}_C)\delta(\omega_A + \omega_B - \omega_C),} \quad (23.56a)$$

fundamental rate for Cerenkov emission of ion-acoustic plasmon by Langmuir plasmon

3. See also Tsytovich (1970, Eq. A.3.12). The rates for many other wave-wave mixing processes are worked out in this book, but beware: it contains a large number of typographical errors.

where

$$R_{AB \leftrightarrow C}(\mathbf{k}_A, \mathbf{k}_B, \mathbf{k}_C) = \frac{8\pi^5 \hbar e^2 (m_p/m_e) \omega_B^3}{(k_B T_e)^2 k_{\text{ia}}^2} (\hat{\mathbf{k}}_A \cdot \hat{\mathbf{k}}_C)^2. \qquad (23.56b)$$

[Here we have written the ion-acoustic plasmon's wave number as k_{ia} instead of k_B, to avoid confusion with Boltzmann's constant; i.e., $k_{\text{ia}} \equiv |\mathbf{k}_B|$.]

This is the analog of the rate (23.45) for Cerenkov emission by an electron. The ion-acoustic occupation number will evolve via an evolution law analogous to Eq. (23.41) with this rate replacing W on the right-hand side, η replaced by the ion-acoustic occupation number η_B, and the electron distribution replaced by the A-mode or C-mode Langmuir occupation number; see Ex. 23.5. Moreover, there will be a similar evolution law for the Langmuir occupation number, involving the same fundamental rate (23.56); Ex. 23.6.

EXERCISES

Exercise 23.5 *Example and Challenge: Three-Wave Mixing—Ion-Acoustic Evolution*

Consider the three-wave processes shown in Fig. 23.5, with A and C Langmuir plasmons and B an ion-acoustic plasmon. The fundamental rate is given by Eqs. (23.56a) and (23.56b).

(a) By summing the rates of forward and backward reactions (Fig. 23.5a,b), show that the occupation number for the ion-acoustic plasmons satisfies the kinetic equation

$$\frac{d\eta_B}{dt} = \int W_{AB \leftrightarrow C}[(1 + \eta_A + \eta_B)\eta_C - \eta_A \eta_B] \frac{d\mathcal{V}_{k_A}}{(2\pi)^3} \frac{d\mathcal{V}_{k_C}}{(2\pi)^3}. \qquad (23.57)$$

[Hint: (i) The rate for $A + B \rightarrow C$ (Fig. 23.5a) will be proportional to $(\eta_C + 1)\eta_A \eta_B$; why? (ii) When you sum the rates for the two diagrams, Fig. 23.5a and b, the terms involving $\eta_A \eta_B \eta_C$ should cancel.]

(b) The ion-acoustic plasmons have far lower frequencies than the Langmuir plasmons, so $\omega_B \ll \omega_A \simeq \omega_C$. Assume that they also have far lower wave numbers, $|\mathbf{k}_B| \ll |\mathbf{k}_A| \simeq |\mathbf{k}_C|$. Assume further (as will typically be the case) that the ion-acoustic plasmons, because of their tiny individual energies, have far larger occupation numbers than the Langmuir plasmons, so $\eta_B \gg \eta_A \sim \eta_C$. Using these approximations, show that the evolution law (23.57) for the ion-acoustic waves reduces to the form

$$\frac{d\eta_B(\mathbf{k})}{dt}$$

$$= \eta_B(\mathbf{k}) \int R_{AB \leftrightarrow C}(\mathbf{k}' - \mathbf{k}, \mathbf{k}, \mathbf{k}') \delta[\omega_B(\mathbf{k}) - \mathbf{k} \cdot \mathbf{V}_{gL}(\mathbf{k}')] \mathbf{k} \cdot \nabla_{\mathbf{k}'} \eta_L(\mathbf{k}') \frac{d\mathcal{V}_{k'}}{(2\pi)^6},$$

$$(23.58)$$

where η_L is the Langmuir (waves A and C) occupation number, $\mathbf{V}_{g\,L}$ is the Langmuir group velocity, and $R_{C \leftrightarrow BA}$ is the fundamental rate (23.56b).

(c) Notice the strong similarities between the evolution equation (23.58) for the ion-acoustic plasmons that are Cerenkov emitted and absorbed by Langmuir plasmons, and the evolution equation (23.44) for Langmuir plasmons that are Cerenkov emitted and absorbed by fast electrons! Discuss the similarities and the physical reasons for them.

(d) *Challenge:* Carry out an explicit classical calculation of the nonlinear interaction between Langmuir waves with wave vectors $\mathbf{k}_A$ and $\mathbf{k}_C$ to produce ion-acoustic waves with wave vector $\mathbf{k}_B = \mathbf{k}_C - \mathbf{k}_A$. Base your calculation on the nonlinear Vlasov equation (23.55) and [for use in relating $\mathbf{E}$ and f_1 in the nonlinear term] the 3-dimensional analog of Eq. (23.13). Assume a spatially independent Maxwellian averaged electron velocity distribution f_0 with temperature T_e (so $\nabla f_0 = 0$). From your result compute, in the random-phase approximation, the evolution of the ion-acoustic energy density $\mathcal{E}_\mathbf{k}$ and thence the evolution of the occupation number $\eta(\mathbf{k})$. Bring that evolution equation into the functional form (23.58). By comparing quantitatively with Eq. (23.58), read off the fundamental rate $R_{C \leftrightarrow BA}$. Your result should be the rate in Eq. (23.56b).

Exercise 23.6 *Example and Challenge: Three-Wave Mixing—Langmuir Evolution*
This exercise continues the analysis of the preceding one.

(a) Derive the kinetic equation for the Langmuir occupation number. [Hint: You will have to sum over four Feynman diagrams, corresponding to the mode of interest playing the role of A and then the role of C in each of the two diagrams in Fig. 23.5.]

(b) Using the approximations outlined in part (b) of Ex. 23.5, show that the Langmuir occupation number evolves in accord with the diffusion equation

$$\frac{d\eta_L(\mathbf{k}')}{dt} = \nabla_{\mathbf{k}'} \cdot [\mathbf{D}(\mathbf{k}') \cdot \nabla_{\mathbf{k}'}\eta_L(\mathbf{k}')], \qquad (23.59a)$$

where the diffusion coefficient is given by the following integral over the ion-acoustic-wave distribution:

$$\mathbf{D}(\mathbf{k}') = \int \eta_B(\mathbf{k})\, \mathbf{k} \otimes \mathbf{k}\, R_{AB \leftrightarrow C}(\mathbf{k}' - \mathbf{k}, \mathbf{k}, \mathbf{k}')\, \delta[\omega_B(\mathbf{k}) - \mathbf{k} \cdot \mathbf{V}_{g\,L}(\mathbf{k}')]\frac{d\mathcal{V}_{\mathbf{k}'}}{(2\pi)^6}.$$

$$(23.59b)$$

(c) Discuss the strong similarity between the evolution law (23.59) for resonant Langmuir plasmons interacting with ion-acoustic waves, and the one for resonant electrons interacting with Langmuir waves [Eqs. (23.29c), (23.29d)]. Why are they so similar?

23.4 Quasilinear Evolution of Unstable Distribution Functions—A Bump in the Tail

In plasma physics, one often encounters a weak beam of electrons passing through a stable Maxwellian plasma, with electron density n_e, beam density n_b, and beam speed v_b much larger than the thermal width σ_e of the background plasma. When the velocity width of the beam σ_b is small compared with v_b, the distribution is known as a *bump-in-tail* distribution (see Fig. 23.6). In this section, we explore the stability and nonlinear evolution of such a distribution.

We focus on the simple case of a 1-dimensional electron distribution function $F_0(v)$ and approximate the beam by a Maxwellian distribution:

$$F_b(v) = \frac{n_b}{(2\pi)^{1/2}\sigma_b} e^{-(v-v_b)^2/(2\sigma_b^2)}, \tag{23.60}$$

where n_b is the beam electron density. For simplicity, we treat the protons as a uniform neutralizing background.

Now, let us suppose that at time $t = 0$, the beam is established, and the Langmuir wave energy density $\mathcal{E}_k$ is very small. The waves will grow fastest when the waves' phase velocity $V_{\text{ph}} = \omega_r/k$ resides where the slope of the distribution function is most positive (i.e., when $V_{\text{ph}} = v_b - \sigma_b$). The associated maximum growth rate as computed from Eq. (23.12b) is

maximum growth rate for Langmuir waves

$$\omega_{i\max} = \left(\frac{\pi}{8e}\right)^{1/2} \left(\frac{v_b}{\sigma_b}\right)^2 \left(\frac{n_b}{n_e}\right) \omega_p, \tag{23.61}$$

where $e = 2.718\ldots$ is not the electron charge. Modes will grow over a range of wave phase velocities $\Delta V_{\text{ph}} \sim \sigma_b$. By using the Bohm-Gross dispersion relation (21.31) rewritten in the form

$$\omega = \omega_p (1 - 3\sigma_e^2/V_{\text{ph}}^2)^{-1/2}, \tag{23.62}$$

we find that the bandwidth of the growing modes is given roughly by

$$\Delta\omega = K\omega_p \frac{\sigma_b}{v_b}, \tag{23.63}$$

where $K = 3(\sigma_e/v_b)^2[1 - 3(\sigma_e/v_b)^2]^{-3/2}$ is a constant typically $\gtrsim 0.1$. Combining Eqs. (23.61) and (23.63), we obtain

$$\frac{\omega_{i\max}}{\Delta\omega} \sim \left(\frac{\pi}{8eK^2}\right)^{1/2} \left(\frac{v_b}{\sigma_b}\right)^3 \left(\frac{n_b}{n_e}\right). \tag{23.64}$$

Dropping constants of order unity, we conclude that the waves' growth time $\sim(\omega_{i\max})^{-1}$ is long compared with their coherence time $\sim(\Delta\omega)^{-1}$, provided that

$$\sigma_b \gtrsim \left(\frac{n_b}{n_e}\right)^{1/3} v_b. \tag{23.65}$$

validity of quasilinear analysis

When inequality (23.65) is satisfied, the waves will take several coherence times to grow, and so we expect that no permanent phase relations will be established in

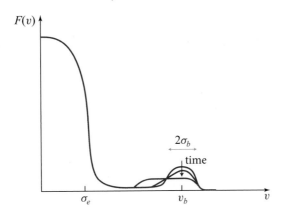

FIGURE 23.6 Evolution of the 1-dimensional electron distribution function from a bump-on-tail shape to a flat distribution function, due to the growth and scattering of electrostatic waves. The direction of evolution is indicated by the downward-pointing arrow at $v = v_b$.

the electric field and that quasilinear theory is an appropriate tool. However, when this inequality is reversed, the instability resembles more the two-stream instability of Chap. 21, and the growth is so rapid as to imprint special phase relations on the waves, so the random-phase approximation fails and quasilinear theory is invalid.

Restricting ourselves to slow growth, we use quasilinear theory to explore the evolution of the wave and particle distributions. We can associate the wave energy density $\mathcal{E}_k$ not just with a given value of k but also with a corresponding value of $V_{\text{ph}} = \omega_r / k$, and thence with the velocities $v = V_{\text{ph}}$ of electrons that resonate with the waves. Using Eq. (23.22a) for the velocity diffusion coefficient and Eq. (23.12b) for the associated wave-growth rate, we can then write the temporal evolution equations for the electron distribution function $F_0(v, t)$ and the wave energy density $\mathcal{E}_k(v, t)$ as

$$\frac{\partial F_0}{\partial t} = \frac{\pi e^2}{m_e^2 \epsilon_0} \frac{\partial}{\partial v} \left(\frac{\mathcal{E}_k}{v} \frac{\partial F_0}{\partial v} \right), \quad \frac{\partial \mathcal{E}_k}{\partial t} = \frac{\pi e^2}{m_e \epsilon_0 \omega_p} v^2 \mathcal{E}_k \frac{\partial F_0}{\partial v}. \tag{23.66}$$

evolution equations for bump-in-tail instability of electrons interacting with Langmuir waves that all propagate in the same direction

Here $v = \omega_r / k$ and, for simplicity, we have assumed a spatially homogeneous distribution of particles and waves so $d/dt \rightarrow \partial/\partial t$.

This pair of nonlinear equations must be solved numerically, but their qualitative behavior can be understood analytically without much effort; see Fig. 23.6. Waves resonant with the rising part of the electron distribution function at first grow exponentially, causing the particles to diffuse and flatten the slope of F_0 and thereby reduce the wave-growth rate. Ultimately, the slope $\partial F_0/\partial v$ diminishes to zero and the wave energy density becomes constant, with its integral, by energy conservation (Ex. 23.2), equal to the total kinetic energy lost by the beam. In this way we see that *a velocity-space irregularity in the distribution function leads to the growth of electrostatic waves, which can react back on the particles in such a way as to saturate the instability.* The

qualitative description of the evolution

net result is a beam of particles with a much-broadened width propagating through the plasma. The waves will ultimately damp through three-wave processes or other damping mechanisms, converting their energy into heat.

Exercise 23.7 *Problem: Stability of Isotropic Distribution Function*
For the bump-in-tail instability, the bump must show up in the 1-dimensional distribution function F_0 (after integrating out the electron velocity components orthogonal to the wave vector $\mathbf{v}$).

Consider an arbitrary, isotropic, 3-dimensional distribution function $f_0 = f_0(|\mathbf{v}|)$ for electron velocities—one that might even have an isotropic bump at large $|\mathbf{v}|$. Show that this distribution is stable against the growth of Langmuir waves (i.e., it produces $\omega_i < 0$ for all wave vectors $\mathbf{k}$).

23.4.1

23.4.1 Instability of Streaming Cosmic Rays

For a simple illustration of this general type of instability (but one dealing with a different wave mode), we return to the issue of the isotropization of galactic cosmic rays, which we introduced in Sec. 19.7. We argued there that cosmic rays propagating through the interstellar medium are effectively scattered by hydromagnetic Alfvén waves. We did not explain where these Alfvén waves originated. Although often there is an independent turbulence spectrum, sometimes these waves are generated by the cosmic rays themselves.

example of bump-in-tail instability: Alfvén waves generated by Cerenkov emission from cosmic rays' motion along interstellar magnetic field

Suppose we have a beam of cosmic rays propagating through the interstellar gas at high speed. The interstellar gas is magnetized, which allows many more types of wave modes to propagate than in the unmagnetized case. It turns out that the particle distribution is unstable to the growth of Alfvén wave modes satisfying the resonance condition (23.53), modified to account for mildly relativistic protons rather than nonrelativistic electrons:

$$\omega - \mathbf{k}_\| \cdot \mathbf{v}_\| = \omega_g. \tag{23.67}$$

Here $\omega_g = \omega_{cp}\sqrt{1 - v^2/c^2}$ is the proton's gyro frequency, equal to its nonrelativistic cyclotron frequency reduced by its Lorentz factor, and we assume that the number of cyclotron (gyro) quanta fed into each emitted Alfvén plasmon is $N = 1$. For pedagogical simplicity, let the wave propagate parallel to the magnetic field so $k = k_\|$. Then, since cosmic rays travel much faster than the Alfvén waves ($v \gg a = \omega/k$), the wave's frequency ω can be neglected in the resonance condition (23.67); equivalently, we can transform to a reference frame that moves with the Alfvén wave, so ω becomes zero, altering $\mathbf{v}_\|$ hardly at all. The resonance condition (23.67) then implies that the distance the particle travels along the magnetic field when making one trip around its Larmor orbit is $v_\|(2\pi/\omega_g) = 2\pi/k$, which is the Alfvén wave's wavelength. This makes sense physically; it enables the wave to resonate with the particle's gyrational motion, absorbing energy from it.

The growth rate of these waves can be studied using a kinetic theory analogous to the one we have just developed for Langmuir waves (Kulsrud, 2005, Chap. 12; Melrose, 1984, Sec. 10.5). Dropping factors of order unity, the growth rate for waves that scatter cosmic rays of energy E is given approximately by

$$\omega_i \simeq \left(\frac{n_{cr}}{n_p}\right) \omega_{cp} \left(\frac{u_{cr}}{a} - 1\right), \tag{23.68}$$

where n_{cr} is the number density of cosmic rays with energy $\sim E$, n_p is the number density of thermal protons in the background plasma, u_{cr} is the mean speed of the cosmic ray protons through the background plasma, and a is the Alfvén speed; see Ex. 23.8. Therefore, if the particles have a mean speed greater than the Alfvén speed, the waves will grow, exponentially at first.

It is observed that the energy density of cosmic rays in our galaxy builds up until it is roughly comparable with that of the thermal plasma. As more cosmic rays are produced, they escape from the galaxy at a sufficient rate to maintain this balance. Therefore, in a steady state, the ratio of the number density of cosmic rays to the thermal proton density is roughly the inverse of their mean-energy ratio. Adopting a mean cosmic ray energy of $\sim 1\,\mathrm{GeV}$ and an ambient temperature in the interstellar medium of $T \sim 10^4\,\mathrm{K}$, this ratio of number densities is $\sim 10^{-9}$. The ion gyro period in the interstellar medium is $\sim 100\,\mathrm{s}$ for a typical field of strength of $\sim 100\,\mathrm{pT}$. Cosmic rays streaming at a few times the Alfvén speed create Alfvén waves in $\sim 10^{10}\,\mathrm{s}$, of order a few hundred years—long before they escape from the galaxy (e.g., Longair, 2011).

The waves then react back on the cosmic rays, scattering them in momentum space [Eq. (23.51a)]. Now, each time a particle is scattered by an Alfvén-wave quantum, the ratio of its energy change to the magnitude of its momentum change must be the same as that in the waves and equal to the Alfvén speed, which is far smaller than the original energy-to-momentum ratio of the particle ($\sim c$) for a mildly relativistic proton. Therefore, the effect of the Alfvén waves is to scatter the particle directions without changing their energies significantly. As the particles are already gyrating around the magnetic field, the effect of the waves is principally to change the angle between their momenta and the field (known as the *pitch angle*), so as to reduce their mean speed along the magnetic field.

Alfvén waves scatter cosmic rays until their speeds along the magnetic field are reduced to Alfvén speed, saturating the bump-in-tail instability

When this mean speed is reduced to a value of order the Alfvén speed, the growth rate diminishes, just like the growth rate of Langmuir waves is diminished after the electron distribution function is flattened. Under a wide variety of conditions, cosmic rays are believed to maintain the requisite energy density in Alfvén-wave turbulence to prevent them from streaming along the magnetic field with a mean speed much faster than the Alfvén speed (which varies between ~ 3 and $\sim 30\,\mathrm{km\,s^{-1}}$). This model of their transport differs from spatial diffusion, which we assumed in Sec. 19.7.3, but the end result is similar, and cosmic rays are confined to our galaxy for more than ~ 10 million years. These processes can be observed directly using spacecraft in the interplanetary medium.

Exercise 23.8 *Challenge: Alfvén Wave Emission by Streaming Cosmic Rays*
Consider a beam of high-energy cosmic ray protons streaming along a uniform background magnetic field in a collisionless plasma. Let the cosmic rays have an isotropic distribution function in a frame that moves along the magnetic field with speed u, and assume that u is large compared with the Alfvén speed but small compared with the speeds of the individual cosmic rays. Adapt our discussion of the emission of Langmuir waves by a bump-on-tail distribution to show that the growth rate is given to order of magnitude by Eq. (23.68).

For a solution using an order-of-magnitude analysis, see Kulsrud (2005, Sec. 12.2). For a detailed analysis using quasilinear theory, and then a quasilinear study of the diffusion of cosmic rays in the Alfvén waves they have generated, see Kulsrud (2005, Secs. 12.3–12.5).

23.5

23.5 Parametric Instabilities; Laser Fusion

laser fusion

One of the approaches to the goal of attaining commercial nuclear fusion (Sec. 19.3.1) is to compress and heat pellets containing a mixture of deuterium and tritium by using powerful lasers. The goal is to produce gas densities and temperatures large enough, and for long enough, that the nuclear energy released exceeds the energy expended in producing the compression (Box 23.2). At these densities the incident laser radiation behaves like a large-amplitude plasma wave and is subject to a new type of instability that may already be familiar from dynamics, namely, a *parametric instability*.

Consider how the incident light is absorbed by the relatively tenuous ionized plasma around the pellet. The critical density at which the incident wave frequency equals the plasma frequency is $\rho \sim 5\lambda_{\mu m}^{-2}$ kg m^{-3}, where $\lambda_{\mu m}$ is the wavelength measured in microns. For an incident wave energy flux $F \sim 10^{18}$ W m^{-2}, the amplitude of the wave's electric field $E \sim [F/(\epsilon_0 c)]^{1/2} \sim 2 \times 10^{10}$ V m^{-1}. The velocity of a free electron oscillating in a wave this strong is $v \sim eE/(m_e\omega) \sim 2{,}000$ km s^{-1}, which is almost 1% of the speed of light. It is therefore not surprising that nonlinear wave processes are important.

stimulated Raman scattering of laser beam parametrically amplifies a Langmuir wave, strengthening the scattering (a parametric instability) and impeding laser fusion

One of the most important such processes is called *stimulated Raman scattering*. In this interaction the coherent electromagnetic wave with frequency ω convects a small preexisting density fluctuation—one associated with a relatively low-frequency Langmuir wave with frequency ω_{pe}—and converts the Langmuir density fluctuation into a current that varies at the beat frequency $\omega - \omega_{pe}$. This creates a new electromagnetic mode with this frequency. The vector sum of the **k** vectors of the two modes must also equal the incident **k** vector. When this condition can first be met, the new **k** is almost antiparallel to that of the incident mode, and so the radiation is backscattered.

The new mode can combine nonlinearly with the original electromagnetic wave to produce a pressure force $\propto \nabla E^2$, which amplifies the original density fluctuation. Provided the growth rate of the wave is faster than the natural damping rates (e.g.,

In the simplest proposed scheme for laser fusion, solid pellets of deuterium and tritium would be compressed and heated to allow the reaction

$$d + t \rightarrow \alpha + n + 17.6\,\text{MeV} \qquad (1)$$

to proceed. (We will meet this reaction again in Sec. 28.4.2.) An individual pellet would have mass $m \sim 3\,\text{mg}$ and initial diameter $r_i \sim 2\,\text{mm}$. The total latent nuclear energy in a single pellet is $\sim 1\,\text{GJ}$. So if, optimistically, the useful energy extraction were $\sim 3\%$ of this, pellets would have to burn at a combined rate of $\sim 10^5\,\text{s}^{-1}$ in reactors all around the world to supply, say, a fifth of the *current* (2016) global power usage $\sim 15\,\text{TW}$.

The largest program designed to accomplish this goal is at the National Ignition Facility in Lawrence Livermore National Laboratory in California. Neodymium-glass pulsed lasers are used to illuminate a small pellet from 192 directions at a wavelength of 351 nm. (The frequency is tripled from the initial infrared frequency using KDP crystals; Box 10.2.) The goal is to deliver a power of $\sim 0.5\,\text{PW}$ for a few nanoseconds. The nominal initial laser energy is $\sim 3\,\text{MJ}$ derived from a capacitor bank storing $\sim 400\,\text{MJ}$ of energy. The energy is delivered to a small metal cylinder, called a *hohlraum*, where a significant fraction of the incident energy is converted to X-rays that illuminate the fuel pellet, a technique known as *indirect drive*. The energy absorbed in the surface layers of the pellet is $\sim 150\,\text{kJ}$. This causes the pellet to implode to roughly a tenth of its initial size and a density of $\sim 10^6\,\text{kg m}^{-3}$; nuclear energy releases of 20 MJ are projected. It has recently become possible to create as much fusion energy as the energy deposited in the pellet (Hurricane et al., 2014). The shot repetition rate is currently about one per day. There is clearly a long road ahead to safe commercial reactors supplying a significant fraction of the global power usage, but progress is being made.

that of Landau damping) there can be a strong backscattering of the incident wave at a density well below the critical density of the incident radiation. (A further condition that must be satisfied is that the bandwidth of the incident wave must also be less than the growth rate. This will generally be true for a laser.) This stimulated Raman scattering is an example of a parametric instability. The incident wave frequency is called the *pump* frequency. One difference between parametric instabilities involving waves as opposed to just oscillations is that it is necessary to match spatial as well as temporal frequencies.

Reflection of the incident radiation by this mechanism reduces the ablation of the pellet and also creates a population of suprathermal electrons, which conduct heat into the interior of the pellet and inhibit compression. Various strategies, including increasing the wave frequency, have been devised to circumvent Raman backscattering (and also a related process called *Brillouin backscattering*, in which the Langmuir mode is replaced by an ion-acoustic mode).

Exercise 23.9 *Example: Ablation vs. Radiation Pressure in Laser Fusion*
Demonstrate that ablation pressure is necessary to compress a pellet to densities at which nuclear reactions can progress efficiently. More specifically, do the following.

(a) Assume that a temperature $T \sim 30 \, \text{keV}$ is necessary for nuclear reactions to proceed. Using the criteria developed in Sec. 20.2.2, evaluate the maximum density before the electrons become degenerate. Estimate the associated pressure and sound speed.

(b) Employ the Lawson criterion (19.25) to estimate the minimum size of a pellet for break-even.

(c) Using the laser performance advertised in Box 23.2, compare the radiation pressure in the laser light with the pressure needed from part (a). Do you think that radiation pressure can create laser fusion?

(d) Assume instead that the incident laser energy is carried off by an outflow. Show that the associated ablation pressure can be orders of magnitude higher than the laser light pressure, and compute an upper bound on the speed of the outflow for laser fusion to be possible.

23.6

23.6 Solitons and Collisionless Shock Waves

In Sec. 21.4.3, we introduced ion-acoustic waves that have a phase speed $V_{\text{ph}} \sim (k_B T_e / m_p)^{1/2}$, determined by a combination of electron pressure and ion inertia. In Sec. 22.3.6, we argued that these waves would be strongly Landau-damped unless the electron temperature greatly exceeded the proton temperature. However, this formalism was only valid for waves of small amplitude so that the linear approximation could be used. In Ex. 21.6, we considered the profile of a nonlinear wave and found a solution for a single ion-acoustic soliton valid when the waves are weakly nonlinear. We now consider this problem in a slightly different way, one valid for a wave amplitude that is strongly nonlinear. However, we restrict our attention to waves that propagate without change of form and so will not generalize the KdV equation.

Once again we use the fluid model and introduce an ion-fluid velocity u. We presume the electrons are highly mobile and so have a local density $n_e \propto \exp[e\phi/(k_B T_e)]$, where ϕ (not Φ as above) is the electrostatic potential. The ions must satisfy equations of continuity and motion:

$$\frac{\partial n}{\partial t} + \frac{\partial}{\partial z}(nu) = 0,$$

$$\frac{\partial u}{\partial t} + u \frac{\partial u}{\partial x} = -\frac{e}{m_p} \frac{\partial \phi}{\partial z}. \tag{23.69}$$

We now seek a solution for an ion-acoustic wave moving with constant speed V through the ambient plasma. In this case, all physical quantities must be functions of a single dependent variable: $\xi = z - Vt$. Using a prime to denote differentiation with respect to ξ, Eqs. (23.69) become

two-fluid theory of a strongly nonlinear ion-acoustic soliton with speed V in a plasma with thermalized electrons: ion speed u and electrostatic potential ϕ as functions of $\xi = z - Vt$

$$(u - V)n' = -nu',$$

$$(u - V)u' = -\frac{e}{m_p}\phi'. \tag{23.70}$$

These two equations can be integrated and combined to obtain an expression for the ion density n in terms of the electrostatic potential:

$$n = n_0[1 - 2e\phi/(m_p V^2)]^{-1/2}, \tag{23.71}$$

where n_0 is the ion density, presumed uniform, long before the wave arrives. The next step is to combine this ion density with the electron density and substitute into Poisson's equation to obtain a nonlinear ordinary differential equation for the potential:

$$\phi'' = -\frac{n_0 e}{\epsilon_0} \left\{ \left(1 - \frac{2e\phi}{m_p V^2}\right)^{-1/2} - e^{e\phi/(k_B T_e)} \right\}. \tag{23.72}$$

Now, the best way to think about this problem is to formulate the equivalent dynamical problem of a particle moving in a 1-dimensional potential well $\Phi(\phi)$, with ϕ the particle's position coordinate and ξ its time coordinate. Then Eq. (23.72) becomes $\phi'' = -d\Phi/d\phi$, whence $\frac{1}{2}\phi'^2 + \Phi(\phi) = \mathcal{E}$, where $\mathcal{E}$ is the particle's conserved energy. Integrating $-d\Phi/d\phi = $ [right-hand side of Eq. (23.72)], and assuming that $\Phi \to 0$ as $\phi \to 0$ (i.e., as $\xi \to \infty$, long before the arrival of the pulse), we obtain

$$\Phi(\phi) = \frac{n_0 k_B T_e}{\epsilon_0} \left\{ \left[1 - (1 - \phi/\phi_o)^{1/2}\right] M^2 - (e^{M^2\phi/(2\phi_o)} - 1) \right\}, \tag{23.73a}$$

where

$$\phi_o = m_p V^2/(2e), \qquad M = [m_p V^2/(k_B T_e)]^{1/2}. \tag{23.73b}$$

We have assumed that $0 < \phi < \phi_o$; when $\phi \to \phi_o$, the proton density $n \to \infty$ [Eq. (23.71)].

The shape of this potential well $\Phi(\phi)$ is sketched in Fig. 23.7; it is determined by the parameter $M = [m_p V^2/(k_B T_e)]^{1/2}$, which is readily recognizable as the ion-acoustic Mach number (i.e., the ratio of the speed of the soliton to the ion-acoustic speed in the undisturbed medium). A solution for the soliton's potential profile $\phi(\xi)$

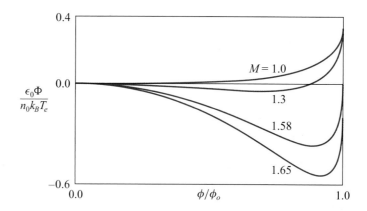

FIGURE 23.7 Potential function $\Phi(\phi)$ used for exhibiting the properties of an ion-acoustic soliton for four different values of the ion-acoustic Mach number M.

corresponds to the trajectory of the particle. Physically, a soliton (solitary wave) must be concentrated in a finite region of space at fixed time; that is, $\phi(\xi) = \phi(z - Vt)$ must go to zero as $\xi \to \pm\infty$. For the particle moving in the potential $\Phi(\phi)$ this is possible only if its total energy $\mathcal{E}$ vanishes, so $\frac{1}{2}\phi'^2 + \Phi(\phi) = 0$. The particle then starts at $\phi = 0$, with zero kinetic energy (i.e., $\phi' = 0$) and then accelerates to a maximum speed near the minimum in the potential before decelerating. If there is a turning point, the particle will come to rest, $\phi(\xi)$ will attain a maximum, and then the particle will return to the origin. This particle trajectory corresponds to a symmetrical soliton, propagating with uniform speed.

Two conditions must be satisfied for a soliton solution. First, the potential well must be attractive. This only happens when $d^2\Phi/d\phi^2(0) < 0$, which implies that $M > 1$. Second, there must be a turning point. This happens if $\Phi(\phi = \phi_o) > 0$. The maximum value of M for which these two conditions are met is a solution of the equation

$$e^{M^2/2} - 1 - M^2 = 0 \qquad (23.74)$$

(i.e., $M = 1.58$). Hence, ion-acoustic-soliton solutions only exist for

$$1 < M < 1.59. \qquad (23.75)$$

Mach number for ion-acoustic soliton

If $M < 1$, the particle's potential Φ acts as a barrier, preventing it from moving into the $\phi > 0$ region. If $M > 1.59$, the plasma's electrons short out the potential; that is, the electron term ($e^{M^2\phi/(2\phi_o)} - 1$) in Eq. (23.73a) becomes so negative that it makes $\Phi(\phi_o)$ negative.

This analogy with particle dynamics is not only helpful in understanding large-amplitude solitons. It also assists us to understand a deep connection between these solitons and laminar shock fronts. The equations that we have been solving so far

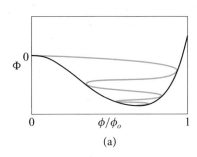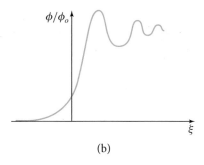

FIGURE 23.8 Ion-acoustic shock wave. (a) Solution in terms of damped particle motion $\phi(\xi)$ (blue curve) in the equivalent potential $\Phi(\phi)$ (black curve). (b) Electrostatic potential profile $\phi(\xi) = \phi(z - Vt)$ in the shock. The proton number-density variation $n(\xi)$ in the shock can be inferred from this $\phi(\xi)$ using Eq. (23.71).

contain the two key ingredients for a soliton: nonlinearity to steepen the wave profile and dispersion to spread it. However, they do not make provision for any form of dissipation, a necessary condition for a shock front, in which the entropy must increase.

In a real collisionless plasma, this dissipation can take on many forms. It may be associated with anomalous resistivity or perhaps with some viscosity associated with the ions. In many circumstances, some ions are reflected by the electrostatic potential barrier and counterstream against the incoming ions, which they eventually heat. Whatever its origin, the net effect of this dissipation will be to cause the equivalent particle to lose its total energy, so that it can never return to its starting point. Given an attractive and bounded potential well, we find that the particle has no alternative except to sink to the bottom of the well. Depending on the strength of the dissipation, the particle may undergo several oscillations before coming to rest. See Fig. 23.8a.

adding dissipation converts soliton into a laminar, collisionless shock

The structure to which this type of solution corresponds is a laminar shock front. Unlike that for a soliton, the wave profile in a shock wave is not symmetric in this case and instead describes a permanent change in the electrostatic potential ϕ (Fig. 23.8b). Repeating the arguments above, we find that a shock wave can only exist when $M > 1$, that is to say, it must be supersonic with respect to the ion-acoustic sound speed. In addition there is a maximum critical Mach number close to $M = 1.6$, above which a laminar shock becomes impossible.

What happens when the critical Mach number is exceeded? In this case there are several possibilities, which include relying on a more rapidly moving wave to form the shock front or appealing to turbulent conditions downstream from the front to enhance the dissipation rate.

This ion-acoustic shock front is the simplest example of a collisionless shock. Essentially every wave mode can be responsible for the formation of a shock. The dissipation in these shocks is still not very well understood, but we can observe them in several different environments in the heliosphere and also in the laboratory. The

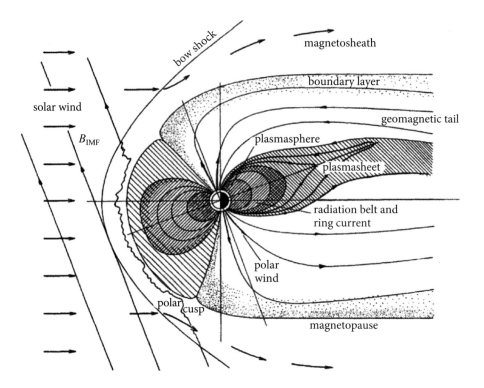

FIGURE 23.9 The form of the collisionless bow shock formed around Earth's magnetosphere. Earth's bow shock has been extensively studied using spacecraft. Alfvén, ion-acoustic, whistler, and Langmuir waves are all generated with large amplitudes in the vicinity of shock fronts by the highly nonthermal particle distributions. Adapted from Parks (2004).

collisionless bow shock at interface of Earth's magnetic field with solar wind

best studied of these shock waves are those based on magnetosonic waves, which were introduced in Sec. 19.7. The solar wind moves with a speed that is typically 5 times the Alfvén speed. It should therefore form a bow shock (one based on the fast magnetosonic mode) when it encounters a planetary magnetosphere. This bow shock forms even though the mean free path of the ions for Coulomb scattering in the solar wind is typically much larger than the thickness of the shock front. The thickness turns out to be a few ion Larmor radii. This is a dramatic illustration of the importance of collective effects in controlling the behavior of essentially collisionless plasmas (Fig. 23.9; see Sagdeev and Kennel, 1991).

EXERCISES

Exercise 23.10 *Derivation: Maximum Mach Number for an Ion-Acoustic Shock Wave*
Verify Eq. (23.73a), and show numerically that the maximum Mach number for a laminar shock front is $M = 1.58$.

Exercise 23.11 *Problem: Solar-Wind Termination Shock*
The solar wind is a quasi-spherical outflow of plasma from the Sun. At the radius of Earth's orbit, the mean proton and electron densities are $n_p \sim n_e \sim 4 \times 10^6 \, \mathrm{m}^{-3}$, their temperatures are $T_p \sim T_e \sim 10^5 \, \mathrm{K}$, and their common radial fluid speed is

~ 400 km s^{-1}. The mean magnetic field strength is ~ 1 nT. Eventually, the radial momentum flux in the solar wind falls to the value of the mean interstellar pressure, $\sim 10^{-13}$ N m^{-2}, and a shock develops.

(a) Estimate the radius where the shock develops.

(b) The solar system moves through the interstellar medium with a speed ~ 30 km s^{-1}. Sketch the likely flow pattern near this radius.

(c) How do you expect the magnetic field to vary with radius in the outflowing solar wind? Estimate its value at the termination shock.

(d) Estimate the electron plasma frequency, the ion-acoustic Mach number, and the proton Larmor radius just ahead of the termination shock front, and comment on the implications of these values for the shock structure.

(e) The Voyager 1 spacecraft was launched in 1977 and is traveling radially away from the Sun with a terminal speed ~ 17 km s^{-1}. It was observed to cross the termination shock in 2004, and in 2012 it passed beyond the limit of the shocked solar wind into the interstellar medium. The Voyager 2 spacecraft passed through the termination shock a few years after Voyager 1. How do these observations compare with your answer to part (a)?

Exercise 23.12 *Example: Diffusive Shock Acceleration of Galactic Cosmic Rays*
Cosmic ray particles with energies between ~ 1 GeV and ~ 1 PeV are believed to be accelerated at the strong shock fronts formed by supernova explosions in the strongly scattering, local, interstellar medium. We explore a simple model of the way in which this happens.

(a) In a reference frame where the shock is at rest, consider the stationary (time-independent) flow of a medium (a plasma) with velocity $u(x)\,\mathbf{e}_x$, and consider relativistic cosmic rays, diffusing through the medium, that have reached a stationary state. Assume that the mean free paths of the cosmic rays are so short that their distribution function $f = dN/d\mathcal{V}_x d\mathcal{V}_p$ is nearly isotropic in the local rest frame of the medium, so $f = f_0(\tilde{p}, x) + f_1(\mathbf{p}, x)$, where $\tilde{p}$ is the magnitude of a cosmic ray's momentum as measured in the medium's local rest frame, $\tilde{p} \equiv |\tilde{\mathbf{p}}|$, $f_0 \equiv \langle f \rangle$ is the average over cosmic-ray propagation direction in the medium's local rest frame, and $|f_1| \ll |f_0|$. Show that in terms of the cosmic ray's momentum $p = |\mathbf{p}|$ and energy $\mathcal{E} = \sqrt{p^2 + m^2}$ as measured in the shock's rest frame, $\tilde{p} = p - u(x)\mathcal{E}\cos\theta$, where θ is the angle between $\mathbf{p}$ and the direction $\mathbf{e}_x$ of the medium's motion. (Here and throughout this exercise we set the speed of light to unity, as in Chap. 2.)

(b) By expanding the full Vlasov equation to second order in the ratio of the scattering mean free path to the scale on which $u(x)$ varies, it can be shown (e.g., Blandford and Eichler, 1987, Sec. 3.5) that, in the shock's frame,

$$u\frac{\partial f}{\partial x} - u\frac{\partial}{\partial x}\left(D\frac{\partial f}{\partial x}\right) = \frac{\tilde{p}}{3}\frac{\partial f}{\partial \tilde{p}}\frac{\partial u}{\partial x}, \qquad (23.76a)$$

where $D(\tilde{p}, x) > 0$ is the spatial diffusion coefficient which arises from f_1. Explore this *convection-diffusion* equation. For example, by integrating it over momentum space in the shock's frame, show that it conserves cosmic ray particles. Also, show that, when the diffusion term is unimportant, f_0 is conserved moving with the medium's flow in the sense that $u\partial f_0/\partial x + (d\tilde{p}/dt)\partial f_0/\partial \tilde{p} = 0$, where $d\tilde{p}/dt$ is the rate of change of cosmic ray momenta due to elastic scattering in the expanding medium (for which a volume element changes as $d \ln \mathcal{V}_x/dt = \partial u/\partial x$).

(c) Argue that the flux of cosmic-ray particles, measured in the shock's frame, is given by $\int_0^\infty F(p, x)4\pi p^2 dp$, where $F(p, x) = -D(\partial f_0/\partial x) - u(p/3)(\partial f/\partial \tilde{p})$; or, to leading order in u/v, where $v = p/\mathcal{E}$ is the cosmic ray speed,

$$F(\tilde{p}, x) = -D\frac{\partial f_0}{\partial x} - u\frac{\tilde{p}}{3}\frac{\partial f}{\partial \tilde{p}} . \qquad (23.76b)$$

This F is often called the flux of particles at momentum $\tilde{p}$.

(d) Idealize the shock as a planar discontinuity at $x = 0$ in the medium's velocity $u(x)$; idealize the velocity as constant before and after the shock, $u = u_- = $ constant for $x < 0$, and $u = u_+ = $ constant for $x > 0$; and denote by $r = u_-/u_+ > 1$ the shock's compression ratio. Show that upstream from the shock, the solution to the convection-diffusion equation (23.76a) is

$$f_0(\tilde{p}, x) = f_-(\tilde{p}) + [f_s(\tilde{p}) - f_-(\tilde{p})] \exp\left[-\int_x^0 \frac{u_- dx'}{D(\tilde{p}, x')}\right] \quad \text{at} \quad x < 0. \quad (23.76c)$$

Here $f_-(\tilde{p}) \equiv f_0(\tilde{p}, x = -\infty)$, and $f_s(\tilde{p}) \equiv f_0(\tilde{p}, x = 0)$ is the value of f_0 at the shock, which must be continuous across the shock. (Why?) Show further that downstream from the shock f_0 cannot depend on x, so

$$f_0(\tilde{p}, x) = f_s(\tilde{p}) \quad \text{at} \quad x > 0 . \qquad (23.76d)$$

(e) By matching the flux of particles at momentum $\tilde{p}$, $F(\tilde{p}, x)$ [Eq. (23.76b)], across the shock, show that the post-shock distribution function is

$$f_s(\tilde{p}) = q\tilde{p}^{-q} \int_0^{\tilde{p}} d\tilde{p}' \, f_-(\tilde{p}') \, \tilde{p}'^{(q-1)} , \qquad (23.76e)$$

where $q = 3r/(r - 1)$.

(f) The fact that this stationary solution to the convection-diffusion equation has a power-law spectrum $f_s \propto \tilde{p}^{-q}$, in accord with observations, suggests that the observed cosmic rays may indeed be accelerated in shock fronts. How, physically, do you think the acceleration occurs and how does this lead to a power-law spectrum? What is the energy source for the acceleration? It may help to explore the manner in which Eq. (23.76c) evolves an arbitrary initial distribution function $f_-(\tilde{p})$ into the power-law distribution (23.76d). [For discussion and details, see, e.g., Blandford and Eichler (1987), and Longair (2011).]

Bibliographic Note

For a concise treatment of the classical quasilinear theory of wave-particle interactions as in Sec. 23.2, see Lifshitz and Pitaevskii (1981, Sec. 49 for quasilinear theory, and Sec. 51 for fluctuations and correlations).

For more detailed and rich, pedagogical, classical treatments of quasilinear theory and some stronger nonlinear effects in plasmas, see Bellan (2006, Chaps. 14, 15), Boyd and Sanderson (2003, Chap. 10), Swanson (2003, Chaps. 7, 8), and Krall and Trivelpiece (1973, Chaps. 10, 11). For a classical treatment of quasilinear theory that is extended to include excitations of magnetized plasmas, see Stix (1992, Chaps. 16–18). For the interactions of cosmic rays with Alfvén waves, see Kulsrud (2005, Chap. 12). For wave-wave coupling parametric instabilities, see Swanson (2003, Sec. 8.4). For ion-acoustic solitons and associated shocks, see Krall and Trivelpiece (1973, Sec. 3.9.4) and Swanson (2003, Sec. 8.2).

A classic text on radiation processes in plasmas is Bekefi (1966). An extensive treatment of nonlinear plasma physics from a quantum viewpoint is contained in Melrose (2008, 2012).

For applications to astrophysical plasmas, see Melrose (1984), Parks (2004), and Kulsrud (2005), and for applications to laser-plasma interactions, see Kruer (1988).

Evolution of Vorticity

(Selected Sections of Chapter 14

of *Modern Classical Physics*)

14.2 Vorticity, Circulation, and Their Evolution

In Sec. 13.5.4, we defined the vorticity as the curl of the velocity field, $\boldsymbol{\omega} = \nabla \times \mathbf{v}$, analogous to defining the magnetic field as the curl of a vector potential. To get insight into vorticity, consider the three simple 2-dimensional flows shown in Fig. 14.1.

Figure 14.1a shows *uniform (rigid) rotation*, with constant angular velocity $\boldsymbol{\Omega} = \Omega \mathbf{e}_z$. The velocity field is $\mathbf{v} = \boldsymbol{\Omega} \times \mathbf{x}$, where $\mathbf{x}$ is measured from the rotation axis. Taking its curl, we discover that $\boldsymbol{\omega} = 2\boldsymbol{\Omega}$ everywhere.

Figure 14.1b shows a flow in which the angular momentum per unit mass $\mathbf{j} = j\mathbf{e}_z$ is constant, because it was approximately conserved as the fluid gradually drifted inward to create this flow. In this case the rotation is *differential* (radially changing angular velocity), with $\mathbf{v} = \mathbf{j} \times \mathbf{x}/\varpi^2$ (where $\varpi = |\mathbf{x}|$ and $\mathbf{j} = $ const). This is the kind of flow that occurs around a bathtub vortex and around a tornado—but outside the vortex's or tornado's core. The vorticity is $\boldsymbol{\omega} = 2\pi j \delta(\mathbf{x})$ (where $\delta(\mathbf{x})$ is the 2-dimensional Dirac delta function in the plane of the flow), so it vanishes everywhere except at the center, $\mathbf{x} = 0$ (or, more precisely, except in the vortex's or tornado's core). Anywhere in the flow, two neighboring fluid elements, separated tangentially, rotate about each other with an angular velocity $+\mathbf{j}/\varpi^2$, but when the two elements are separated radially, their relative angular velocity is $-\mathbf{j}/\varpi^2$; see Ex. 14.1. The average of these two angular velocities vanishes, which seems reasonable, since the vorticity vanishes.

vortex without vorticity except at its center

The vanishing vorticity in this case is an illustration of a simple geometrical description of vorticity in any 2-dimensional flow (Ex. 13.15): If we orient the $\mathbf{e}_z$ axis of a Cartesian coordinate system along the vorticity, then

$$\omega = \left(\frac{\partial v_y}{\partial x} - \frac{\partial v_x}{\partial y} \right) \mathbf{e}_z. \tag{14.1}$$

This expression implies that the vorticity at a point is the sum of the angular velocities of a pair of mutually perpendicular, infinitesimal lines passing through that point (one

---

This appendix is extracted from Volume 3 of *Modern Classical Physics*. It consists of portions of Sec. 14.2 of Chapter 14, "Vorticity," with no changes aside from pagination.

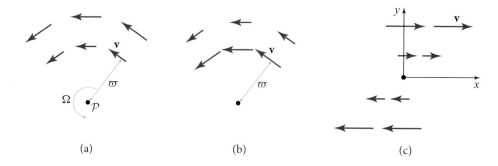

FIGURE 14.1 Vorticity in three 2-dimensional flows. The vorticity vector points in the z direction (orthogonal to the plane of the flow) and so can be thought of as a scalar ($\omega = \omega_z$). (a) Constant angular velocity Ω. If we measure radius ϖ from the center $\mathcal{P}$, the circular velocity satisfies $v = \Omega\varpi$. This flow has vorticity $\omega = 2\Omega$ everywhere. (b) Constant angular momentum per unit mass j, with $v = j/\varpi$. This flow has zero vorticity except at its center, $\omega = 2\pi j\delta(\mathbf{x})$. (c) Shearing flow in a laminar boundary layer, $v_x = -\omega y$ with $\omega < 0$. The vorticity is $\omega = -v_x/y$, and the rate of shear is $\sigma_{xy} = \sigma_{yx} = -\frac{1}{2}\omega$.

vorticity measured by a vane with orthogonal fins

along the x direction, the other along the y direction) and moving with the fluid; for example, these lines could be thin straws suspended in the fluid. If we float a little vane with orthogonal fins in the flow, with the vane parallel to $\boldsymbol{\omega}$, then the vane will rotate with an angular velocity that is the average of the flow's angular velocities at its fins, which is half the vorticity. Equivalently, the vorticity is twice the rotation rate of the vane. In the case of constant-angular-momentum flow in Fig. 14.1b, the average of the two angular velocities is zero, the vane doesn't rotate, and the vorticity vanishes.

Figure 14.1c shows the flow in a plane-parallel shear layer. In this case, a line in the flow along the x direction does not rotate, while a line along the y direction rotates with angular velocity ω. The sum of these two angular velocities, $0 + \omega = \omega$, is the vorticity. Evidently, curved streamlines are not a necessary condition for vorticity.

EXERCISES

Exercise 14.1 *Practice: Constant-Angular-Momentum Flow—Relative Motion of Fluid Elements*
Verify that for the constant-angular-momentum flow of Fig. 14.1b, with $\mathbf{v} = \mathbf{j} \times \mathbf{x}/\varpi^2$, two neighboring fluid elements move around each other with angular velocity $+j/\varpi^2$ when separated tangentially and $-j/\varpi^2$ when separated radially. [Hint: If the fluid elements' separation vector is $\boldsymbol{\xi}$, then their relative velocity is $\nabla_{\boldsymbol{\xi}}\mathbf{v} = \boldsymbol{\xi} \cdot \nabla\mathbf{v}$. Why?]

Exercise 14.2 *Practice: Vorticity and Incompressibility*
Sketch the streamlines for the following stationary 2-dimensional flows, determine whether the flow is compressible, and evaluate its vorticity. The coordinates are Cartesian in parts (a) and (b), and are circular polar with orthonormal bases $\{\mathbf{e}_\varpi, \mathbf{e}_\phi\}$ in (c) and (d).

(a) $v_x = 2xy, \quad v_y = x^2,$

(b) $v_x = x^2, \quad v_y = -2xy,$

(c) $v_\varpi = 0, \quad v_\phi = \varpi,$

(d) $v_\varpi = 0, \quad v_\phi = \varpi^{-1}.$

Exercise 14.3 **Example: Rotating Superfluids*

At low temperatures certain fluids undergo a phase transition to a superfluid state. A good example is ^{4}He, for which the transition temperature is 2.2 K. As a superfluid has no viscosity, it cannot develop vorticity. How then can it rotate? The answer (e.g., Feynman 1972, Chap. 11) is that not all the fluid is in a superfluid state; some of it is normal and can have vorticity. When the fluid rotates, all the vorticity is concentrated in microscopic vortex cores of normal fluid that are parallel to the rotation axis and have quantized circulations $\Gamma = h/m$, where m is the mass of the atoms and h is Planck's constant. The fluid external to these vortex cores is irrotational (has vanishing vorticity). These normal fluid vortices may be pinned at the walls of the container.

(a) Explain, using a diagram, how the vorticity of the macroscopic velocity field, averaged over many vortex cores, is twice the mean angular velocity of the fluid.

(b) Make an order-of-magnitude estimate of the spacing between these vortex cores in a beaker of superfluid helium on a turntable rotating at 10 rpm.

(c) Repeat this estimate for a neutron star, which mostly comprises superfluid neutron pairs at the density of nuclear matter and spins with a period of order a millisecond. (The mass of the star is roughly 3×10^{30} kg.)

14.2.1 Vorticity Evolution

By analogy with magnetic field lines, we define a flow's *vortex lines* to be parallel to the vorticity vector $\boldsymbol{\omega}$ and to have a line density proportional to $\omega = |\boldsymbol{\omega}|$. These vortex lines are always continuous throughout the fluid, because the vorticity field, like the magnetic field, is a curl and therefore is necessarily solenoidal ($\nabla \cdot \boldsymbol{\omega} = 0$). However, vortex lines can begin and end on solid surfaces, as the equations of fluid dynamics no longer apply there. Figure 14.2 shows an example: vortex lines that emerge from the wingtip of a flying airplane.

Vorticity and its vortex lines depend on the velocity field at a particular instant and evolve with time as the velocity field evolves. We can determine how by manipulating the Navier-Stokes equation.

In this chapter and the next one, we restrict ourselves to flows that are incompressible in the sense that $\nabla \cdot \mathbf{v} = 0$. As we saw in Sec. 13.6, this is the case when the flow is substantially subsonic and gravitational potential differences are not too extreme. We also require (as is almost always the case) that the shear viscosity vary spatially far more slowly than the shear itself. These restrictions allow us to write the Navier-Stokes equation in its simplest form:

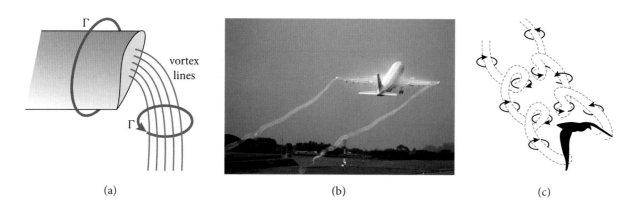

(a)   (b)   (c)

FIGURE 14.2 (a) Sketch of the wing of a flying airplane and the vortex lines that emerge from the wing tip and sweep backward behind the plane. The lines are concentrated in a region with small cross section, a vortex of whirling air. The closed red curves encircle the wing and the vortex; the integral of the velocity field around these curves, $\Gamma = \int \mathbf{v} \cdot d\mathbf{x}$, is the circulation contained in the wing and its boundary layers, and in the vortex; see Sec. 14.2.4 and especially Ex. 14.8. (b) Photograph of the two vortices emerging from the wingtips of an Airbus, made visible by light scattering off water droplets in the vortex cores (Ex. 14.6). Photo © Daniel Umaña. (c) Sketch of vortex lines (dashed) in the wingtip vortices of a flying bird and the flow lines (solid) of air around them. Sketch from Vogel, 1994, Fig. 12.7c, reprinted by permission.

$$\frac{d\mathbf{v}}{dt} \equiv \frac{\partial \mathbf{v}}{\partial t} + (\mathbf{v} \cdot \boldsymbol{\nabla})\mathbf{v} = -\frac{\boldsymbol{\nabla} P}{\rho} - \boldsymbol{\nabla}\Phi + \nu\nabla^2\mathbf{v} \tag{14.2}$$

[Eq. (13.70) with $\mathbf{g} = -\boldsymbol{\nabla}\Phi$].

To derive the desired evolution equation for vorticity, we take the curl of Eq. (14.2) and use the vector identity $(\mathbf{v} \cdot \boldsymbol{\nabla})\mathbf{v} = \boldsymbol{\nabla}(v^2)/2 - \mathbf{v} \times \boldsymbol{\omega}$ (easily derivable using the Levi-Civita tensor and index notation) to obtain

$$\frac{\partial \boldsymbol{\omega}}{\partial t} = \boldsymbol{\nabla} \times (\mathbf{v} \times \boldsymbol{\omega}) - \frac{\boldsymbol{\nabla} P \times \boldsymbol{\nabla} \rho}{\rho^2} + \nu\nabla^2\boldsymbol{\omega}. \tag{14.3}$$

Although the flow is assumed incompressible, $\boldsymbol{\nabla} \cdot \mathbf{v} = 0$, the density can vary spatially due to a varying chemical composition (e.g., some regions might be oil and others water) or varying temperature and associated thermal expansion. Therefore, we must not omit the $\boldsymbol{\nabla} P \times \boldsymbol{\nabla} \rho$ term.

It is convenient to rewrite the vorticity evolution equation (14.3) with the aid of the relation (again derivable using the Levi-Civita tensor)

$$\boldsymbol{\nabla} \times (\mathbf{v} \times \boldsymbol{\omega}) = (\boldsymbol{\omega} \cdot \boldsymbol{\nabla})\mathbf{v} + \mathbf{v}(\boldsymbol{\nabla} \cdot \boldsymbol{\omega}) - \boldsymbol{\omega}(\boldsymbol{\nabla} \cdot \mathbf{v}) - (\mathbf{v} \cdot \boldsymbol{\nabla})\boldsymbol{\omega}. \tag{14.4}$$

Inserting this into Eq. (14.3), using $\boldsymbol{\nabla} \cdot \boldsymbol{\omega} = 0$ and $\boldsymbol{\nabla} \cdot \mathbf{v} = 0$, and introducing a new type of time derivative[2]

---

2. The combination of spatial derivatives appearing here is called the *Lie derivative* and is denoted $\mathcal{L}_{\mathbf{v}}\boldsymbol{\omega} \equiv (\mathbf{v} \cdot \boldsymbol{\nabla})\boldsymbol{\omega} - (\boldsymbol{\omega} \cdot \boldsymbol{\nabla})\mathbf{v}$; it is also the *commutator* of $\mathbf{v}$ and $\boldsymbol{\omega}$ and is denoted $[\mathbf{v}, \boldsymbol{\omega}]$. It is often encountered in differential geometry.

$$\boxed{\frac{D\omega}{Dt} \equiv \frac{\partial\omega}{\partial t} + (\mathbf{v}\cdot\nabla)\omega - (\omega\cdot\nabla)\mathbf{v} = \frac{d\omega}{dt} - (\omega\cdot\nabla)\mathbf{v},}$$ (14.5)

we bring Eq. (14.3) into the following form:

$$\boxed{\frac{D\omega}{Dt} = -\frac{\nabla P \times \nabla\rho}{\rho^2} + \nu\nabla^2\omega.}$$ (14.6)

vorticity evolution equation for incompressible flow

This is our favorite form for the vorticity evolution equation for an incompressible flow, $\nabla\cdot\mathbf{v} = 0$. If there are additional accelerations acting on the fluid, then their curls must be added to the right-hand side. The most important examples are the Coriolis acceleration $-2\mathbf{\Omega}\times\mathbf{v}$ in a reference frame that rotates rigidly with angular velocity $\mathbf{\Omega}$ (Sec. 14.5.1), and the Lorentz-force acceleration $\mathbf{j}\times\mathbf{B}/\rho$ when the fluid has an internal electric current density $\mathbf{j}$ and an immersed magnetic field $\mathbf{B}$ (Sec. 19.2.1); then Eq. (14.6) becomes

$$\frac{D\omega}{Dt} = -\frac{\nabla P \times \nabla\rho}{\rho^2} + \nu\nabla^2\omega - 2\nabla\times(\mathbf{\Omega}\times\mathbf{v}) + \nabla\times(\mathbf{j}\times\mathbf{B}/\rho).$$ (14.7)

influence of Coriolis and Lorentz forces on vorticity

In the remainder of Sec. 14.2, we explore the predictions of our favorite form [Eq. (14.6)] of the vorticity evolution equation.

The operator D/Dt [defined by Eq. (14.5) when acting on a vector and by $D/Dt = d/dt$ when acting on a scalar] is called the *fluid derivative*. (Warning: The notation D/Dt is used in some older texts for the convective derivative d/dt.) The geometrical meaning of the fluid derivative can be understood from Fig. 14.3. Denote by $\Delta\mathbf{x}(t)$ the vector connecting two points $\mathcal{P}$ and $\mathcal{Q}$ that are moving with the fluid. Then the figure shows that the convective derivative $d\Delta\mathbf{x}/dt$ is the relative velocity of these two points, namely $(\Delta\mathbf{x}\cdot\nabla)\mathbf{v}$. Therefore, by the second equality in Eq. (14.5), the fluid derivative of $\Delta\mathbf{x}$ vanishes:

$$\frac{D\Delta\mathbf{x}}{Dt} = 0.$$ (14.8)

Correspondingly, *the fluid derivative of any vector is its rate of change relative to a vector, such as $\Delta\mathbf{x}$, whose tail and head move with the fluid.*

meaning of the fluid derivative D/Dt

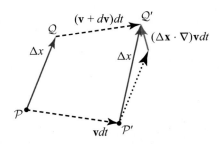

FIGURE 14.3 Equation of motion for an infinitesimal vector $\Delta\mathbf{x}$ connecting two fluid elements. As the fluid elements at $\mathcal{P}$ and $\mathcal{Q}$ move to $\mathcal{P}'$ and $\mathcal{Q}'$ in a time interval dt, the vector changes by $(\Delta\mathbf{x}\cdot\nabla)\mathbf{v}dt$.

14.2.2 Barotropic, Inviscid, Compressible Flows: Vortex Lines Frozen into Fluid

To understand the vorticity evolution law (14.6) physically, we explore various special cases in this and the next few subsections.

Here we specialize to a barotropic $[P = P(\rho)]$, inviscid ($\nu = 0$) fluid flow. (This kind of flow often occurs in Earth's atmosphere and oceans, well away from solid boundaries.) Then the right-hand side of Eq. (14.6) vanishes, leaving $D\boldsymbol{\omega}/Dt = 0$.

For generality, we temporarily (this subsection only) abandon our restriction to incompressible flow, $\boldsymbol{\nabla} \cdot \mathbf{v} = 0$, but keep the flow barotropic and inviscid. Then it is straightforward to deduce, from the curl of the Euler equation (13.44), that

$$\frac{D\boldsymbol{\omega}}{Dt} = -\boldsymbol{\omega}\boldsymbol{\nabla} \cdot \mathbf{v} \tag{14.9}$$

(Ex. 14.4). This equation shows that the vorticity has a fluid derivative parallel to itself: the fluid slides along its vortex lines, or equivalently, the vortex lines are frozen into (i.e., are carried by) the moving fluid. The wingtip vortex lines of Fig. 14.2 are an example. They are carried backward by the air that flowed over the wingtips, and they endow that air with vorticity that emerges from a wingtip.

We can actually make the fluid derivative vanish by substituting $\boldsymbol{\nabla} \cdot \mathbf{v} = -\rho^{-1}d\rho/dt$ (the equation of mass conservation) into Eq. (14.9); the result is

$$\boxed{\frac{D}{Dt}\left(\frac{\boldsymbol{\omega}}{\rho}\right) = 0 \quad \text{for barotropic, inviscid flow.}} \tag{14.10}$$

Therefore, the quantity $\boldsymbol{\omega}/\rho$ evolves according to the same equation as the separation $\Delta\mathbf{x}$ of two points in the fluid. To see what this implies, consider a small cylindrical fluid element whose symmetry axis is parallel to $\boldsymbol{\omega}$ (Fig. 14.4). Denote its vectorial length by $\Delta\mathbf{x}$, its vectorial cross sectional area by $\Delta\boldsymbol{\Sigma}$, and its conserved mass by $\Delta M = \rho\Delta\mathbf{x} \cdot \Delta\boldsymbol{\Sigma}$. Then, since $\boldsymbol{\omega}/\rho$ points along $\Delta\mathbf{x}$ and both are frozen into the fluid, it must be that $\boldsymbol{\omega}/\rho = \text{const} \times \Delta\mathbf{x}$. Therefore, the fluid element's conserved mass is $\Delta M = \rho\Delta\mathbf{x} \cdot \Delta\boldsymbol{\Sigma} = \text{const} \times \boldsymbol{\omega} \cdot \Delta\boldsymbol{\Sigma}$, so $\boldsymbol{\omega} \cdot \Delta\boldsymbol{\Sigma}$ is conserved as the cylindrical fluid element moves and deforms. We thereby conclude that the fluid's vortex lines, with number per unit area proportional to $|\boldsymbol{\omega}|$, are convected by our barotropic, inviscid fluid, without being created or destroyed.

for barotropic, inviscid flow: vortex lines are convected (frozen into the fluid)

Now return to an incompressible flow, $\boldsymbol{\nabla} \cdot \mathbf{v} = 0$ (which includes, of course, Earth's oceans and atmosphere), so the vorticity evolution equation becomes $D\boldsymbol{\omega}/Dt = 0$. Suppose that the flow is 2 dimensional (as it commonly is to moderate accuracy when averaged over transverse scales large compared to the thickness of the atmosphere and oceans), so $\mathbf{v}$ is in the x and y directions and is independent of z. Then $\boldsymbol{\omega} = \omega\mathbf{e}_z$, and we can regard the vorticity as the scalar ω. Then Eq. (14.5) with $(\boldsymbol{\omega} \cdot \boldsymbol{\nabla})\mathbf{v} = 0$ implies that the vorticity obeys the simple propagation law

$$\frac{d\omega}{dt} = 0. \tag{14.11}$$

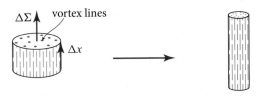

FIGURE 14.4 Simple demonstration of the kinematics of vorticity propagation in a compressible, barotropic, inviscid flow. A short, thick cylindrical fluid element with generators parallel to the local vorticity is deformed, by the flow, into a long, slender cylinder. By virtue of Eq. (14.10), we can think of the vortex lines as being convected with the fluid, with no creation of new lines or destruction of old ones, so that the number of vortex lines passing through the cylinder (through its end surface $\Delta\mathbf{\Sigma}$) remains constant.

Thus, *in a 2-dimensional, incompressible, barotropic, inviscid flow, the scalar vorticity is conserved when convected,* just like entropy per unit mass in an adiabatic fluid.

EXERCISES

Exercise 14.4 *Derivation: Vorticity Evolution in a Compressible, Barotropic, Inviscid Flow*
By taking the curl of the Euler equation (13.44), derive the vorticity evolution equation (14.9) for a compressible, barotropic, inviscid flow.

Geometric Optics

(Selected Sections of Chapter 7

of *Modern Classical Physics*)

7.2 Waves in a Homogeneous Medium

7.2.1 Monochromatic Plane Waves; Dispersion Relation

Consider a monochromatic plane wave propagating through a homogeneous medium. Independently of the physical nature of the wave, it can be described mathematically by

$$\psi = A e^{i(\mathbf{k}\cdot\mathbf{x}-\omega t)} \equiv A e^{i\varphi}, \tag{7.1}$$

where ψ is any oscillatory physical quantity associated with the wave, for example, the y component of the magnetic field associated with an electromagnetic wave. If, as is usually the case, the physical quantity is real (not complex), then we must take the real part of Eq. (7.1). In Eq. (7.1), A is the wave's *complex amplitude*; $\varphi = \mathbf{k}\cdot\mathbf{x} - \omega t$ is the wave's *phase*; t and $\mathbf{x}$ are time and location in space; $\omega = 2\pi f$ is the wave's *angular frequency*; and $\mathbf{k}$ is its *wave vector* (with $k \equiv |\mathbf{k}|$ its *wave number*, $\lambda = 2\pi/k$ its *wavelength*, $\lambdabar = \lambda/(2\pi)$ its *reduced wavelength*, and $\hat{\mathbf{k}} \equiv \mathbf{k}/k$ its *propagation direction*). Surfaces of constant phase φ are orthogonal to the propagation direction $\hat{\mathbf{k}}$ and move in the $\hat{\mathbf{k}}$ direction with the *phase velocity*

> plane wave: complex amplitude, phase, angular frequency, wave vector, wavelength, and propagation direction

$$\mathbf{V}_{\text{ph}} \equiv \left(\frac{\partial \mathbf{x}}{\partial t}\right)_{\varphi} = \frac{\omega}{k}\hat{\mathbf{k}} \tag{7.2}$$

> phase velocity

(cf. Fig. 7.1). The frequency ω is determined by the wave vector $\mathbf{k}$ in a manner that depends on the wave's physical nature; the functional relationship

$$\omega = \Omega(\mathbf{k}) \tag{7.3}$$

> dispersion relation

is called the wave's *dispersion relation*, because (as we shall see in Ex. 7.2) it governs the dispersion (spreading) of a wave packet that is constructed by superposing plane waves.

This appendix is extracted from Volume 2 of *Modern Classical Physics*. It consists of Secs. 7.2 and 7.3 of Chapter 7, "Geometric Optics," with no changes aside from pagination.

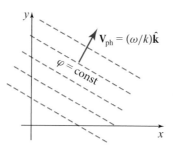

FIGURE 7.1 A monochromatic plane wave in a homogeneous medium.

examples:

Some examples of plane waves that we study in this book are:

electromagnetic waves

1. Electromagnetic waves propagating through an isotropic dielectric medium with index of refraction n [Eq. 10.20)], for which ψ could be any Cartesian component of the electric or magnetic field or vector potential and the dispersion relation is

$$\omega = \Omega(\mathbf{k}) = Ck \equiv C|\mathbf{k}|, \tag{7.4}$$

with $C = c/n$ the phase speed and c the speed of light in vacuum.

sound waves

2. Sound waves propagating through a solid (Sec. 12.2.3) or fluid (liquid or vapor; Secs. 7.3.1 and 16.5), for which ψ could be the pressure or density perturbation produced by the sound wave (or it could be a potential whose gradient is the velocity perturbation), and the dispersion relation is the same as for electromagnetic waves, Eq. (7.4), but with C now the sound speed.

water waves

3. Waves on the surface of a deep body of water (depth $\gg \lambda$; Sec. 16.2.1), for which ψ could be the height of the water above equilibrium, and the dispersion relation is [Eq. (16.9)]:

$$\omega = \Omega(\mathbf{k}) = \sqrt{gk} = \sqrt{g|\mathbf{k}|}, \tag{7.5}$$

with g the acceleration of gravity.

flexural waves

4. Flexural waves on a stiff beam or rod (Sec. 12.3.4), for which ψ could be the transverse displacement of the beam from equilibrium, and the dispersion relation is

$$\omega = \Omega(\mathbf{k}) = \sqrt{\frac{D}{\Lambda}} k^2 = \sqrt{\frac{D}{\Lambda}} \mathbf{k} \cdot \mathbf{k}, \tag{7.6}$$

with Λ the rod's mass per unit length and D its "flexural rigidity" [Eq. (12.33)].

5. Alfvén waves in a magnetized, nonrelativistic plasma (bending waves of the plasma-laden magnetic field lines; Sec. 19.7.2), for which ψ could be the transverse displacement of the field and plasma, and the dispersion relation is [Eq. (19.75)]

Alfvén waves

$$\omega = \Omega(\mathbf{k}) = \mathbf{a} \cdot \mathbf{k}, \tag{7.7}$$

with $\mathbf{a} = \mathbf{B}/\sqrt{\mu_o \rho}$, $[= \mathbf{B}/\sqrt{4\pi\rho}]^1$ the Alfvén speed, $\mathbf{B}$ the (homogeneous) magnetic field, μ_o the magnetic permittivity of the vacuum, and ρ the plasma mass density.

6. Gravitational waves propagating across the universe, for which ψ can be a component of the waves' metric perturbation which describes the waves' stretching and squeezing of space; these waves propagate nondispersively at the speed of light, so their dispersion relation is Eq. (7.4) with C replaced by the vacuum light speed c.

gravitational waves

In general, one can derive the dispersion relation $\omega = \Omega(\mathbf{k})$ by inserting the plane-wave ansatz (7.1) into the dynamical equations that govern one's physical system [e.g., Maxwell's equations, the equations of elastodynamics (Chap. 12), or the equations for a magnetized plasma (Part VI)]. We shall do so time and again in this book.

7.2.2 Wave Packets

7.2.2

Waves in the real world are not precisely monochromatic and planar. Instead, they occupy wave packets that are somewhat localized in space and time. Such wave packets can be constructed as superpositions of plane waves:

$$\psi(\mathbf{x}, t) = \int A(\mathbf{k}) e^{i\alpha(\mathbf{k})} e^{i(\mathbf{k}\cdot\mathbf{x} - \omega t)} \frac{d^3 k}{(2\pi)^3}, \tag{7.8a}$$
where $A(\mathbf{k})$ is concentrated around some $\mathbf{k} = \mathbf{k}_o$.

wave packet

Here A and α (both real) are the modulus and phase of the complex amplitude $Ae^{i\alpha}$, and the integration element is $d^3 k \equiv d\mathcal{V}_k \equiv dk_x dk_y dk_z$ in terms of components of $\mathbf{k}$ along Cartesian axes x, y, and z. In the integral (7.8a), the contributions from adjacent $\mathbf{k}$s will tend to cancel each other except in that region of space and time where the oscillatory phase factor changes little with changing $\mathbf{k}$ (when $\mathbf{k}$ is near $\mathbf{k}_o$). This is the spacetime region in which the wave packet is concentrated, and its center is where $\nabla_{\mathbf{k}}$(phase factor) $= 0$:

$$\left(\frac{\partial \alpha}{\partial k_j} + \frac{\partial}{\partial k_j}(\mathbf{k} \cdot \mathbf{x} - \omega t) \right)_{\mathbf{k}=\mathbf{k}_o} = 0. \tag{7.8b}$$

---

1. Gaussian unit equivalents will be given with square brackets.

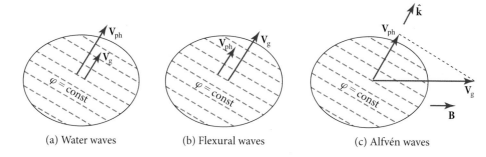

(a) Water waves	(b) Flexural waves	(c) Alfvén waves

FIGURE 7.2 (a) A wave packet of waves on a deep body of water. The packet is localized in the spatial region bounded by the ellipse. The packet's (ellipse's) center moves with the group velocity $\mathbf{V}_g$. The ellipse expands slowly due to wave-packet dispersion (spreading; Ex. 7.2). The surfaces of constant phase (the wave's oscillations) move twice as fast as the ellipse and in the same direction, $\mathbf{V}_{\mathrm{ph}} = 2\mathbf{V}_g$ [Eq. (7.11)]. This means that the wave's oscillations arise at the back of the packet and move forward through the packet, disappearing at the front. The wavelength of these oscillations is $\lambda = 2\pi/k_o$, where $k_o = |\mathbf{k}_o|$ is the wave number about which the wave packet is concentrated [Eq. (7.8a) and associated discussion]. (b) A flexural wave packet on a beam, for which $\mathbf{V}_{\mathrm{ph}} = \frac{1}{2}\mathbf{V}_g$ [Eq. (7.12)], so the wave's oscillations arise at the packet's front and, traveling more slowly than the packet, disappear at its back. (c) An Alfvén wave packet. Its center moves with a group velocity $\mathbf{V}_g$ that points along the direction of the background magnetic field [Eq. (7.13)], and its surfaces of constant phase (the wave's oscillations) move with a phase velocity $\mathbf{V}_{\mathrm{ph}}$ that can be in any direction $\hat{\mathbf{k}}$. The phase speed is the projection of the group velocity onto the phase propagation direction, $|\mathbf{V}_{\mathrm{ph}}| = \mathbf{V}_g \cdot \hat{\mathbf{k}}$ [Eq. (7.13)], which implies that the wave's oscillations remain fixed inside the packet as the packet moves; their pattern inside the ellipse does not change. (An even more striking example is provided by the Rossby wave, discussed in Sec. 16.4, in which the group velocity is equal and oppositely directed to the phase velocity.)

Evaluating the derivative with the aid of the wave's dispersion relation $\omega = \Omega(\mathbf{k})$, we obtain for the location of the wave packet's center

$$x_j - \left(\frac{\partial \Omega}{\partial k_j}\right)_{\mathbf{k}=\mathbf{k}_o} t = -\left(\frac{\partial \alpha}{\partial k_j}\right)_{\mathbf{k}=\mathbf{k}_o} = \text{const.} \tag{7.8c}$$

This tells us that the *wave packet* moves with the *group velocity*

group velocity

$$\boxed{\mathbf{V}_g = \nabla_{\mathbf{k}}\Omega, \quad \text{i.e.,} \quad V_{g\,j} = \left(\frac{\partial \Omega}{\partial k_j}\right)_{\mathbf{k}=\mathbf{k}_o}.} \tag{7.9}$$

When, as for electromagnetic waves in a dielectric medium or sound waves in a solid or fluid, the dispersion relation has the simple form of Eq. (7.4), $\omega = \Omega(\mathbf{k}) = Ck$ with $k \equiv |\mathbf{k}|$, then the group and phase velocities are the same,

$$\mathbf{V}_g = \mathbf{V}_{\mathrm{ph}} = C\hat{\mathbf{k}}, \tag{7.10}$$

and the waves are said to be *dispersionless*. If the dispersion relation has any other form, then the group and phase velocities are different, and the wave is said to exhibit *dispersion;* cf. Ex. 7.2. Examples are (see Fig. 7.2 and the list in Sec. 7.2.1, from which our numbering is taken):

3. Waves on a deep body of water [dispersion relation (7.5); Fig. 7.2a], for which

$$\mathbf{V}_g = \frac{1}{2}\mathbf{V}_{ph} = \frac{1}{2}\sqrt{\frac{g}{k}}\,\hat{\mathbf{k}};\tag{7.11}$$

4. Flexural waves on a stiff beam or rod [dispersion relation (7.6); Fig. 7.2b], for which

$$\mathbf{V}_g = 2\mathbf{V}_{ph} = 2\sqrt{\frac{D}{\Lambda}}\,k\hat{\mathbf{k}};\tag{7.12}$$

5. Alfvén waves in a magnetized and nonrelativistic plasma [dispersion relation (7.7); Fig. 7.2c], for which

$$\mathbf{V}_g = \mathbf{a}, \qquad \mathbf{V}_{ph} = (\mathbf{a}\cdot\hat{\mathbf{k}})\hat{\mathbf{k}}.\tag{7.13}$$

Notice that, depending on the dispersion relation, the group speed $|\mathbf{V}_g|$ can be less than or greater than the phase speed, and if the homogeneous medium is anisotropic (e.g., for a magnetized plasma), the group velocity can point in a different direction than the phase velocity.

Physically, it should be obvious that the energy contained in a wave packet must remain always with the packet and cannot move into the region outside the packet where the wave amplitude vanishes. Correspondingly, the wave packet's energy must propagate with the group velocity $\mathbf{V}_g$ and not with the phase velocity $\mathbf{V}_{ph}$. When one examines the wave packet from a quantum mechanical viewpoint, its quanta must move with the group velocity $\mathbf{V}_g$. Since we have required that the wave packet have its wave vectors concentrated around $\mathbf{k}_o$, the energy and momentum of each of the packet's quanta are given by the standard quantum mechanical relations:

$$\boxed{\mathcal{E} = \hbar\Omega(\mathbf{k}_o), \quad \text{and} \quad \mathbf{p} = \hbar\mathbf{k}_o.}\tag{7.14}$$

EXERCISES

Exercise 7.1 *Practice: Group and Phase Velocities*
Derive the group and phase velocities (7.10)–(7.13) from the dispersion relations (7.4)–(7.7).

Exercise 7.2 **Example: Gaussian Wave Packet and Its Dispersion*
Consider a 1-dimensional wave packet, $\psi(x,t) = \int A(k)e^{i\alpha(k)}e^{i(kx-\omega t)}dk/(2\pi)$, with dispersion relation $\omega = \Omega(k)$. For concreteness, let $A(k)$ be a narrow Gaussian peaked around k_o: $A \propto \exp[-\kappa^2/(2(\Delta k)^2)]$, where $\kappa = k - k_o$.

(a) Expand α as $\alpha(k) = \alpha_o - x_o\kappa$ with x_o a constant, and assume for simplicity that higher order terms are negligible. Similarly, expand $\omega \equiv \Omega(k)$ to quadratic order, and explain why the coefficients are related to the group velocity V_g at $k = k_o$ by $\Omega = \omega_o + V_g\kappa + (dV_g/dk)\kappa^2/2$.

(b) Show that the wave packet is given by

$$\psi \propto \exp[i(\alpha_o + k_o x - \omega_o t)] \int_{-\infty}^{+\infty} \exp[i\kappa(x - x_o - V_g t)] \tag{7.15a}$$

$$\times \exp\left[-\frac{\kappa^2}{2}\left(\frac{1}{(\Delta k)^2} + i\frac{dV_g}{dk}t\right)\right] d\kappa.$$

The term in front of the integral describes the phase evolution of the waves inside the packet; cf. Fig. 7.2.

(c) Evaluate the integral analytically (with the help of a computer, if you wish). From your answer, show that the modulus of ψ satisfies

$$\boxed{|\psi| \propto \frac{1}{L^{1/2}} \exp\left[-\frac{(x - x_o - V_g t)^2}{2L^2}\right], \quad \text{where } L = \frac{1}{\Delta k}\sqrt{1 + \left(\frac{dV_g}{dk}(\Delta k)^2 t\right)^2}}$$

$$\tag{7.15b}$$

is the packet's half-width.

(d) Discuss the relationship of this result at time $t = 0$ to the uncertainty principle for the localization of the packet's quanta.

(e) Equation (7.15b) shows that the wave packet spreads (i.e., disperses) due to its containing a range of group velocities [Eq. (7.11)]. How long does it take for the packet to enlarge by a factor 2? For what range of initial half-widths can a water wave on the ocean spread by less than a factor 2 while traveling from Hawaii to California?

7.3

7.3 Waves in an Inhomogeneous, Time-Varying Medium: The Eikonal Approximation and Geometric Optics

Suppose that the medium in which the waves propagate is spatially inhomogeneous and varies with time. If the lengthscale $\mathcal{L}$ and timescale $\mathcal{T}$ for substantial variations are long compared to the waves' reduced wavelength and period,

$$\mathcal{L} \gg \lambdabar = 1/k, \qquad \mathcal{T} \gg 1/\omega, \tag{7.16}$$

then the waves can be regarded locally as planar and monochromatic. The medium's inhomogeneities and time variations may produce variations in the wave vector **k** and frequency ω, but those variations should be substantial only on scales $\gtrsim \mathcal{L} \gg 1/k$ and $\gtrsim \mathcal{T} \gg 1/\omega$. This intuitively obvious fact can be proved rigorously using a two-lengthscale expansion (i.e., an expansion of the wave equation in powers of $\lambdabar/\mathcal{L} = 1/k\mathcal{L}$ and $1/\omega\mathcal{T}$). Such an expansion, in this context of wave propagation, is called

two-lengthscale expansion

the *geometric-optics approximation* or the *eikonal approximation* (after the Greek word εικων, meaning image). When the waves are those of elementary quantum mechanics, it is called the *WKB approximation*.[2] The eikonal approximation converts the laws of wave propagation into a remarkably simple form, in which the waves' amplitude is transported along trajectories in spacetime called *rays*. In the language of quantum mechanics, these rays are the world lines of the wave's quanta (photons for light, phonons for sound, plasmons for Alfvén waves, and gravitons for gravitational waves), and the law by which the wave amplitude is transported along the rays is one that conserves quanta. These ray-based propagation laws are called the laws of *geometric optics*.

eikonal approximation

WKB approximation

In this section we develop and study the eikonal approximation and its resulting laws of geometric optics. We begin in Sec. 7.3.1 with a full development of the eikonal approximation and its geometric-optics consequences for a prototypical dispersion-free wave equation that represents, for example, sound waves in a weakly inhomogeneous fluid. In Sec. 7.3.3, we extend our analysis to cover all other types of waves. In Sec. 7.3.4 and a number of exercises we explore examples of geometric-optics waves, and in Sec. 7.3.5 we discuss conditions under which the eikonal approximation breaks down and some non-geometric-optics phenomena that result from the breakdown. Finally, in Sec. 7.3.6 we return to nondispersive light and sound waves, deduce Fermat's principle, and explore some of its consequences.

7.3.1 Geometric Optics for a Prototypical Wave Equation

7.3.1

Our prototypical wave equation is

$$\frac{\partial}{\partial t}\left(W\frac{\partial\psi}{\partial t}\right) - \boldsymbol{\nabla}\cdot(WC^2\boldsymbol{\nabla}\psi) = 0. \tag{7.17}$$

prototypical wave equation in a slowly varying medium

Here $\psi(\mathbf{x}, t)$ is the quantity that oscillates (the *wave field*), $C(\mathbf{x}, t)$ will turn out to be the wave's slowly varying *propagation speed*, and $W(\mathbf{x}, t)$ is a slowly varying *weighting function* that depends on the properties of the medium through which the wave propagates. As we shall see, W has no influence on the wave's dispersion relation or on its geometric-optics rays, but it does influence the law of transport for the waves' amplitude.

The wave equation (7.17) describes sound waves propagating through a static, isentropic, inhomogeneous fluid (Ex. 16.13), in which case ψ is the wave's pressure perturbation δP, $C(\mathbf{x}) = \sqrt{(\partial P/\partial\rho)_s}$ is the adabiatic sound speed, and the weighting function is $W(\mathbf{x}) = 1/(\rho C^2)$, with ρ the fluid's unperturbed density. This wave equation also describes waves on the surface of a lake or pond or the ocean, in the limit that the slowly varying depth of the undisturbed water $h_o(\mathbf{x})$ is small compared

---

2. Sometimes called "JWKB," adding Jeffreys to the attribution, though Carlini, Liouville, and Green used it a century earlier.

to the wavelength (shallow-water waves; e.g., tsunamis); see Ex. 16.3. In this case ψ is the perturbation of the water's depth, $W = 1$, and $C = \sqrt{gh_o}$ with g the acceleration of gravity. In both cases—sound waves in a fluid and shallow-water waves—if we turn on a slow time dependence in the unperturbed fluid, then additional terms enter the wave equation (7.17). For pedagogical simplicity we leave those terms out, but in the analysis below we do allow W and C to be slowly varying in time, as well as in space: $W = W(\mathbf{x}, t)$ and $C = C(\mathbf{x}, t)$.

Associated with the wave equation (7.17) are an energy density $U(\mathbf{x}, t)$ and energy flux $\mathbf{F}(\mathbf{x}, t)$ given by

energy density and flux

$$U = W\left[\frac{1}{2}\left(\frac{\partial \psi}{\partial t}\right)^2 + \frac{1}{2}C^2(\nabla\psi)^2\right], \qquad \mathbf{F} = -WC^2\frac{\partial \psi}{\partial t}\nabla\psi; \qquad (7.18)$$

see Ex. 7.4. It is straightforward to verify that, if C and W are independent of time t, then the scalar wave equation (7.17) guarantees that the U and $\mathbf{F}$ of Eq. (7.18) satisfy the law of energy conservation:

$$\frac{\partial U}{\partial t} + \nabla \cdot \mathbf{F} = 0; \qquad (7.19)$$

cf. Ex. 7.4.[3]

We now specialize to a weakly inhomogeneous and slowly time-varying fluid and to nearly plane waves, and we seek a solution of the wave equation (7.17) that locally has approximately the plane-wave form $\psi \simeq Ae^{i\mathbf{k}\cdot\mathbf{x}-\omega t}$. Motivated by this plane-wave form, (i) we express the waves in the eikonal approximation as the product of a real amplitude $A(\mathbf{x}, t)$ that varies slowly on the length- and timescales $\mathcal{L}$ and $\mathcal{T}$, and the exponential of a complex phase $\varphi(\mathbf{x}, t)$ that varies rapidly on the timescale $1/\omega$ and lengthscale λ:

eikonal approximated wave: amplitude, phase, wave vector, and angular frequency

$$\psi(\mathbf{x}, t) = A(\mathbf{x}, t)e^{i\varphi(\mathbf{x},t)}; \qquad (7.20)$$

and (ii) we define the wave vector (field) and angular frequency (field) by

$$\mathbf{k}(\mathbf{x}, t) \equiv \nabla\varphi, \qquad \omega(\mathbf{x}, t) \equiv -\partial\varphi/\partial t. \qquad (7.21)$$

In addition to our two-lengthscale requirement, $\mathcal{L} \gg 1/k$ and $\mathcal{T} \gg 1/\omega$, we also require that A, $\mathbf{k}$, and ω vary slowly (i.e., vary on lengthscales $\mathcal{R}$ and timescales $\mathcal{T}'$ long compared to $\lambda = 1/k$ and $1/\omega$).[4] This requirement guarantees that the waves are locally planar, $\varphi \simeq \mathbf{k} \cdot x - \omega t + \text{constant}$.

---

3. Alternatively, one can observe that a stationary medium will not perform work.

4. Note that these variations can arise both (i) from the influence of the medium's inhomogeneity (which puts limits $\mathcal{R} \lesssim \mathcal{L}$ and $\mathcal{T}' \lesssim \mathcal{T}$ on the wave's variations) and (ii) from the chosen form of the wave. For example, the wave might be traveling outward from a source and so have nearly spherical phase fronts with radii of curvature $r \simeq$ (distance from source); then $\mathcal{R} = \min(r, \mathcal{L})$.

We now insert the eikonal-approximated wave field (7.20) into the wave equation (7.17), perform the differentiations with the aid of Eqs. (7.21), and collect terms in a manner dictated by a two-lengthscale expansion (see Box 7.2):

$$0 = \frac{\partial}{\partial t}\left(W\frac{\partial\psi}{\partial t}\right) - \nabla\cdot(WC^2\nabla\psi) \tag{7.22}$$

$$= \left(-\omega^2 + C^2 k^2\right)W\psi$$

$$+ i\left[-2\left(\omega\frac{\partial A}{\partial t} + C^2 k_j A_{,j}\right)W - \frac{\partial(W\omega)}{\partial t}A - (WC^2 k_j)_{,j}A\right]e^{i\varphi} + \cdots.$$

The first term on the right-hand side, $(-\omega^2 + C^2 k^2)W\psi$, scales as λ^{-2} when we make the reduced wavelength λ shorter and shorter while holding the macroscopic lengthscales $\mathcal{L}$ and $\mathcal{R}$ fixed; the second term (in square brackets) scales as λ^{-1}; and the omitted terms scale as λ^0. This is what we mean by "collecting terms in a manner dictated by a two-lengthscale expansion." Because of their different scaling, the first, second, and omitted terms must vanish separately; they cannot possibly cancel one another.

dispersion relation

The vanishing of the first term in the eikonal-approximated wave equation (7.22) implies that the waves' frequency field $\omega(\mathbf{x}, t) \equiv -\partial\varphi/\partial t$ and wave-vector field $\mathbf{k} \equiv \nabla\varphi$ satisfy the dispersionless dispersion relation,

$$\omega = \Omega(\mathbf{k}, \mathbf{x}, t) \equiv C(\mathbf{x}, t)k, \tag{7.23}$$

where (as throughout this chapter) $k \equiv |\mathbf{k}|$. Notice that, as promised, this dispersion relation is independent of the weighting function W in the wave equation. Notice further that this dispersion relation is identical to that for a precisely plane wave in a homogeneous medium, Eq. (7.4), except that the propagation speed C is now a slowly varying function of space and time. This will always be so.

One can always deduce the geometric-optics dispersion relation by (i) considering a precisely plane, monochromatic wave in a precisely homogeneous, time-independent medium and deducing $\omega = \Omega(\mathbf{k})$ in a functional form that involves the medium's properties (e.g., density) and then (ii) allowing the properties to be slowly varying functions of $\mathbf{x}$ and t. The resulting dispersion relation [e.g., Eq. (7.23)] then acquires its $\mathbf{x}$ and t dependence from the properties of the medium.

The vanishing of the second term in the eikonal-approximated wave equation (7.22) dictates that the wave's real amplitude A is transported with the group velocity $\mathbf{V}_g = C\hat{\mathbf{k}}$ in the following manner:

propagation law for amplitude

$$\frac{dA}{dt} \equiv \left(\frac{\partial}{\partial t} + \mathbf{V}_g \cdot \nabla\right) A = -\frac{1}{2W\omega}\left[\frac{\partial(W\omega)}{\partial t} + \nabla \cdot (WC^2\mathbf{k})\right] A. \tag{7.24}$$

This propagation law, by contrast with the dispersion relation, does depend on the weighting function W. We return to this propagation law shortly and shall understand more deeply its dependence on W, but first we must investigate in detail the directions along which A is transported.

rays

The time derivative $d/dt = \partial/\partial t + \mathbf{V}_g \cdot \nabla$ appearing in the propagation law (7.24) is similar to the derivative with respect to proper time along a world line in special relativity, $d/d\tau = u^0\partial/\partial t + \mathbf{u} \cdot \nabla$ (with u^α the world line's 4-velocity). This analogy tells us that the waves' amplitude A is being propagated along some sort of world lines (trajectories). Those world lines (the waves' rays), in fact, are governed by Hamilton's equations of particle mechanics with the dispersion relation $\Omega(\mathbf{x}, t, \mathbf{k})$ playing the role of the hamiltonian and $\mathbf{k}$ playing the role of momentum:

Hamilton's equations for rays

$$\boxed{\frac{dx_j}{dt} = \left(\frac{\partial\Omega}{\partial k_j}\right)_{\mathbf{x},t} \equiv V_{g\,j},} \quad \boxed{\frac{dk_j}{dt} = -\left(\frac{\partial\Omega}{\partial x_j}\right)_{\mathbf{k},t},} \quad \boxed{\frac{d\omega}{dt} = \left(\frac{\partial\Omega}{\partial t}\right)_{\mathbf{x},\mathbf{k}}.} \tag{7.25}$$

The first of these Hamilton equations is just our definition of the group velocity, with which [according to Eq. (7.24)] the amplitude is transported. The second tells us how the wave vector $\mathbf{k}$ changes along a ray, and together with our knowledge of $C(\mathbf{x}, t)$, it tells us how the group velocity $\mathbf{V}_g = C\hat{\mathbf{k}}$ for our dispersionless waves changes along a ray, and thence defines the ray itself. The third tells us how the waves' frequency changes along a ray.

To deduce the second and third of these Hamilton equations, we begin by inserting the definitions $\omega = -\partial\varphi/\partial t$ and $\mathbf{k} = \nabla\varphi$ [Eqs. (7.21)] into the dispersion relation $\omega = \Omega(\mathbf{x}, t; \mathbf{k})$ for an arbitrary wave, thereby obtaining

$$\boxed{\frac{\partial\varphi}{\partial t} + \Omega(\mathbf{x}, t; \nabla\varphi) = 0.} \tag{7.26a}$$

This equation is known in optics as the *eikonal equation*. It is formally the same as the Hamilton-Jacobi equation of classical mechanics (see, e.g., Goldstein, Poole, and Safko, 2002), if we identify Ω with the hamiltonian and φ with Hamilton's principal function (cf. Ex. 7.9). This suggests that, to derive the second and third of Eqs. (7.25), we can follow the same procedure as is used to derive Hamilton's equations of motion. We take the gradient of Eq. (7.26a) to obtain

eikonal equation and
Hamilton-Jacobi equation

$$\frac{\partial^2\varphi}{\partial t\,\partial x_j} + \frac{\partial\Omega}{\partial k_l}\frac{\partial^2\varphi}{\partial x_l\,\partial x_j} + \frac{\partial\Omega}{\partial x_j} = 0, \tag{7.26b}$$

where the partial derivatives of Ω are with respect to its arguments $(\mathbf{x}, t; \mathbf{k})$; we then use $\partial\varphi/\partial x_j = k_j$ and $\partial\Omega/\partial k_l = V_{g\,l}$ to write Eq. (7.26b) as $dk_j/dt = -\partial\Omega/\partial x_j$. This is the second of Hamilton's equations (7.25), and it tells us how the wave vector changes along a ray. The third Hamilton equation, $d\omega/dt = \partial\Omega/\partial t$ [Eq. (7.25)], is obtained by taking the time derivative of the eikonal equation (7.26a).

Not only is the waves' amplitude A propagated along the rays, so also is their phase:

$$\frac{d\varphi}{dt} = \frac{\partial\varphi}{\partial t} + \mathbf{V}_g \cdot \nabla\varphi = -\omega + \mathbf{V}_g \cdot \mathbf{k}. \tag{7.27}$$

propagation equation for
phase

Since our dispersionless waves have $\omega = Ck$ and $\mathbf{V}_g = C\hat{\mathbf{k}}$, this vanishes. Therefore, for the special case of dispersionless waves (e.g., sound waves in a fluid and electromagnetic waves in an isotropic dielectric medium), the phase is constant along each ray:

$$\boxed{d\varphi/dt = 0.} \tag{7.28}$$

7.3.2 Connection of Geometric Optics to Quantum Theory

7.3.2

Although the waves $\psi = Ae^{i\varphi}$ are classical and our analysis is classical, their propagation laws in the eikonal approximation can be described most nicely in quantum mechanical language.[5] Quantum mechanics insists that, associated with any wave in

---

5. This is intimately related to the fact that quantum mechanics underlies classical mechanics; the classical world is an approximation to the quantum world, often a very good approximation.

the geometric-optics regime, there are real quanta: the wave's quantum mechanical particles. If the wave is electromagnetic, the quanta are photons; if it is gravitational, they are gravitons; if it is sound, they are phonons; if it is a plasma wave (e.g., Alfvén), they are plasmons. When we multiply the wave's $\mathbf{k}$ and ω by $\hbar$, we obtain the particles' momentum and energy:

momentum and energy of quanta

$$\boxed{\mathbf{p} = \hbar \mathbf{k}, \qquad \mathcal{E} = \hbar \omega.} \tag{7.29}$$

Although the originators of the nineteenth-century theory of classical waves were unaware of these quanta, once quantum mechanics had been formulated, the quanta became a powerful conceptual tool for thinking about classical waves.

In particular, we can regard the rays as the world lines of the quanta, and by multiplying the dispersion relation by $\hbar$, we can obtain the hamiltonian for the quanta's world lines:

hamiltonian for quanta

$$\boxed{H(\mathbf{x}, t; \mathbf{p}) = \hbar \Omega(\mathbf{x}, t; \mathbf{k} = \mathbf{p}/\hbar).} \tag{7.30}$$

Hamilton's equations (7.25) for the rays then immediately become Hamilton's equations for the quanta: $dx_j/dt = \partial H/\partial p_j$, $dp_j/dt = -\partial H/\partial x_j$, and $d\mathcal{E}/dt = \partial H/\partial t$.

Return now to the propagation law (7.24) for the waves' amplitude, and examine its consequences for the waves' energy. By inserting the ansatz $\psi = \Re(A e^{i\varphi}) = A \cos(\varphi)$ into Eqs. (7.18) for the energy density U and energy flux $\mathbf{F}$ and averaging over a wavelength and wave period (so $\overline{\cos^2 \varphi} = \overline{\sin^2 \varphi} = 1/2$), we find that

$$U = \frac{1}{2} W C^2 k^2 A^2 = \frac{1}{2} W \omega^2 A^2, \qquad \mathbf{F} = U(C\hat{\mathbf{k}}) = U \mathbf{V}_{\mathrm{g}}. \tag{7.31}$$

Inserting these into the expression $\partial U/\partial t + \boldsymbol{\nabla} \cdot \mathbf{F}$ for the rate at which energy (per unit volume) fails to be conserved and using the propagation law (7.24) for A, we obtain

$$\frac{\partial U}{\partial t} + \boldsymbol{\nabla} \cdot \mathbf{F} = U \frac{\partial \ln C}{\partial t}. \tag{7.32}$$

Thus, as the propagation speed C slowly changes at a fixed location in space due to a slow change in the medium's properties, the medium slowly pumps energy into the waves or removes it from them at a rate per unit volume of $U \partial \ln C/\partial t$.

This slow energy change can be understood more deeply using quantum concepts. The number density and number flux of quanta are

number density and flux for quanta

$$\boxed{n = \frac{U}{\hbar \omega}, \qquad \mathbf{S} = \frac{\mathbf{F}}{\hbar \omega} = n \mathbf{V}_{\mathrm{g}}.} \tag{7.33}$$

By combining these equations with the energy (non)conservation equation (7.32), we obtain

$$\frac{\partial n}{\partial t} + \boldsymbol{\nabla} \cdot \mathbf{S} = n \left[\frac{\partial \ln C}{\partial t} - \frac{d \ln \omega}{dt} \right]. \tag{7.34}$$

The third Hamilton equation (7.25) tells us that

$$d\omega/dt = (\partial\Omega/\partial t)_{x,k} = [\partial(Ck)/\partial t]_{x,k} = k\partial C/\partial t,$$

whence $d\ln\omega/dt = \partial\ln C/\partial t$, which, when inserted into Eq. (7.34), implies that the quanta are conserved:

$$\boxed{\frac{\partial n}{\partial t} + \mathbf{\nabla}\cdot\mathbf{S} = 0.} \qquad (7.35a)$$

conservation of quanta

Since $\mathbf{S} = n\mathbf{V}_g$ and $d/dt = \partial/\partial t + \mathbf{V}_g\cdot\mathbf{\nabla}$, we can rewrite this conservation law as a propagation law for the number density of quanta:

$$\boxed{\frac{dn}{dt} + n\mathbf{\nabla}\cdot\mathbf{V}_g = 0.} \qquad (7.35b)$$

The propagation law for the waves' amplitude, Eq. (7.24), can now be understood much more deeply: *The amplitude propagation law is nothing but the law of conservation of quanta in a slowly varying medium, rewritten in terms of the amplitude. This is true quite generally, for any kind of wave (Sec. 7.3.3); and the quickest route to the amplitude propagation law is often to express the wave's energy density U in terms of the amplitude and then invoke conservation of quanta,* Eq. (7.35b).

In Ex. 7.3 we show that the conservation law (7.35b) is equivalent to

$$\boxed{\frac{d(nC\mathcal{A})}{dt} = 0, \quad \text{i.e., } nC\mathcal{A} \text{ is a constant along each ray.}} \qquad (7.35c)$$

Here $\mathcal{A}$ is the cross sectional area of a bundle of rays surrounding the ray along which the wave is propagating. Equivalently, by virtue of Eqs. (7.33) and (7.31) for the number density of quanta in terms of the wave amplitude A, we have

$$\frac{d}{dt}A\sqrt{CW\omega\mathcal{A}} = 0, \quad \text{i.e., } A\sqrt{CW\omega\mathcal{A}} \text{ is a constant along each ray.} \quad (7.35d)$$

In Eqs. (7.33) and (7.35), we have boxed those equations that are completely general (because they embody conservation of quanta) and have not boxed those that are specialized to our prototypical wave equation.

EXERCISES

Exercise 7.3 ** *Derivation and Example: Amplitude Propagation for Dispersionless Waves Expressed as Constancy of Something along a Ray*

(a) In connection with Eq. (7.35b), explain why $\mathbf{\nabla}\cdot\mathbf{V}_g = d\ln\mathcal{V}/dt$, where $\mathcal{V}$ is the tiny volume occupied by a collection of the wave's quanta.

(b) Choose for the collection of quanta those that occupy a cross sectional area $\mathcal{A}$ orthogonal to a chosen ray, and a longitudinal length Δs along the ray, so $\mathcal{V} = \mathcal{A}\Delta s$. Show that $d\ln\Delta s/dt = d\ln C/dt$ and correspondingly, $d\ln\mathcal{V}/dt = d\ln(C\mathcal{A})/dt$.

(c) Given part (b), show that the conservation law (7.35b) is equivalent to the constancy of $nC\mathcal{A}$ along a ray, Eq. (7.35c).

(d) From the results of part (c), derive the constancy of $A\sqrt{CW\omega\mathcal{A}}$ along a ray (where A is the wave's amplitude), Eq. (7.35d).

Exercise 7.4 **Example: Energy Density and Flux, and Adiabatic Invariant,
for a Dispersionless Wave*

(a) Show that the prototypical scalar wave equation (7.17) follows from the variational principle

$$\delta \int \mathcal{L} dt\, d^3x = 0, \qquad (7.36a)$$

where $\mathcal{L}$ is the lagrangian density

$$\mathcal{L} = W\left[\frac{1}{2}\left(\frac{\partial\psi}{\partial t}\right)^2 - \frac{1}{2}C^2(\boldsymbol{\nabla}\psi)^2\right] \qquad (7.36b)$$

(not to be confused with the lengthscale $\mathcal{L}$ of inhomogeneities in the medium).

(b) For any scalar-field lagrangian density $\mathcal{L}(\psi, \partial\psi/\partial t, \boldsymbol{\nabla}\psi, \mathbf{x}, t)$, the energy density and energy flux can be expressed in terms of the lagrangian, in Cartesian coordinates, as

$$U(\mathbf{x}, t) = \frac{\partial\psi}{\partial t}\frac{\partial\mathcal{L}}{\partial\psi/\partial t} - \mathcal{L}, \qquad F_j = \frac{\partial\psi}{\partial t}\frac{\partial\mathcal{L}}{\partial\psi/\partial x_j}. \qquad (7.36c)$$

(Goldstein, Poole, and Safko, 2002, Sec. 13.3). Show, from the Euler-Lagrange equations for $\mathcal{L}$, that these expressions satisfy energy conservation, $\partial U/\partial t + \boldsymbol{\nabla}\cdot\mathbf{F} = 0$, *if* $\mathcal{L}$ has no explicit time dependence [e.g., for the lagrangian (7.36b) if $C = C(\mathbf{x})$ and $W = W(\mathbf{x})$ do not depend on time t].

(c) Show that expression (7.36c) for the field's energy density U and its energy flux F_j agree with Eqs. (7.18).

(d) Now, regard the wave amplitude ψ as a generalized (field) coordinate. Use the lagrangian $L = \int \mathcal{L} d^3x$ to define a field momentum Π conjugate to this ψ, and then compute a *wave action*,

$$J \equiv \int_0^{2\pi/\omega}\int \Pi(\partial\psi/\partial t)d^3x\, dt, \qquad (7.36d)$$

which is the continuum analog of Eq. (7.43) in Sec. 7.3.6. The temporal integral is over one wave period. Show that this J is proportional to the wave energy divided by the frequency and thence to the number of quanta in the wave.

It is shown in standard texts on classical mechanics that, for approximately periodic oscillations, the particle action (7.43), with the integral limited to one period of oscillation of q, is an *adiabatic invariant*. By the extension of that proof to continuum physics, the wave action (7.36d) is also an adiabatic invariant. This means that the wave action and hence the number of quanta in the waves are conserved when the medium [in our case the index of refraction $n(\mathbf{x})$] changes very slowly in time—a result asserted in the text, and one that also follows from quantum mechanics. We study the particle version (7.43) of this adiabatic invariant in detail when we analyze charged-particle motion in a slowly varying magnetic field in Sec. 20.7.4.

Exercise 7.5 *Problem: Propagation of Sound Waves in a Wind*
Consider sound waves propagating in an atmosphere with a horizontal wind. Assume that the sound speed C, as measured in the air's local rest frame, is constant. Let the wind velocity $\mathbf{u} = u_x \mathbf{e}_x$ increase linearly with height z above the ground: $u_x = Sz$, where S is the constant shearing rate. Consider only rays in the x-z plane.

(a) Give an expression for the dispersion relation $\omega = \Omega(\mathbf{x}, t; \mathbf{k})$. [Hint: In the local rest frame of the air, Ω should have its standard sound-wave form.]

(b) Show that k_x is constant along a ray path, and then demonstrate that sound waves will not propagate when

$$\left| \frac{\omega}{k_x} - u_x(z) \right| < C. \tag{7.37}$$

(c) Consider sound rays generated on the ground that make an angle θ to the horizontal initially. Derive the equations describing the rays, and use them to sketch the rays, distinguishing values of θ both less than and greater than $\pi/2$. (You might like to perform this exercise numerically.)

7.3.3 Geometric Optics for a General Wave

With the simple case of nondispersive sound waves (Secs. 7.3.1 and 7.3.2) as our model, we now study an arbitrary kind of wave in a weakly inhomogeneous and slowly time varying medium (e.g., any of the examples in Sec. 7.2.1: light waves in a dielectric medium, deep water waves, flexural waves on a stiff beam, or Alfvén waves). Whatever the wave may be, we seek a solution to its wave equation using the eikonal approximation $\psi = Ae^{i\varphi}$ with slowly varying amplitude A and rapidly varying phase φ. Depending on the nature of the wave, ψ and A might be a scalar (e.g., sound waves), a vector (e.g., light waves), or a tensor (e.g., gravitational waves).

When we insert the ansatz $\psi = Ae^{i\varphi}$ into the wave equation and collect terms in the manner dictated by our two-lengthscale expansion [as in Eq. (7.22) and Box 7.2], the leading-order term will arise from letting every temporal or spatial derivative act on the $e^{i\varphi}$. This is precisely where the derivatives would operate in the case of a plane wave in a homogeneous medium, and here, as there, the result of each differentiation is $\partial e^{i\varphi}/\partial t = -i\omega e^{i\varphi}$ or $\partial e^{i\varphi}/\partial x_j = ik_j e^{i\varphi}$. Correspondingly, the leading-order terms in the wave equation here will be identical to those in the homogeneous plane wave case: they will be the dispersion relation multiplied by something times the wave,

$$[-\omega^2 + \Omega^2(\mathbf{x}, t; \mathbf{k})] \times (\text{something}) A e^{i\varphi} = 0, \tag{7.38a}$$

dispersion relation for general wave

with the spatial and temporal dependence of Ω^2 entering through the medium's properties. This guarantees that (as we claimed in Sec. 7.3.1) the dispersion relation can be obtained by analyzing a plane, monochromatic wave in a homogeneous, time-independent medium and then letting the medium's properties, in the dispersion relation, vary slowly with $\mathbf{x}$ and t.

Each next-order ("subleading") term in the wave equation will entail just one of the wave operator's derivatives acting on a slowly varying quantity (A, a medium property, ω, or $\mathbf{k}$) and all the other derivatives acting on $e^{i\varphi}$. The subleading terms that interest us, for the moment, are those in which the one derivative acts on A, thereby propagating it. Therefore, the subleading terms can be deduced from the leading-order terms (7.38a) by replacing just one $i\omega A e^{i\varphi} = -A(e^{i\varphi})_{,t}$ by $-A_{,t} e^{i\varphi}$, and replacing just one $ik_j A e^{i\varphi} = A(e^{i\varphi})_{,j}$ by $A_{,j} e^{i\varphi}$ (where the subscript commas denote partial derivatives in Cartesian coordinates). A little thought then reveals that the equation for the vanishing of the subleading terms must take the form [deducible from the leading terms (7.38a)]:

$$-2i\omega \frac{\partial A}{\partial t} - 2i\Omega(\mathbf{k}, \mathbf{x}, t) \frac{\partial \Omega(\mathbf{k}, \mathbf{x}, t)}{\partial k_j} \frac{\partial A}{\partial x_j} = \text{terms proportional to } A. \tag{7.38b}$$

Using the dispersion relation $\omega = \Omega(\mathbf{x}, t; \mathbf{k})$ and the group velocity (first Hamilton equation) $\partial \Omega / \partial k_j = V_{g\,j}$, we bring this into the "propagate A along a ray" form:

$$\frac{dA}{dt} \equiv \frac{\partial A}{\partial t} + \mathbf{V}_g \cdot \nabla A = \text{terms proportional to } A. \tag{7.38c}$$

Let us return to the leading-order terms (7.38a) in the wave equation [i.e., to the dispersion relation $\omega = \Omega(\mathbf{x}, t; k)$]. For our general wave, as for the prototypical dispersionless wave of the previous two sections, the argument embodied in Eqs. (7.26) shows that the rays are determined by Hamilton's equations (7.25),

Hamilton's equations for general wave

$$\boxed{\frac{dx_j}{dt} = \left(\frac{\partial \Omega}{\partial k_j}\right)_{\mathbf{x},t} \equiv V_{g\,j}, \quad \frac{dk_j}{dt} = -\left(\frac{\partial \Omega}{\partial x_j}\right)_{\mathbf{k},t}, \quad \frac{d\omega}{dt} = \left(\frac{\partial \Omega}{\partial t}\right)_{\mathbf{x},\mathbf{k}},} \tag{7.39}$$

but using the general wave's dispersion relation $\Omega(\mathbf{k}, \mathbf{x}, t)$ rather than $\Omega = C(\mathbf{x}, t)k$. These Hamilton equations include propagation laws for $\omega = -\partial\varphi/\partial t$ and $k_j = \partial\varphi/\partial x_j$, from which we can deduce the propagation law (7.27) for φ along the rays:

propagation law for phase of general wave

$$\boxed{\frac{d\varphi}{dt} = -\omega + \mathbf{V}_g \cdot \mathbf{k}.} \tag{7.40}$$

For waves with dispersion, by contrast with sound in a fluid and other waves that have $\Omega = Ck$, φ will not be constant along a ray.

For our general wave, as for dispersionless waves, the Hamilton equations for the rays can be reinterpreted as Hamilton's equations for the world lines of the waves' quanta [Eq. (7.30) and associated discussion]. And for our general wave, as for dispersionless waves, the medium's slow variations are incapable of creating or destroying wave quanta.[6] Correspondingly, if one knows the relationship between the waves' energy density U and their amplitude A, and thence the relationship between the waves' quantum number density $n = U/\hbar\omega$ and A, then *from the quantum conservation law* [boxed Eqs. (7.35)]

$$\boxed{\frac{\partial n}{\partial t} + \mathbf{\nabla} \cdot (n\mathbf{V}_{\mathrm{g}}) = 0, \quad \frac{dn}{dt} + n\mathbf{\nabla} \cdot \mathbf{V}_{\mathrm{g}} = 0, \quad \text{or} \quad \frac{d(nC\mathcal{A})}{dt} = 0,} \tag{7.41}$$

conservation of quanta and propagation of amplitude for general wave

one can deduce the propagation law for A—and the result must be the same propagation law as one obtains from the subleading terms in the eikonal approximation.

7.3.4 Examples of Geometric-Optics Wave Propagation

7.3.4

SPHERICAL SOUND WAVES

As a simple example of these geometric-optics propagation laws, consider a sound wave propagating radially outward through a homogeneous fluid from a spherical source (e.g., a radially oscillating ball; cf. Sec. 16.5.3). The dispersion relation is Eq. (7.4): $\Omega = Ck$. It is straightforward (Ex. 7.6) to integrate Hamilton's equations and learn that the rays have the simple form $\{r = Ct + \text{constant}, \theta = \text{constant}, \phi = \text{constant}, \mathbf{k} = (\omega/C)\mathbf{e}_r\}$ in spherical polar coordinates, with $\mathbf{e}_r$ the unit radial vector. Because the wave is dispersionless, its phase φ must be conserved along a ray [Eq. (7.28)], so φ must be a function of $Ct - r, \theta$, and ϕ. For the waves to propagate radially, it is essential that $\mathbf{k} = \mathbf{\nabla}\varphi$ point very nearly radially, which implies that φ must be a rapidly varying function of $Ct - r$ and a slowly varying one of θ and ϕ. The law of conservation of quanta in this case reduces to the propagation law $d(rA)/dt = 0$ (Ex. 7.6), so rA is also a constant along the ray; we call it $\mathcal{B}$. Putting this all together, we conclude that the sound waves' pressure perturbation $\psi = \delta P$ has the form

$$\psi = \frac{\mathcal{B}(Ct - r, \theta, \phi)}{r}e^{i\varphi(Ct-r,\theta,\phi)}, \tag{7.42}$$

where the phase φ is rapidly varying in $Ct - r$ and slowly varying in the angles, and the amplitude $\mathcal{B}$ is slowly varying in $Ct - r$ and the angles.

FLEXURAL WAVES

As another example of the geometric-optics propagation laws, consider flexural waves on a spacecraft's tapering antenna. The dispersion relation is $\Omega = k^2\sqrt{D/\Lambda}$ [Eq. (7.6)] with $D/\Lambda \propto h^2$, where h is the antenna's thickness in its direction of bend (or the

---

6. This is a general feature of quantum theory; creation and destruction of quanta require imposed oscillations at the high frequency and short wavelength of the waves themselves, or at some submultiple of them (in the case of nonlinear creation and annihilation processes; Chap. 10).

antenna's diameter, if it has a circular cross section); cf. Eq. (12.33). Since Ω is independent of t, as the waves propagate from the spacecraft to the antenna's tip, their frequency ω is conserved [third of Eqs. (7.39)], which implies by the dispersion relation that $k = (D/\Lambda)^{-1/4}\omega^{1/2} \propto h^{-1/2}$; hence the wavelength decreases as $h^{1/2}$. The group velocity is $V_g = 2(D/\Lambda)^{1/4}\omega^{1/2} \propto h^{1/2}$. Since the energy per quantum $\hbar\omega$ is constant, particle conservation implies that the waves' energy must be conserved, which in this 1-dimensional problem means that the energy flowing through a segment of the antenna per unit time must be constant along the antenna. On physical grounds this constant energy flow rate must be proportional to $A^2 V_g h^2$, which means that the amplitude A must increase $\propto h^{-5/4}$ as the flexural waves approach the antenna's end. A qualitatively similar phenomenon is seen in the cracking of a bullwhip (where the speed of the end can become supersonic).

LIGHT THROUGH A LENS AND ALFVÉN WAVES

Figure 7.3 sketches two other examples: light propagating through a lens and Alfvén waves propagating in the magnetosphere of a planet. In Sec. 7.3.6 and the exercises we explore a variety of other applications, but first we describe how the geometric-optics propagation laws can fail (Sec. 7.3.5).

EXERCISES

Exercise 7.6 *Derivation and Practice: Quasi-Spherical Solution to Vacuum Scalar Wave Equation*
Derive the quasi-spherical solution (7.42) of the vacuum scalar wave equation $-\partial^2\psi/\partial t^2 + \nabla^2\psi = 0$ from the geometric-optics laws by the procedure sketched in the text.

7.3.5

7.3.5 Relation to Wave Packets; Limitations of the Eikonal Approximation and Geometric Optics

The form $\psi = Ae^{i\varphi}$ of the waves in the eikonal approximation is remarkably general. At some initial moment of time, A and φ can have any form whatsoever, so long as the two-lengthscale constraints are satisfied [A, $\omega \equiv -\partial\varphi/\partial t$, $\mathbf{k} \equiv \boldsymbol{\nabla}\varphi$, and dispersion relation $\Omega(\mathbf{k}; \mathbf{x}, t)$ all vary on lengthscales long compared to $\lambdabar = 1/k$ and on timescales long compared to $1/\omega$]. For example, ψ could be as nearly planar as is allowed by the inhomogeneities of the dispersion relation. At the other extreme, ψ could be a moderately narrow wave packet, confined initially to a small region of space (though not too small; its size must be large compared to its mean reduced wavelength). In either case, the evolution will be governed by the above propagation laws.

Of course, the eikonal approximation is an approximation. Its propagation laws make errors, though when the two-lengthscale constraints are well satisfied, the errors will be small for sufficiently short propagation times. Wave packets provide an important example. Dispersion (different group velocities for different wave vectors) causes wave packets to spread (disperse) as they propagate; see Ex. 7.2. This spreading

phenomena missed by geometric optics

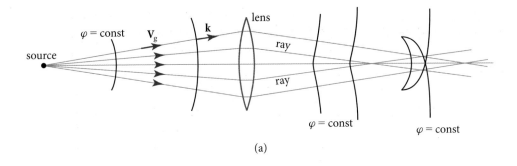

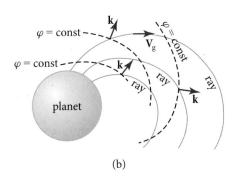

FIGURE 7.3 (a) The rays and the surfaces of constant phase φ at a fixed time for light passing through a converging lens [dispersion relation $\Omega = ck/\mathfrak{n}(\mathbf{x})$, where $\mathfrak{n}$ is the index of refraction]. In this case the rays (which always point along $\mathbf{V}_g$) are parallel to the wave vector $\mathbf{k} = \nabla\varphi$ and thus are also parallel to the phase velocity $\mathbf{V}_{ph}$, and the waves propagate along the rays with a speed $V_g = V_{ph} = c/\mathfrak{n}$ that is independent of wavelength. The strange self-intersecting shape of the last phase front is due to caustics; see Sec. 7.5. (b) The rays and surfaces of constant phase for Alfvén waves in the magnetosphere of a planet [dispersion relation $\Omega = \mathbf{a}(\mathbf{x}) \cdot \mathbf{k}$]. In this case, because $\mathbf{V}_g = \mathbf{a} \equiv \mathbf{B}/\sqrt{\mu_0\rho}$, the rays are parallel to the magnetic field lines and are not parallel to the wave vector, and the waves propagate along the field lines with speeds V_g that are independent of wavelength; cf. Fig. 7.2c. As a consequence, if some electric discharge excites Alfvén waves on the planetary surface, then they will be observable by a spacecraft when it passes magnetic field lines on which the discharge occurred. As the waves propagate, because $\mathbf{B}$ and ρ are time independent and hence $\partial\Omega/\partial t = 0$, the frequency ω and energy $\hbar\omega$ of each quantum is conserved, and conservation of quanta implies conservation of wave energy. Because the Alfvén speed generally diminishes with increasing distance from the planet, conservation of wave energy typically requires the waves' energy density and amplitude to increase as they climb upward.

is not included in the geometric-optics propagation laws; it is a fundamentally wave-based phenomenon and is lost when one goes to the particle-motion regime. In the limit that the wave packet becomes very large compared to its wavelength or that the packet propagates for only a short time, the spreading is small (Ex. 7.2). This is the geometric-optics regime, and geometric optics ignores the spreading.

Many other wave phenomena are missed by geometric optics. Examples are diffraction (e.g., at a geometric-optics caustic; Secs. 7.5 and 8.6), nonlinear wave-wave coupling (Chaps. 10 and 23, and Sec. 16.3), and parametric amplification of waves by rapid time variations of the medium (Sec. 10.7.3)—which shows up in quantum mechanics as particle production (i.e., a breakdown of the law of conservation of quanta). In Sec. 28.7.1 , we will encounter such particle production in inflationary models of the early universe.

7.3.6 Fermat's Principle

Hamilton's equations of optics allow us to solve for the paths of rays in media that vary both spatially and temporally. When the medium is time independent, the rays $\mathbf{x}(t)$ can be computed from a variational principle due to Fermat. This is the optical **principle of least action** analog of the classical dynamics principle of least action,[7] which states that, when a particle moves from one point to another through a time-independent potential (so its energy, the hamiltonian, is conserved), then the path $\mathbf{q}(t)$ that it follows is one that extremizes the action

$$J = \int \mathbf{p} \cdot d\mathbf{q} \qquad (7.43)$$

(where $\mathbf{q}$ and $\mathbf{p}$ are the particle's generalized coordinates and momentum), subject to the constraint that the paths have a fixed starting point, a fixed endpoint, and constant energy. The proof (e.g., Goldstein, Poole, and Safko, 2002, Sec. 8.6) carries over directly to optics when we replace the hamiltonian by Ω, $\mathbf{q}$ by $\mathbf{x}$, and $\mathbf{p}$ by $\mathbf{k}$. The resulting Fermat principle, stated with some care, has the following form.

Fermat's principle
Consider waves whose hamiltonian $\Omega(\mathbf{k}, \mathbf{x})$ is independent of time. Choose an initial location $\mathbf{x}_{\text{initial}}$ and a final location $\mathbf{x}_{\text{final}}$ in space, and consider the rays $\mathbf{x}(t)$ that connect these two points. The rays (usually only one) are those paths that satisfy the variational principle

$$\boxed{\delta \int \mathbf{k} \cdot d\mathbf{x} = 0.} \qquad (7.44)$$

In this variational principle, $\mathbf{k}$ must be expressed in terms of the trial path $\mathbf{x}(t)$ using Hamilton's equation $dx^j/dt = -\partial\Omega/\partial k_j$; the rate that the trial path is traversed (i.e., the magnitude of the group velocity) must be adjusted to keep Ω constant along the trial path (which means that the total time taken to go from $\mathbf{x}_{\text{initial}}$ to $\mathbf{x}_{\text{final}}$ can differ from one trial path to another). And of course, the trial paths must all begin at $\mathbf{x}_{\text{initial}}$ and end at $\mathbf{x}_{\text{final}}$.

PATH INTEGRALS

Notice that, once a ray has been identified by this action principle, it has $\mathbf{k} = \nabla\varphi$, and therefore the extremal value of the action $\int \mathbf{k} \cdot d\mathbf{x}$ along the ray is equal to the waves'

---

7. This is commonly attributed to Maupertuis, though others, including Leibniz and Euler, understood it earlier or better. This "action" and the rules for its variation are different from those in play in Hamilton's principle.

phase difference $\Delta\varphi$ between $\mathbf{x}_{\text{initial}}$ and $\mathbf{x}_{\text{final}}$. Correspondingly, for any trial path, we can think of the action as a phase difference along that path,

$$\Delta\varphi = \int \mathbf{k} \cdot d\mathbf{x}, \tag{7.45a}$$

and we can think of Fermat's principle as saying that the particle travels along a path of extremal phase difference $\Delta\varphi$. This can be reexpressed in a form closely related to *Feynman's path-integral formulation of quantum mechanics* (Feynman, 1966). We can regard all the trial paths as being followed with equal probability. For each path, we are to construct a probability amplitude $e^{i\Delta\varphi}$, and we must then add together these amplitudes,

$$\sum_{\text{all paths}} e^{i\Delta\varphi}, \tag{7.45b}$$

to get the net complex amplitude for quanta associated with the waves to travel from $\mathbf{x}_{\text{initial}}$ to $\mathbf{x}_{\text{final}}$. The contributions from almost all neighboring paths will interfere destructively. The only exceptions are those paths whose neighbors have the same values of $\Delta\varphi$, to first order in the path difference. These are the paths that extremize the action (7.44): they are the wave's rays, the actual paths of the quanta.

SPECIALIZATION TO $\Omega = C(\mathbf{x})k$

Fermat's principle takes on an especially simple form when not only is the hamiltonian $\Omega(\mathbf{k}, \mathbf{x})$ time independent, but it also has the simple dispersion-free form $\Omega = C(\mathbf{x})k$—a form valid for the propagation of light through a time-independent dielectric, and sound waves through a time-independent, inhomogeneous fluid, and electromagnetic or gravitational waves through a time-independent, Newtonian gravitational field (Sec. 7.6). In this $\Omega = C(\mathbf{x})k$ case, the hamiltonian dictates that for each trial path, $\mathbf{k}$ is parallel to $d\mathbf{x}$, and therefore $\mathbf{k} \cdot d\mathbf{x} = k\,ds$, where s is distance along the path. Using the dispersion relation $k = \Omega/C$ and noting that Hamilton's equation $dx^j/dt = \partial\Omega/\partial k_j$ implies $ds/dt = C$ for the rate of traversal of the trial path, we see that $\mathbf{k} \cdot d\mathbf{x} = k\,ds = \Omega\,dt$. Since the trial paths are constrained to have Ω constant, Fermat's principle (7.44) becomes a principle of extremal time: The rays between $\mathbf{x}_{\text{initial}}$ and $\mathbf{x}_{\text{final}}$ are those paths along which

principle of extreme time for dispersionless wave

$$\boxed{\int dt = \int \frac{ds}{C(\mathbf{x})} = \int \frac{n(\mathbf{x})}{c}ds} \tag{7.46}$$

is extremal. In the last expression we have adopted the convention used for light in a dielectric medium, that $C(\mathbf{x}) = c/n(\mathbf{x})$, where c is the speed of light in vacuum, and n is the medium's index of refraction. Since c is constant, the rays are paths of extremal optical path length $\int n(\mathbf{x})ds$.

index of refraction

We can use Fermat's principle to demonstrate that, if the medium contains no opaque objects, then there will always be at least one ray connecting any two points. This is because there is a lower bound on the optical path between any two points, given by $n_{min}L$, where n_{min} is the lowest value of the refractive index anywhere in the medium, and L is the distance between the two points. This means that for some path the optical path length must be a minimum, and that path is then a ray connecting the two points.

From the principle of extremal time, we can derive the Euler-Lagrange differential equation for the ray. For ease of derivation, we write the action principle in the form

$$\delta \int n(\mathbf{x}) \sqrt{\frac{d\mathbf{x}}{d\mathbf{s}} \cdot \frac{d\mathbf{x}}{d\mathbf{s}}} \, ds, \tag{7.47}$$

where the quantity in the square root is identically one. Performing a variation in the usual manner then gives

ray equation for dispersionless wave

$$\boxed{\frac{d}{ds}\left(n\frac{d\mathbf{x}}{ds}\right) = \nabla n, \quad \text{i.e.,} \quad \frac{d}{ds}\left(\frac{1}{C}\frac{d\mathbf{x}}{ds}\right) = \nabla\left(\frac{1}{C}\right).} \tag{7.48}$$

This is equivalent to Hamilton's equations for the ray, as one can readily verify using the hamiltonian $\Omega = kc/n$ (Ex. 7.7).

Equation (7.48) is a second-order differential equation requiring two boundary conditions to define a solution. We can either choose these to be the location of the start of the ray and its starting direction, or the start and end of the ray. A simple case arises when the medium is stratified [i.e., when $n = n(z)$, where (x, y, z) are Cartesian coordinates]. Projecting Eq. (7.48) perpendicular to $\mathbf{e}_z$, we discover that $n\, dy/ds$ and $n\, dx/ds$ are constant, which implies

Snell's law

$$\boxed{n \sin\theta = \text{constant},} \tag{7.49}$$

where θ is the angle between the ray and $\mathbf{e}_z$. This is a variant of Snell's law of refraction. Snell's law is just a mathematical statement that the rays are normal to surfaces (wavefronts) on which the eikonal (phase) φ is constant (cf. Fig. 7.4).[8] Snell's law is valid not only when $n(\mathbf{x})$ varies slowly but also when it jumps discontinuously, despite the assumptions underlying geometric optics failing at a discontinuity.

EXERCISES

Exercise 7.7 *Derivation: Hamilton's Equations for Dispersionless Waves; Fermat's Principle*

Show that Hamilton's equations for the standard dispersionless dispersion relation (7.4) imply the same ray equation (7.48) as we derived using Fermat's principle.

---

8. Another important application of this general principle is to the design of optical instruments, where it is known as the *Abbé condition*. See, e.g., Born and Wolf (1999).

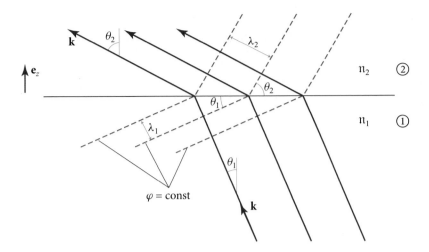

FIGURE 7.4 Illustration of Snell's law of refraction at the interface between two media, for which the refractive indices are n_1 and n_2 (assumed less than n_1). As the wavefronts must be continuous across the interface, simple geometry tells us that $\lambda_1/\sin\theta_1 = \lambda_2/\sin\theta_2$. This and the fact that the wavelengths are inversely proportional to the refractive index, $\lambda_j \propto 1/n_j$, imply that $n_1\sin\theta_1 = n_2\sin\theta_2$, in agreement with Eq. (7.49).

Exercise 7.8 *Example: Self-Focusing Optical Fibers*

Optical fibers in which the refractive index varies with radius are commonly used to transport optical signals. When the diameter of the fiber is many wavelengths, we can use geometric optics. Let the refractive index be

$$n = n_0(1 - \alpha^2 r^2)^{1/2}, \tag{7.50a}$$

where n_0 and α are constants, and r is radial distance from the fiber's axis.

(a) Consider a ray that leaves the axis of the fiber along a direction that makes an angle β to the axis. Solve the ray-transport equation (7.48) to show that the radius of the ray is given by

$$r = \frac{\sin\beta}{\alpha}\left|\sin\left(\frac{\alpha z}{\cos\beta}\right)\right|, \tag{7.50b}$$

where z measures distance along the fiber.

(b) Next consider the propagation time T for a light pulse propagating along the ray with $\beta \ll 1$, down a long length L of fiber. Show that

$$T = \frac{n_0 L}{C}[1 + O(\beta^4)], \tag{7.50c}$$

and comment on the implications of this result for the use of fiber optics for communication.

Exercise 7.9 **Example: Geometric Optics for the Schrödinger Equation*

Consider the nonrelativistic Schrödinger equation for a particle moving in a time-dependent, 3-dimensional potential well:

$$-\frac{\hbar}{i}\frac{\partial \psi}{\partial t} = \left[\frac{1}{2m}\left(\frac{\hbar}{i}\nabla\right)^2 + V(\mathbf{x}, t)\right]\psi. \tag{7.51}$$

(a) Seek a geometric-optics solution to this equation with the form $\psi = Ae^{iS/\hbar}$, where A and V are assumed to vary on a lengthscale $\mathcal{L}$ and timescale $\mathcal{T}$ long compared to those, $1/k$ and $1/\omega$, on which S varies. Show that the leading-order terms in the two-lengthscale expansion of the Schrödinger equation give the Hamilton-Jacobi equation

$$\frac{\partial S}{\partial t} + \frac{1}{2m}(\nabla S)^2 + V = 0. \tag{7.52a}$$

Our notation $\varphi \equiv S/\hbar$ for the phase φ of the wave function ψ is motivated by the fact that the geometric-optics limit of quantum mechanics is classical mechanics, and the function $S = \hbar\varphi$ becomes, in that limit, "Hamilton's principal function," which obeys the Hamilton-Jacobi equation (see, e.g., Goldstein, Poole, and Safko, 2002, Chap. 10). [Hint: Use a formal parameter σ to keep track of orders (Box 7.2), and argue that terms proportional to $\hbar^n$ are of order σ^n. This means there must be factors of σ in the Schrödinger equation (7.51) itself.]

(b) From Eq. (7.52a) derive the equation of motion for the rays (which of course is identical to the equation of motion for a wave packet and therefore is also the equation of motion for a classical particle):

$$\frac{d\mathbf{x}}{dt} = \frac{\mathbf{p}}{m}, \qquad \frac{d\mathbf{p}}{dt} = -\nabla V, \tag{7.52b}$$

where $\mathbf{p} = \nabla S$.

(c) Derive the propagation equation for the wave amplitude A and show that it implies

$$\frac{d|A|^2}{dt} + |A|^2\frac{\nabla \cdot \mathbf{p}}{m} = 0. \tag{7.52c}$$

Interpret this equation quantum mechanically.

Distribution Function and Mean Occupation Number

(Selected Sections of Chapter 3
of *Modern Classical Physics*)

3.2 Phase Space and Distribution Function

3.2.1 Newtonian Number Density in Phase Space, $\mathcal{N}$

In Newtonian, 3-dimensional space (*physical space*), consider a particle with rest mass m that moves along a path $\mathbf{x}(t)$ as universal time t passes (Fig. 3.1a). The particle's time-varying velocity and momentum are $\mathbf{v}(t) = d\mathbf{x}/dt$ and $\mathbf{p}(t) = m\mathbf{v}$. The path $\mathbf{x}(t)$ is a curve in the physical space, and the momentum $\mathbf{p}(t)$ is a time-varying, coordinate-independent vector in the physical space.

It is useful to introduce an auxiliary 3-dimensional space, called *momentum space*, in which we place the tail of $\mathbf{p}(t)$ at the origin. As time passes, the tip of $\mathbf{p}(t)$ sweeps out a curve in momentum space (Fig. 3.1b). This momentum space is "secondary" in the sense that it relies for its existence on the physical space of Fig. 3.1a. Any Cartesian coordinate system of physical space, in which the location $\mathbf{x}(t)$ of the particle has coordinates $\{x, y, z\}$, induces in momentum space a corresponding coordinate system $\{p_x, p_y, p_z\}$. The 3-dimensional physical space and 3-dimensional momentum space together constitute a 6-dimensional *phase space*, with coordinates $\{x, y, z, p_x, p_y, p_z\}$.

In this chapter, we study a collection of a very large number of identical particles (all with the same rest mass m).[1] As tools for this study, consider a tiny 3-dimensional volume $d\mathcal{V}_x$ centered on some location $\mathbf{x}$ in physical space and a tiny 3-dimensional volume $d\mathcal{V}_p$ centered on location $\mathbf{p}$ in momentum space. Together these make up a tiny 6-dimensional volume

$$d^2\mathcal{V} \equiv d\mathcal{V}_x d\mathcal{V}_p. \qquad (3.1)$$

In any Cartesian coordinate system, we can think of $d\mathcal{V}_x$ as being a tiny cube located at (x, y, z) and having edge lengths dx, dy, dz, and similarly for $d\mathcal{V}_p$. Then, as

---

This appendix is extracted from Volume 1 of *Modern Classical Physics*. It consists of Secs. 3.2.1, 3.2.3, and 3.2.5 of Chapter 3, "Kinetic Theory," with no changes aside from pagination.

1. In Ex. 3.2 and Box 3.2, we extend kinetic theory to particles with a range of rest masses.

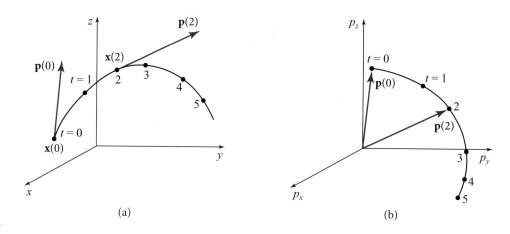

FIGURE 3.1 (a) Euclidean physical space, in which a particle moves along a curve $\mathbf{x}(t)$ that is parameterized by universal time t. In this space, the particle's momentum $\mathbf{p}(t)$ is a vector tangent to the curve. (b) Momentum space, in which the particle's momentum vector $\mathbf{p}$ is placed, unchanged, with its tail at the origin. As time passes, the momentum's tip sweeps out the indicated curve $\mathbf{p}(t)$.

computed in this coordinate system, these tiny volumes are

$$dV_x = dx\,dy\,dz, \quad dV_p = dp_x\,dp_y\,dp_z, \quad d^2V = dx\,dy\,dz\,dp_x\,dp_y\,dp_z. \quad (3.2)$$

Denote by dN the number of particles (all with rest mass m) that reside inside d^2V in phase space (at some moment of time t). Stated more fully: dN is the number of particles that, at time t, are located in the 3-volume dV_x centered on the location $\mathbf{x}$ in physical space and that also have momentum vectors whose tips at time t lie in the 3-volume dV_p centered on location $\mathbf{p}$ in momentum space. Denote by

distribution function

$$\boxed{\mathcal{N}(\mathbf{x}, \mathbf{p}, t) \equiv \frac{dN}{d^2V} = \frac{dN}{dV_x\,dV_p}} \quad (3.3)$$

the *number density of particles at location* $(\mathbf{x}, \mathbf{p})$ *in phase space at time t*. This is also called the *distribution function*.

This distribution function is kinetic theory's principal tool for describing any collection of a large number of identical particles.

In Newtonian theory, the volumes dV_x and dV_p occupied by our collection of dN particles are independent of the reference frame that we use to view them. Not so in relativity theory: dV_x undergoes a Lorentz contraction when one views it from a moving frame, and dV_p also changes; but (as we shall see in Sec. 3.2.2) their product $d^2V = dV_x dV_p$ is the same in all frames. Therefore, in both Newtonian theory and relativity theory, the distribution function $\mathcal{N} = dN/d^2V$ is independent of reference frame, and also, of course, independent of any choice of coordinates. It is a coordinate-independent scalar in phase space.

The normalization that one uses for the distribution function is arbitrary: renormalize $\mathcal{N}$ by multiplying with any constant, and $\mathcal{N}$ will still be a geometric, coordinate-independent, and frame-independent quantity and will still contain the same information as before. In this book, we use several renormalized versions of $\mathcal{N}$, depending on the situation. We now introduce them, beginning with the version used in plasma physics.

In Part VI, when dealing with nonrelativistic plasmas (collections of electrons and ions that have speeds small compared to light), we regard the distribution function as depending on time t, location $\mathbf{x}$ in Euclidean space, and velocity $\mathbf{v}$ (instead of momentum $\mathbf{p} = m\mathbf{v}$), and we denote it by[2]

$$f(t, \mathbf{x}, \mathbf{v}) \equiv \frac{dN}{d\mathcal{V}_x \, d\mathcal{V}_v} = \frac{dN}{dx\,dy\,dz \, dv_x \, dv_y \, dv_z} = m^3 \mathcal{N}. \qquad (3.10)$$

plasma distribution function

(This change of viewpoint and notation when transitioning to plasma physics is typical of this textbook. When presenting any subfield of physics, we usually adopt the conventions, notation, and also the system of units that are generally used in that subfield.)

3.2.5 Mean Occupation Number η

3.2.5

Although this book is about classical physics, we cannot avoid making frequent contact with quantum theory. The reason is that modern classical physics rests on a quantum mechanical foundation. Classical physics is an approximation to quantum physics, not conversely. Classical physics is derivable from quantum physics, not conversely.

In statistical physics, the classical theory cannot fully shake itself free from its quantum roots; it must rely on them in crucial ways that we shall meet in this chapter and the next. Therefore, rather than try to free it from its roots, we expose these roots and profit from them by introducing a quantum mechanics-based normalization for the distribution function: the mean occupation number η.

As an aid in defining the mean occupation number, we introduce the concept of the *density of states:* Consider a particle of mass m, described quantum mechanically. Suppose that the particle is known to be located in a volume $d\mathcal{V}_x$ (as observed in a specific inertial reference frame) and to have a spatial momentum in the region $d\mathcal{V}_p$ centered on $\mathbf{p}$. Suppose, further, that *the particle does not interact with any other particles or fields;* for example, ignore Coulomb interactions. (In portions of Chaps. 4 and 5, we include interactions.) Then how many single-particle quantum mechanical states[3] are available to the free particle? This question is answered most easily by

density of states

---

2. The generalization to relativistic plasmas is straightforward; see, e.g., Ex. 23.12.
3. A quantum mechanical state for a single particle is called an "orbital" in the chemistry literature (where the particle is an election) and in the classic thermal physics textbook by Kittel and Kroemer (1980);

constructing (in some arbitrary inertial frame) a complete set of wave functions for the particle's spatial degrees of freedom, with the wave functions (i) confined to be eigenfunctions of the momentum operator and (ii) confined to satisfy the standard periodic boundary conditions on the walls of a box with volume $d\mathcal{V}_x$. For simplicity, let the box have edge length L along each of the three spatial axes of the Cartesian spatial coordinates, so $d\mathcal{V}_x = L^3$. (This L is arbitrary and will drop out of our analysis shortly.) Then a complete set of wave functions satisfying (i) and (ii) is the set $\{\psi_{j,k,l}\}$ with

$$\psi_{j,k,l}(x, y, z) = \frac{1}{L^{3/2}} e^{i(2\pi/L)(jx+ky+lz)} e^{-i\omega t} \tag{3.14a}$$

[cf., e.g., Cohen-Tannoudji, Diu, and Laloë (1977, pp. 1440–1442), especially the Comment at the end of this page range]. Here the demand that the wave function take on the same values at the left and right faces of the box ($x = -L/2$ and $x = +L/2$), at the front and back faces, and at the top and bottom faces (the demand for periodic boundary conditions) dictates that the quantum numbers j, k, and l be integers. The basis states (3.14a) are eigenfunctions of the momentum operator $(\hbar/i)\nabla$ with momentum eigenvalues

$$p_x = \frac{2\pi\hbar}{L} j, \quad p_y = \frac{2\pi\hbar}{L} k, \quad p_z = \frac{2\pi\hbar}{L} l; \tag{3.14b}$$

correspondingly, the wave function's frequency ω has the following values in Newtonian theory $\boxed{\text{N}}$ and relativity $\boxed{\text{R}}$:

$$\boxed{\text{N}} \quad \hbar\omega = E = \frac{\mathbf{p}^2}{2m} = \frac{1}{2m}\left(\frac{2\pi\hbar}{L}\right)^2 (j^2 + k^2 + l^2); \tag{3.14c}$$

$$\boxed{\text{R}} \quad \hbar\omega = \mathcal{E} = \sqrt{m^2 + \mathbf{p}^2} \rightarrow m + E \text{ in the Newtonian limit.} \tag{3.14d}$$

Equations (3.14b) tell us that the allowed values of the momentum are confined to lattice sites in 3-momentum space with one site in each cube of side $2\pi\hbar/L$. Correspondingly, the total number of states in the region $d\mathcal{V}_x d\mathcal{V}_p$ of phase space is the number of cubes of side $2\pi\hbar/L$ in the region $d\mathcal{V}_p$ of momentum space:

$$dN_{\text{states}} = \frac{d\mathcal{V}_p}{(2\pi\hbar/L)^3} = \frac{L^3 d\mathcal{V}_p}{(2\pi\hbar)^3} = \frac{d\mathcal{V}_x d\mathcal{V}_p}{h^3}. \tag{3.15}$$

This is true no matter how relativistic or nonrelativistic the particle may be.

Thus far we have considered only the particle's spatial degrees of freedom. Particles can also have an internal degree of freedom called "spin." For a particle with spin s, the number of independent spin states is

---

we shall use physicists' more conventional but cumbersome phrase "single-particle quantum state," and also, sometimes, "mode."

$$g_s = \begin{cases} 2s + 1 & \text{if } m \neq 0 \text{ (e.g., an electron, proton, or atomic nucleus)} \\ 2 & \text{if } m = 0 \text{ and } s > 0 \text{ [e.g., a photon } (s = 1) \text{ or graviton } (s = 2)] \\ 1 & \text{if } m = 0 \text{ and } s = 0 \text{ (i.e., a hypothetical massless scalar particle)} \end{cases}$$

$$(3.16)$$

A notable exception is each species of neutrino or antineutrino, which has nonzero rest mass and spin 1/2, but $g_s = 1$ rather than $g_s = 2s + 1 = 2$.[4] We call this number of internal spin states g_s the particle's *multiplicity*. [It will turn out to play a crucial role in computing the entropy of a system of particles (Chap. 4); thus, it places the imprint of quantum theory on the entropy of even a highly classical system.]

particle's multiplicity

Taking account of both the particle's spatial degrees of freedom and its spin degree of freedom, we conclude that the total number of independent quantum states available in the region $d\mathcal{V}_x d\mathcal{V}_p \equiv d^2\mathcal{V}$ of phase space is $dN_{\text{states}} = (g_s/h^3)d^2\mathcal{V}$, and correspondingly the *number density of states in phase space* is

$$\mathcal{N}_{\text{states}} \equiv \frac{dN_{\text{states}}}{d^2\mathcal{V}} = \frac{g_s}{h^3}. \qquad (3.17)$$

density of states in phase space

[Relativistic remark: Note that, although we derived this number density of states using a specific inertial frame, it is a frame-independent quantity, with a numerical value depending only on Planck's constant and (through g_s) the particle's rest mass m and spin s.]

The ratio of the number density of particles to the number density of quantum states is obviously the number of particles in each state (the state's *occupation number*) averaged over many neighboring states—but few enough that the averaging region is small by macroscopic standards. In other words, this ratio is the quantum states' *mean occupation number* η:

$$\eta = \frac{\mathcal{N}}{\mathcal{N}_{\text{states}}} = \frac{h^3}{g_s}\mathcal{N}; \quad \text{i.e.,} \quad \boxed{\mathcal{N} = \mathcal{N}_{\text{states}}\eta = \frac{g_s}{h^3}\eta.} \qquad (3.18)$$

mean occupation number

The mean occupation number η plays an important role in quantum statistical mechanics, and its quantum roots have a profound impact on classical statistical physics.

From quantum theory we learn that the allowed values of the occupation number for a quantum state depend on whether the state is that of a *fermion* (a particle with spin 1/2, 3/2, 5/2, . . .) or that of a *boson* (a particle with spin 0, 1, 2, . . .). For fermions, no two particles can occupy the same quantum state, so the occupation number can only take on the eigenvalues 0 and 1. For bosons, one can shove any number of particles one wishes into the same quantum state, so the occupation number can take on the eigenvalues 0, 1, 2, 3, ... Correspondingly, the mean occupation numbers must lie in the ranges

fermions and bosons

$$0 \leq \eta \leq 1 \text{ for fermions}, \qquad 0 \leq \eta < \infty \text{ for bosons}. \qquad (3.19)$$

---

4. The reason for the exception is the particle's fixed chirality: -1 for neutrinos and $+1$ for antineutrinos; to have $g_s = 2$, a spin-1/2 particle must admit both chiralities.

Quantum theory also teaches us that, when $\eta \ll 1$, the particles, whether fermions or bosons, behave like classical, discrete, distinguishable particles; and when $\eta \gg 1$ (possible only for bosons), the particles behave like a classical wave—if the particles are photons ($s = 1$), like a classical electromagnetic wave; and if they are gravitons ($s = 2$), like a classical gravitational wave. This role of η in revealing the particles' physical behavior will motivate us frequently to use η as our distribution function instead of $\mathcal{N}$.

Of course η, like $\mathcal{N}$, is a function of location in phase space, $\eta(\mathcal{P}, \vec{p})$ in relativity with no inertial frame chosen; or $\eta(t, \mathbf{x}, \mathbf{p})$ in both relativity and Newtonian theory when an inertial frame is in use.

REFERENCES

Alfvén, H. (1970). Plasma physics, space research and the origin of the solar system. Nobel lecture. Available at http://www.nobelprize.org/nobel_prizes/physics/laureates/1970/alfven-lecture.pdf. Chapter 19 epigraph reprinted with permission of The Nobel Foundation.

Allis, W. P., S. J. Buchsbaum, and A. Bers (1963). *Waves in Anisotropic Plasmas*. Cambridge, Mass.: MIT Press.

Arfken, G. B., H. J. Weber, and F. E. Harris (2013). *Mathematical Methods for Physicists*. Amsterdam: Elsevier.

Bateman, G. (1978). *MHD Instabilities*. Cambridge, Mass.: MIT Press.

Bekefi, G. (1966). *Radiation Processes in Plasmas*. New York: Wiley.

Bellan, P. M. (2000). *Spheromaks*. London: Imperial College Press.

Bellan, P. M. (2006). *Fundamentals of Plasma Physics*. Cambridge: Cambridge University Press.

Bernstein, I. B., J. M. Greene, and M. D. Kruskal (1957). Exact nonlinear plasma oscillations. *Physical Review* **108**, 546–550.

Bernstein, I. B., E. A. Frieman, M. D. Kruskal, and R. M. Kulsrud (1958). An energy principle for hydromagnetic stability problems. *Proceedings of the Royal Society A* **244**, 17–40.

Binney, J. J., and S. Tremaine (2003). *Galactic Dynamics*. Princeton, N.J.: Princeton University Press.

Birn, J., and E. Priest (2007). *Reconnection of Magnetic Fields*. Cambridge: Cambridge University Press.

Bittencourt, J. A. (2004). *Fundamentals of Plasma Physics*. Berlin: Springer-Verlag.

Blandford, R. D., and D. Eichler (1987). Particle acceleration at astrophysical shocks: A theory of cosmic ray origin. *Physics Reports* **154**, 1–75.

Bogolyubov, N. N. (1962). Problems of a dynamical theory in statistical physics. In J. de Boer and G. E. Uhlenbeck (eds.), *Studies in Statistical Mechanics*, p. 1. Amsterdam: North Holland.

Born, M., and H. S. Green (1949). *A General Kinetic Theory of Liquids*. Cambridge: Cambridge University Press.

Born, M., and E. Wolf (1999). *Principles of Optics: Electromagnetic Theory of Propagation, Interference and Diffraction of Light*. Cambridge: Cambridge University Press.

Boyd, T. J. M., and J. J. Sanderson (2003). *The Physics of Plasmas*. Cambridge: Cambridge University Press.

Brown, L., and G. Gabrielse (1986). Geonium theory: Physics of a single electron or ion in a Penning trap. *Reviews of Modern Physics* **58**, 233–311.

Chandrasekhar, S. (1961). *Hydrodynamics and Hydromagnetic Stability*. Oxford: Oxford University Press.

Chen, F. F. (1974). *Introduction to Plasma Physics*. New York: Plenum Press.

Chen, F. F. (2016). *Introduction to Plasma Physics and Controlled Fusion*, third edition. Heidelberg: Springer.

Chew, G. F., M. L. Goldberger, and F. E. Low (1956). The Boltzmann equation and the one-fluid hydromagnetic equations in the absence of particle collisions. *Proceedings of the Royal Society A* **236**, 112–118.

Clemmow, P. C., and J. P. Dougherty (1969). *Electrodynamics of Particles and Plasmas*. New York: Addison-Wesley.

Cohen-Tannoudji, C., B. Diu, and F. Laloë (1977). *Quantum Mechanics*. New York: Wiley.

Copson, E. T. (1935). *An Introduction to the Theory of Functions of a Complex Variable*. Oxford: Oxford University Press.

Crookes, W. (1879). The Bakerian lecture: On the illumination of lines of molecular pressure, and the trajectory of molecules. *Philosophical Transactions of the Royal Society* **170**, 135–164.

Davidson, P. A. (2001). *An Introduction to Magnetohydrodynamics*. Cambridge: Cambridge University Press.

Davidson, R. C. (1972). *Methods in Nonlinear Plasma Theory*. New York: Academic Press.

Dorf, R. C., and R. H. Bishop (2012). *Modern Control Systems*. Upper Saddle River, N.J.: Pearson.

Dyson, F. (1986). Quoted in T. A. Heppenheimer, After the Sun dies. *Omni* **8**, no. 11, 38. Chapter 23 epigraph reprinted with permission of Freeman Dyson.

Feynman, R. P. (1966). *The Character of Physical Law*. Cambridge, Mass.: MIT Press.

Feynman, R. P. (1972). *Statistical Mechanics*. New York: Benjamin.

Forbes, T., and E. Priest (2007). *Magnetic Reconnection*. Cambridge: Cambridge University Press.

Franklin, G. F., J. D. Powell, and A. Emami-Naeini (2005). *Feedback Control of Dynamic Systems*. Upper Saddle River, N.J.: Pearson.

Goedbloed, J. P., R. Keppens, and S. Poedts (2010). *Advanced Magnetohydrodynamics*. Cambridge: Cambridge University Press.

Goedbloed, J. P., and S. Poedts (2004). *Principles of Magnetohydrodynamics, with Applications to Laboratory and Astrophysical Plasmas*. Cambridge: Cambridge University Press.

Goldstein, H., C. Poole, and J. Safko (2002). *Classical Mechanics*. New York: Addison-Wesley.

Gurnett, D. A. and Bhattarcharjee, A. (2017) *Introduction to Plasma Physics*. Cambridge: Cambridge University Press.

Hassani, S. (2013). *Mathematical Physics: A Modern Introduction to Its Foundations*. Cham, Switzerland: Springer.

Heaviside, O. (1912). *Electromagnetic Theory*, Volume III, p. 1. London: "The Electrician" Printing and Publishing.

Hénon, M. (1982). Vlasov equation? *Astronomy and Astrophysics* **114**, 211–212.

Hurricane, O. A., D. A. Callahan, D. T. Casey, P. M. Cellers, et al. (2014). Fuel gain exceeding unity in an inertially confined fusion explosion. *Nature* **506**, 343–348.

Jackson, J. D. (1999). *Classical Electrodynamics*. New York: Wiley.

Jeans, J. H. (1929). *Astronomy and Cosmogony*, second edition. Cambridge: Cambridge University Press.

Jeffrey, A. and T. Taniuti, eds. (1966). *Magnetohydrodynamic Stability and Thermonuclear Confinement: A Collection of Reprints*. New York: Academic Press.

Keilhacker, M., and the JET Team (1998). Fusion physics progress on JET. *Fusion Engineering and Design* **46**, 273–290.

Kirkwood, J. G. (1946). Statistical mechanical theory of transport processes. I. General theory. *Journal of Chemical Physics* **14**, 180–201.

Kittel, C. (2004). *Elementary Statistical Physics*. Mineola, N.Y.: Courier Dover Publications.

Kittel, C., and H. Kroemer (1980). *Thermal Physics*. London: Macmillan.

Krall, N. A., and A. W. Trivelpiece (1973). *Principles of Plasma Physics*. New York: McGraw-Hill.

Kruer, W. L. (1988). *The Physics of Laser-Plasma Interactions*. New York: Addison-Wesley.

Kulsrud, R. M. (2005). *Plasma Physics for Astrophysics*. Princeton, N.J.: Princeton University Press.

Landau, L. D. (1946). On the vibrations of the electronic plasma. *Journal of Physics USSR* **10**, 25–37.

Landau, L. D., L. P. Pitaevskii, and E. M. Lifshitz (1979). *Electrodynamics of Continuous Media*. Oxford: Butterworth-Heinemann.

Langmuir, I. (1928). Oscillations in Ionized Gases. *Proceedings of the National Academy of Sciences* **14**, 627–637. Chapter 22 epigraph reprinted with permission of the publisher.

Lifshitz, E. M., and L. P. Pitaevskii (1981). *Physical Kinetics*. Oxford: Pergamon.

Longair, M. S. (2011). *High Energy Astrophysics*. Cambridge: Cambridge University Press.

Martin, R. F. (1986). Chaotic particle dynamics near a two-dimensional neutral point with application to the geomagnetic tail. *Journal of Geophysical Research* **91**, 11985–11992.

Mathews, J., and R. L. Walker (1970). *Mathematical Methods of Physics*. New York: Benjamin.

Melrose, D. B. (1980). *Plasma Astrophysics*. New York: Gordon and Breach.

Melrose, D. B. (1984). *Instabilities in Space and Laboratory Plasmas*. Cambridge: Cambridge University Press.

Melrose, D. B. (2008). *Quantum Plasmadynamics, Vol 1: Unmagnetized Plasmas*. Cham, Switzerland: Springer.

Melrose, D. B. (2012). *Quantum Plasmadynamics, Vol 2: Magnetized Plasmas*. Cham, Switzerland: Springer.

Mikhailovskii, A. B. (1998). *Instabilities in a Confined Plasma*. Bristol and Philadelphia: Institute of Physics Publishing.

Northrop, T. (1963). *Adiabatic Motion of Charged Particles*. New York: Interscience.

Parker, E. N. (1979). *Cosmical Magnetic Fields*. Oxford: Clarendon Press.

Parks, G. K. (2004). *Physics of Space Plasmas: An Introduction*. Boulder: Westview Press.

Penrose, O. (1960). Electrostatic instabilities of a uniform non-Maxwellian plasma. *Physics of Fluids* **3**, 258–265.

Piel. A. (2017) *Plasma Physics*. Switzerland: Springer International Publishing AG.

Pines, D., and J. R. Schrieffer (1962). Approach to equilibrium of electrons, plasmons and phonons in quantum and classical plasmas. *Physical Review* **125**, 804–812.

Reif, F. (2008). *Fundamentals of Statistical and Thermal Physics*. Long Grove, Ill.: Waveland Press.

Rosenbluth, M. N., M. MacDonald, and D. L. Judd (1957). Fokker-Planck equation for an inverse square force. *Physical Review* **107**, 1–6.

Sagdeev, R. Z., and C. F. Kennel (1991). Collisionless shock waves. *Scientific American* **264**, April issue, 106–113.

Schmidt, G. (1979). *Physics of High Temperature Plasmas*. New York: Academic Press.

Shapiro, S. L., and S. A. Teukolsky (1983). *Black Holes, White Dwarfs and Neutron Stars: The Physics of Compact Objects*. New York: Wiley.

Shercliff, J. A. (1965). National Committee for Fluid Mechanics Films movie: Magnetohydro-dynamics.

Shkarofsky, I. P., T. W. Johnston, and M. P. Bachynski (1966). *The Particle Kinetics of Plasmas*. New York: Addison-Wesley.

Spitzer, Jr., L. (1962). *Physics of Fully Ionized Gases*. New York: Interscience. Chapter 20 epigraph reprinted with permission of the publisher.

Spitzer, Jr., L., and R. Harm (1953). Transport phenomena in a completely ionized gas. *Physical Review* **89**, 977–981.

Stix, T. H. (1992). *Waves in Plasmas*. New York: American Institute of Physics.

Sturrock, P. A. (1994). *Plasma Physics: An Introduction to the Theory of Astrophysical, Geophysical and Laboratory Plasmas*. Cambridge: Cambridge University Press.

Swanson, D. G. (2003). *Plasma Waves*. Bristol and Philadelphia: Institute of Physics Publishing.

Tsytovich, V. (1970). *Nonlinear Effects in Plasma*. New York: Plenum.

Vogel, S. (1994). *Life In Moving Fluids: The Physical Biology Of Flow,* 2nd Edition, Revised and Expanded by Steven Vogel, Illustrated by Susan Tanner Beety and the Author.

Yvon, J. (1935). *La Théorie des Fluides et l'Équation d'État*. Paris: Hermann.

NAME INDEX

Page numbers for entries in boxes are followed by "b," those for epigraphs at the beginning of a chapter by "e," those for figures by "f," and those for notes by "n."

SUBJECT INDEX

Second and third level entries are not ordered alphabetically. Instead, the most important or general entries come first, followed by less important or less general ones, with specific applications last.

Page numbers for entries in boxes are followed by "b," those for epigraphs at the beginning of a chapter by "e," those for figures by "f," for notes by "n," and for tables by "t."

for geometric optics, 1164–1165, 1166–1167

for quasi-linear theory in plasma physics, 1113–1114

two-point correlation function, 1104–1107

for Coulomb corrections to pressure in a plasma, 1107–1108

for electron antibunching in a plasma, 1104–1107

two-stream instability

in two-fluid formalism, 1065–1068

in kinetic theory, 1079–1080, 1137

van Allen belts, 1028, 1029f

virial theorems

for any system obeying momentum conservation, 982–984

for self-gravitating systems, 984

MHD application, 982, 984

Vlasov equation, in plasma kinetic theory, 1071–1072

solution via Jeans' theorem, 1075

volume in phase space, 1183

vortex. *See also* vortex lines; vorticity

above a water drain, 1151

wingtip vortex, 1154f

tornado, 1151

vortex cores, in superfluid, 1153

vortex lines, 1153–1154, 1154f

frozen into fluid, for barotropic, inviscid flows, 1155–1157, 1157f

vorticity, 1151–1152, 1152f

measured by a vane with orthogonal fins, 1151–1152

evolution equations for, 1153–1157

frozen into an inviscid, barotropic flow, 1155–1157

interaction with magnetic field, 957–958

delta-function: constant-angular-momentum flow, 1151–1152

Voyager spacecraft, 1147

water waves. *See* gravity waves on water; sound waves in a fluid

wave equations

prototypical, 1165

lagrangian, energy density flux, and adiabatic invariant for, 1172–1173

algebratized, 1037

for electromagnetic waves. *See* electromagnetic waves

wave packet, 1161–1163

Gaussian, 1163–1164

spreading of (dispersion), 1163–1164

wave-normal surface, 1062–1063, 1062f, 1064f

wave-wave mixing

in plasmas, 1132–1135

waves, monochromatic in homogeneous medium. *See also* sound waves in a fluid; gravity waves on water; flexural waves on a beam or rod; gravitational waves; Alfvén waves

dispersion relation, 1159

group velocity, 1162

phase velocity, 1159

plane, 1159

whistler wave in plasma, 1053f, 1054–1055, 1062f, 1146f

winds, propagation of sound waves in, 1173

wingtip vortices, 1154f

WKB approximation, as example of eikonal approximation, 1165

zero point energy, 1002

Z-pinch for plasma confinement, 960–962, 961f

stability of, 975–978

T2 Track Two; see page xvi

PREFACE TO *MODERN CLASSICAL PHYSICS*

The study of physics (including astronomy) is one of the oldest academic enterprises. Remarkable surges in inquiry occurred in equally remarkable societies—in Greece and Egypt, in Mesopotamia, India and China—and especially in Western Europe from the late sixteenth century onward. Independent, rational inquiry flourished at the expense of ignorance, superstition, and obeisance to authority.

Physics is a constructive and progressive discipline, so these surges left behind layers of understanding derived from careful observation and experiment, organized by fundamental principles and laws that provide the foundation of the discipline today. Meanwhile the detritus of bad data and wrong ideas has washed away. The laws themselves were so general and reliable that they provided foundations for investigation far beyond the traditional frontiers of physics, and for the growth of technology.

The start of the twentieth century marked a watershed in the history of physics, when attention turned to the small and the fast. Although rightly associated with the names of Planck and Einstein, this turning point was only reached through the curiosity and industry of their many forerunners. The resulting quantum mechanics and relativity occupied physicists for much of the succeeding century and today are viewed very differently from each other. Quantum mechanics is perceived as an abrupt departure from the tacit assumptions of the past, while relativity—though no less radical conceptually—is seen as a logical continuation of the physics of Galileo, Newton, and Maxwell. There is no better illustration of this than Einstein's growing special relativity into the general theory and his famous resistance to the quantum mechanics of the 1920s, which others were developing.

This is a book about classical physics—a name intended to capture the pre-quantum scientific ideas, augmented by general relativity. Operationally, it is physics in the limit that Planck's constant $h \to 0$. Classical physics is sometimes used, pejoratively, to suggest that "classical" ideas were discarded and replaced by new principles and laws. Nothing could be further from the truth. The majority of applications of physics today are still essentially classical. This does not imply that physicists or others working in these areas are ignorant or dismissive of quantum physics. It is simply that the issues with which they are confronted are mostly addressed classically. Furthermore, classical physics has not stood still while the quantum world was being explored. In scope and in practice, it has exploded on many fronts and would now be quite unrecognizable to a Helmholtz, a Rayleigh, or a Gibbs. In this book, we have tried to emphasize these contemporary developments and applications at the expense of historical choices, and this is the reason for our seemingly oxymoronic title, *Modern Classical Physics.*

This book is ambitious in scope, but to make it bindable and portable (and so the authors could spend some time with their families), we do not develop classical mechanics, electromagnetic theory, or elementary thermodynamics. We assume the reader has already learned these topics elsewhere, perhaps as part of an undergraduate curriculum. We also assume a normal undergraduate facility with applied mathematics. This allows us to focus on those topics that are less frequently taught in undergraduate and graduate courses.

Another important exclusion is numerical methods and simulation. High-performance computing has transformed modern research and enabled investigations that were formerly hamstrung by the limitations of special functions and artificially imposed symmetries. To do justice to the range of numerical techniques that have been developed—partial differential equation solvers, finite element methods, Monte Carlo approaches, graphics, and so on—would have more than doubled the scope and size of the book. Nonetheless, because numerical evaluations are crucial for physical insight, the book includes many applications and exercises in which user-friendly numerical packages (such as Maple, Mathematica, and Matlab) can be used to produce interesting numerical results without too much effort. We hope that, via this pathway from fundamental principle to computable outcome, our book will bring readers not only physical insight but also enthusiasm for computational physics.

Classical physics as we develop it emphasizes physical phenomena on macroscopic scales: scales where the particulate natures of matter and radiation are secondary to their behavior in bulk; scales where particles' statistical—as opposed to individual—properties are important, and where matter's inherent graininess can be smoothed over.

In this book, we take a journey through spacetime and phase space; through statistical and continuum mechanics (including solids, fluids, and plasmas); and through optics and relativity, both special and general. In our journey, we seek to comprehend the fundamental laws of classical physics in their own terms, and also in relation to quantum physics. And, using carefully chosen examples, we show how the classical laws are applied to important, contemporary, twenty-first-century problems and to everyday phenomena; and we also uncover some deep relationships among the various fundamental laws and connections among the practical techniques that are used in different subfields of physics.

Geometry is a deep theme throughout this book and a very important connector. We shall see how a few geometrical considerations dictate or strongly limit the basic principles of classical physics. Geometry illuminates the character of the classical principles and also helps relate them to the corresponding principles of quantum physics. Geometrical methods can also obviate lengthy analytical calculations. Despite this, long, routine algebraic manipulations are sometimes unavoidable; in such cases, we occasionally save space by invoking modern computational symbol manipulation programs, such as Maple, Mathematica, and Matlab.

This book is the outgrowth of courses that the authors have taught at Caltech and Stanford beginning 37 years ago. Our goal was then and remains now to fill what we saw as a large hole in the traditional physics curriculum, at least in the United States:

- We believe that every masters-level or PhD physicist should be familiar with the basic concepts of all the major branches of classical physics and should have had some experience in applying them to real-world phenomena; this book is designed to facilitate this goal.

- Many physics, astronomy, and engineering graduate students in the United States and around the world use classical physics extensively in their research, and even more of them go on to careers in which classical physics is an essential component; this book is designed to expedite their efforts.

- Many professional physicists and engineers discover, in mid-career, that they need an understanding of areas of classical physics that they had not previously mastered. This book is designed to help them fill in the gaps and see the relationship to already familiar topics.

In pursuit of this goal, we seek, in this book, to *give the reader a clear understanding of the basic concepts and principles of classical physics*. We present these principles in the language of modern physics (not nineteenth-century applied mathematics), and we present them primarily for physicists—though we have tried hard to make the content interesting, useful, and accessible to a much larger community including engineers, mathematicians, chemists, biologists, and so on. As far as possible, we emphasize theory that involves general principles which extend well beyond the particular topics we use to illustrate them.

In this book, we also seek to *teach the reader how to apply the ideas of classical physics*. We do so by presenting contemporary applications from a variety of fields, such as

- fundamental physics, experimental physics, and applied physics;
- astrophysics and cosmology;
- geophysics, oceanography, and meteorology;
- biophysics and chemical physics; and

- engineering, optical science and technology, radio science and technology, and information science and technology.

Why is the range of applications so wide? Because we believe that physicists should have enough understanding of general principles to attack problems that arise in unfamiliar environments. In the modern era, a large fraction of physics students will go on to careers outside the core of fundamental physics. For such students, a broad exposure to non-core applications can be of great value. For those who wind up in the core, such an exposure is of value culturally, and also because ideas from other fields often turn out to have impact back in the core of physics. Our examples illustrate how basic concepts and problem-solving techniques are freely interchanged across disciplines.

We strongly believe that classical physics should *not* be studied in isolation from quantum mechanics and its modern applications. Our reasons are simple:

- Quantum mechanics has primacy over classical physics. Classical physics is an approximation—often excellent, sometimes poor—to quantum mechanics.

- In recent decades, many concepts and mathematical techniques developed for quantum mechanics have been imported into classical physics and there used to enlarge our classical understanding and enhance our computational capability. An example that we shall study is nonlinearly interacting plasma waves, which are best treated as quanta ("plasmons"), despite their being solutions of classical field equations.

- Ideas developed initially for classical problems are frequently adapted for application to avowedly quantum mechanical subjects; examples (not discussed in this book) are found in supersymmetric string theory and in the liquid drop model of the atomic nucleus.

Because of these intimate connections between quantum and classical physics, quantum physics appears frequently in this book.

The amount and variety of material covered in this book may seem overwhelming. If so, keep in mind the key goals of the book: to teach the fundamental concepts, which are not so extensive that they should overwhelm, and to illustrate those concepts. Our goal is not to provide a mastery of the many illustrative applications contained in the book, but rather to convey the spirit of how to apply the basic concepts of classical physics. To help students and readers who feel overwhelmed, we have labeled as "Track Two" sections that can be skipped on a first reading, or skipped entirely—but are sufficiently interesting that many readers may choose to browse or study them. Track-Two sections are labeled by the symbol **T2** . To keep Track One manageable for a one-year course, the Track-One portion of each chapter is rarely longer than 40 pages (including many pages of exercises) and is often somewhat shorter. Track One is designed for a full-year course at the first-year graduate level; that is how we have mostly used it. (Many final-year undergraduates have taken our course successfully, but rarely easily.)

The book is divided into seven parts:

I. **Foundations**—which introduces our book's powerful *geometric* point of view on the laws of physics and brings readers up to speed on some concepts and mathematical tools that we shall need. Many readers will already have mastered most or all of the material in Part I and might find that they can understand most of the rest of the book without adopting our avowedly geometric viewpoint. Nevertheless, we encourage such readers to browse Part I, at least briefly, before moving on, so as to become familiar with this viewpoint. We believe the investment will be repaid. Part I is split into two chapters, Chap. 1 on Newtonian physics and Chap. 2 on special relativity. Since nearly all of Parts II–VI is Newtonian, readers may choose to skip Chap. 2 and the occasional special relativity sections of subsequent chapters, until they are ready to launch into Part VII, General Relativity. Accordingly, Chap. 2 is labeled Track Two, though it becomes Track One when readers embark on Part VII.

II. **Statistical Physics**—including kinetic theory, statistical mechanics, statistical thermodynamics, and the theory of random processes. These subjects underlie some portions of the rest of the book, especially plasma physics and fluid mechanics.

III. **Optics**—by which we mean classical waves of all sorts: light waves, radio waves, sound waves, water waves, waves in plasmas, and gravitational waves. The major concepts we develop for dealing with all these waves include geometric optics, diffraction, interference, and nonlinear wave-wave mixing.

IV. **Elasticity**—elastic deformations, both static and dynamic, of solids. Here we develop the use of tensors to describe continuum mechanics.

V. **Fluid Dynamics**—with flows ranging from the traditional ones of air and water to more modern cosmic and biological environments. We introduce vorticity, viscosity, turbulence, boundary layers, heat transport, sound waves, shock waves, magnetohydrodynamics, and more.

VI. **Plasma Physics**—including plasmas in Earth-bound laboratories and in technological (e.g., controlled-fusion) devices, Earth's ionosphere, and cosmic environments. In addition to magnetohydrodynamics (treated in Part V), we develop two-fluid and kinetic approaches, and techniques of nonlinear plasma physics.

VII. **General Relativity**—the physics of curved spacetime. Here we show how the physical laws that we have discussed in flat spacetime are modified to account for curvature. We also explain how energy and momentum generate this curvature. These ideas are developed for their principal classical applications to neutron stars, black holes, gravitational radiation, and cosmology.

It should be possible to read and teach these parts independently, provided one is prepared to use the cross-references to access some concepts, tools, and results developed in earlier parts.

Five of the seven parts (II, III, V, VI, and VII) conclude with chapters that focus on applications where there is much current research activity and, consequently, there are many opportunities for physicists.

Exercises are a major component of this book. There are five types of exercises:

1. *Practice.* Exercises that provide practice at mathematical manipulations (e.g., of tensors).

2. *Derivation.* Exercises that fill in details of arguments skipped over in the text.

3. *Example.* Exercises that lead the reader step by step through the details of some important extension or application of the material in the text.

4. *Problem.* Exercises with few, if any, hints, in which the task of figuring out how to set up the calculation and get started on it often is as difficult as doing the calculation itself.

5. *Challenge.* Especially difficult exercises whose solution may require reading other books or articles as a foundation for getting started.

We urge readers to try working many of the exercises, especially the examples, which should be regarded as continuations of the text and which contain many of the most illuminating applications. Exercises that we regard as especially important are designated by **.

A few words on units and conventions. In this book we deal with practical matters and frequently need to have a quantitative understanding of the magnitudes of various physical quantities. This requires us to adopt a particular unit system. Physicists use both Gaussian and SI units; units that lie outside both formal systems are also commonly used in many subdisciplines. Both Gaussian and SI units provide a complete and internally consistent set for all of physics, and it is an often-debated issue as to which system is more convenient or aesthetically appealing. We will not enter this debate! One's choice of units should not matter, and a mature physicist should be able to change from one system to another with little thought. However, when learning new concepts, having to figure out "where the 2πs and 4πs go" is a genuine impediment to progress. Our solution to this problem is as follows. For each physics subfield that we study, we consistently use the set of units that seem most natural or that, we judge, constitute the majority usage by researchers in that subfield. We do not pedantically convert cm to m or vice versa at every juncture; we trust that the reader can easily make whatever translation is necessary. However, where the equations are actually different—primarily in electromagnetic theory—we occasionally provide, in brackets or footnotes, the equivalent equations in the other unit system and enough information for the reader to proceed in his or her preferred scheme.

We encourage readers to consult this book's website, http://press.princeton.edu/titles/MCP.html, for information, errata, and various resources relevant to the book.

A large number of people have influenced this book and our viewpoint on the material in it. We list many of them and express our thanks in the Acknowledgments. Many misconceptions and errors have been caught and corrected. However, in a book of this size and scope, others will remain, and for these we take full responsibility. We would be delighted to learn of these from readers and will post corrections and explanations on this book's website when we judge them to be especially important and helpful.

Above all, we are grateful for the support of our wives, Carolee and Liz—and especially for their forbearance in epochs when our enterprise seemed like a mad and vain pursuit of an unreachable goal, a pursuit that we juggled with huge numbers of other obligations, while Liz and Carolee, in the midst of their own careers, gave us the love and encouragement that were crucial in keeping us going.

ACKNOWLEDGMENTS FOR *MODERN CLASSICAL PHYSICS*

This book evolved gradually from notes written in 1980–81, through improved notes, then sparse prose, and on into text that ultimately morphed into what you see today. Over these three decades and more, courses based on our evolving notes and text were taught by us and by many of our colleagues at Caltech, Stanford, and elsewhere. From those teachers and their students, and from readers who found our evolving text on the web and dove into it, we have received an extraordinary volume of feedback,[1] and also patient correction of errors and misconceptions as well as help with translating passages that were correct but impenetrable into more lucid and accessible treatments. For all this feedback and to all who gave it, we are extremely grateful. We wish that we had kept better records; the heartfelt thanks that we offer all these colleagues, students, and readers, named and unnamed, are deeply sincere.

Teachers who taught courses based on our evolving notes and text, and gave invaluable feedback, include Professors Richard Blade, Yanbei Chen, Michael Cross, Steven Frautschi, Peter Goldreich, Steve Koonin, Christian Ott, Sterl Phinney, David Politzer, John Preskill, John Schwarz, and David Stevenson at Caltech; Professors Tom Abel, Seb Doniach, Bob Wagoner, and the late Shoucheng Zhang at Stanford; and Professor Sandor Kovacs at Washington University in St. Louis.

Our teaching assistants, who gave us invaluable feedback on the text, improvements of exercises, and insights into the difficulty of the material for the students, include Jeffrey Atwell, Nate Bode, Yu Cao, Yi-Yuh Chen, Jane Dai, Alexei Dvoretsky, Fernando Echeverria, Jiyu Feng, Eanna Flanagan, Marc Goroff, Dan Grin, Arun Gupta, Alexandr Ikriannikov, Anton Kapustin, Kihong Kim, Hee-Won Lee, Geoffrey Lovelace, Miloje Makivic, Draza Markovic, Keith Matthews, Eric Morganson, Mike Morris, Chung-Yi Mou, Rob Owen, Yi Pan, Jaemo Park, Apoorva Patel, Alexander Putilin, Shuyan Qi, Soo Jong Rey, Fintan Ryan, Bonnie Shoemaker, Paul Simeon,

---

1. Specific applications that were originated by others, to the best of our memory, are acknowledged in the text.

Hidenori Sinoda, Matthew Stevenson, Wai Mo Suen, Marcus Teague, Guodang Wang, Xinkai Wu, Huan Yang, Jimmy Yee, Piljin Yi, Chen Zheng, and perhaps others of whom we have lost track!

Among the students and readers of our notes and text, who have corresponded with us, sending important suggestions and errata, are Bram Achterberg, Mustafa Amin, Richard Anantua, Alborz Bejnood, Edward Blandford, Jonathan Blandford, Dick Bond, Phil Bucksbaum, James Camparo, Conrado Cano, U Lei Chan, Vernon Chaplin, Mina Cho, Ann Marie Cody, Sandro Commandè, Kevin Fiedler, Krzysztof Findeisen, Jeff Graham, Casey Handmer, John Hannay, Ted Jacobson, Matt Kellner, Deepak Kumar, Andrew McClung, Yuki Moon, Evan O'Connor, Jeffrey Oishi, Keith Olive, Zhen Pan, Eric Peterson, Laurence Perreault Levasseur, Rob Phillips, Vahbod Pourahmad, Andreas Reisenegger, David Reis, Pavlin Savov, Janet Scheel, Yuki Taka-hashi, Clifford Will, Fun Lim Yee, Yajie Yuan, and Aaron Zimmerman.

For computational advice or assistance, we thank Edward Campbell, Mark Scheel, Chris Mach, and Elizabeth Wood.

Academic support staff who were crucial to our work on this book include Christine Aguilar, JoAnn Boyd, Jennifer Formicelli, and Shirley Hampton.

The editorial and production professionals at Princeton University Press (Peter Dougherty, Karen Fortgang, Ingrid Gnerlich, Eric Henney, and Arthur Werneck) and at Princeton Editorial Associates (Peter Strupp and his freelance associates Paul Anagnostopoulos, Laurel Muller, MaryEllen Oliver, Joe Snowden, and Cyd Westmoreland) have been magnificent, helping us plan and design this book, and transforming our raw prose and primitive figures into a visually appealing volume, with sustained attention to detail, courtesy, and patience as we missed deadline after deadline.

Of course, we the authors take full responsibility for all the errors of judgment, bad choices, and mistakes that remain.

Roger Blandford thanks his many supportive colleagues at Caltech, Stanford University, and the Kavli Institute for Particle Astrophysics and Cosmology. He also acknowledges the Humboldt Foundation, the Miller Institute, the National Science Foundation, and the Simons Foundation for generous support during the completion of this book. And he also thanks the Berkeley Astronomy Department; Caltech; the Institute of Astronomy, Cambridge; and the Max Planck Institute for Astrophysics, Garching, for hospitality.

Kip Thorne is grateful to Caltech—the administration, faculty, students, and staff—for the supportive environment that made possible his work on this book, work that occupied a significant portion of his academic career.